Cereals and Nutraceuticals

Joginder Singh • Sawinder Kaur
Prasad Rasane • Jyoti Singh
Editors

Cereals and Nutraceuticals

Springer

Editors
Joginder Singh
Department of Botany
Nagaland University
Lumami, Nagaland, India

Prasad Rasane
Department of Food Technology
and Nutrition
Lovely Professional University
Phagwara, Punjab, India

Sawinder Kaur
Department of Food Technology
and Nutrition
Lovely Professional University
Phagwara, Punjab, India

Jyoti Singh
Department of Food Technology
and Nutrition
Lovely Professional University
Phagwara, Punjab, India

ISBN 978-981-97-2541-0 ISBN 978-981-97-2542-7 (eBook)
https://doi.org/10.1007/978-981-97-2542-7

This Springer imprint is published by the registered company Springer Nature Singapore Pte Ltd.
The registered company address is: 152 Beach Road, #21-01/04 Gateway East, Singapore 189721, Singapore

Preface

Nutricereals more commonly known as coarse cereals have been part of our consumption habits since ancient times. With the advent of green revolution, however, the consumption pattern dramatically shifted towards a more refined form of diet that constituted of limited number of cereals such as rice and wheat, replacing the vast aura of grains consumed. These cereals were easier to domesticate and cultivate in large quantity in a mono-cropping system. Although this brought great accolades in terms of managing hunger around the world, it had drastic and long-lasting effect on our health. The cereal grains that were left behind were termed as coarse grains for a long time that made it unappealing to a mass consumer. This, however, has changed recently as the general consumer has turned its attention to a diet that should be a rich blend of foods that would provide it much needed functional value for growing concerns of health issues that are becoming prevalent in recent past and present. These issues are mostly pertaining to problems such as obesity, hypertension, cardiovascular diseases, and similar such issues that are profoundly related to bad diet that is largely attributed to more refined food that humans gladly embraced so far. Researchers and scientists around the world have now turned their attention to these problems and found in nutricereal a cheaper and most important solution to these problems. Medical professionals and nutritionists have emphasized on the variety in diet and an unrefined, coarser diet for general health and wellness for long. Nutricereals seem to be just the food to provide that much needed value to the food. Therefore, many studies have since then focused on inclusion of nutricereals in diet as well as various health implications associated with its consumption. Moreover, stress is being given on advanced cultivation practices, production, and productivity of these nutricereals. Processing of these cereals is a matter of advanced research in recent times to make it suitable and palatable for human consumption and taste and hence high emphasis is given on processing and value addition of these cereals.

The book focuses on all these aspects that include production, cultivation, processing, composition, nutritional value, value addition, health implications, limitations, and safety of these nutricereals. Special attentions were given to the functional significance and potential of these nutricereals. Chapters are written by active researchers and presented in an accessible way to the public beyond those specialists in the topic dealt with. This book serves as an important basis for knowledge on nutricereals for student, research scholar, food technologists and scientists, nutritionists, and dieticians for paving their way towards health and nourishment and future research on product development and malnutrition.

We are honored that the leading experts with extensive, in-depth experience and expertise in diverse microbial technologies for their time and effort to develop outstanding chapters.

We thank Springer team for their generous assistance, constant support, and patience in initializing the book. Authors are also grateful to our esteemed friends, well-wishers, and faculty colleagues.

Lumami, Nagaland, India	Joginder Singh
Phagwara, Punjab, India	Sawinder Kaur
Phagwara, Punjab, India	Prasad Rasane
Phagwara, Punjab, India	Jyoti Singh

Contents

Editors and Contributors

About the Editors

Joginder Singh is a Professor at the Department of Botany, Nagaland University, Lumami, Nagaland, India. Previously, he worked as a Professor in the school of Bioengineering and Biosciences, Lovely Professional University and also as a Young Scientist at Microbial Biotechnology and Biofertilizer Laboratory, Department of Botany, Jai Narain Vyas University in research project funded by the Department of Science and Technology, Govt. of India. He is an active member of various scientific societies and organizations. He has more than 165 research and review articles in the peer-reviewed journals, edited 25 books, and authored/co-authored 100 chapters in edited books. He serves as a reviewer for many prestigious journals. Dr. Singh attended several international and national seminars, symposia, conferences, and chaired technical sessions and presented papers in them.

Sawinder Kaur is presently serving as the Professor, Department of Food Technology and Nutrition and Associate Dean, School of Agriculture, Lovely Professional University, Punjab, India. She is recipient of Gold Medal at both the graduation and post-graduation levels and also has received Research Excellence award by Lovely Professional University, two times in a row (2018, 2019). Dr. Kaur has working experience in both Food Industry and Academia for more than 15 years. She has more than 109 peer-reviewed international publications and also authored various international book chapters. She also has seven Indian patents filed in her name. She has organized and participated in numerous workshops and seminar and served on the organizing committees of conferences. She is an esteemed reviewer for numerous peer-reviewed international publications and volunteers to review books, book proposals, and projects.

Prasad Rasane is a doctorate in Food Science and Technology from the Banaras Hindu University, Varanasi, Uttar Pradesh, India, and is recipient of the prestigious INSPIRE fellowship Award from the Department of Science and Technology, Government of India, the Merit Award by the Jawaharlal Nehru Memorial Fund, New Delhi, India, and Research Excellence Award at Lovely Professional University, Punjab, India. He is currently working as Associate Professor, Head of he department, Food Technology and Nutrition, School of Agriculture at Lovely Professional

University, Phagwara, Punjab, India, since past 10 years. He has more than 121 high impact peer-reviewed international publications, several international book chapters, and has filed 14 Indian patents in his name. Dr. Rasane also serves as Vice President, Association of Food Scientists and Technologist of India-Jalandhar Chapter. He regularly volunteers as reviewer of scientific publications for international peer-reviewed publications and numerous projects and also serves on the editorial board of various national and international peer-reviewed journals. He has organized various workshops and served on the organizing committee of conferences and seminars.

Jyoti Singh is currently working as Associate Professor (Nutrition and Dietetics) in the Department of Food Technology and Nutrition, Lovely Professional University, Punjab, India. She is recipient of academic honor during her post-graduation and was also awarded research appreciation award (2019 and 2023) at Lovely Professional University. Dr. Jyoti Singh has authored more than 82 peer-reviewed international publications including high impact and Scopus rated research and review papers and book chapters. She also has seven Indian patents filed in her name. Dr. Jyoti Singh is member of the Association of Food Scientists and Technologist of India and Indian Science Congress. Dr. Jyoti Singh regularly participates in various international and national conferences, seminars, and workshops and has presented seminars on various topics like food safety, product development, and functional food research. She regularly volunteers as reviewer of scientific publications for international peer-reviewed publications.

Contributors

Ali Abdullahi Department of Agricultural Economics and Extension, Bayero University, Kano, Nigeria

Amanjyoti Department of Food Technology and Nutrition, School of Agriculture, Lovely Professional University, Phagwara, Punjab, India

Amine Assouguem Department of Plant Protection and Environment, National School of Agriculture, Meknes, Morocco

Aarti Bains Department of Microbiology, Lovely Professional University, Phagwara, Punjab, India

Vishesh Bhadariya School of Chemical Engineering, Oklahoma State University, Stillwater, OK, USA

Ananya Bhardwaj Department of Dairy Science and Food Technology, Institute of Agricultural Sciences, Banaras Hindu University, Varanasi, Uttar Pradesh, India

Prince Chawla Department of Food Technology and Nutrition, Lovely Professional University, Phagwara, Punjab, India

Sezai Ercisli Department of Horticulture, Faculty of Agriculture, Ataturk University, Erzurum, Turkey

Sunakshi Gautam School of Bioengineering and Food Technology, Shoolini University, Solan, Himachal Pradesh, India

Mahendra Gunjal Department of Food Technology and Nutrition, School of Agriculture, Lovely Professional University, Phagwara, Punjab, India

Hamid Food Technology and Nutrition, Lovely Professional University, Phagwara, Punjab, India

Adams Ovie Iyiola Department of Fisheries and Aquatic Resources Management, Faculty of Renewable Natural Resources Management, College of Agriculture and Renewable Natural Resources, Osun State University, Osogbo, Nigeria

Amit K. Jaiswal School of Food Science and Environmental Health, Technological University Dublin-City Campus, Dublin, Ireland

Environmental Sustainability and Health Institute (ESHI), Technological University Dublin-City Campus, Dublin, Ireland

Sapna Jarial Department of Agricultural Economics and Extension, School of Agriculture, Lovely Professional University, Phagwara, Punjab, India

Sakshi Kasodhan Department of Food Technology and Nutrition, Lovely Professional University, Phagwara, Punjab, India

Diksha Kataria Department of Food Technology and Nutrition, School of Agriculture, Lovely Professional University, Phagwara, Punjab, India

Deepika Kathuria Department of Food Science and Technology, Graphic Era University, Dehradun, Uttarakhand, India

Jaspreet Kaur Department of Food Technology and Nutrition, School of Agriculture, Lovely Professional University, Phagwara, Punjab, India

Sawinder Kaur Department of Food Technology and Nutrition, School of Agriculture, Lovely Professional University, Phagwara, Punjab, India

Deepak Kaushik Department of Biotechnology, Shoolini University, Solan, Himachal Pradesh, India

Rupesh Kaushik Department of Agricultural Economics and Extension, School of Agriculture, Lovely Professional University, Phagwara, Punjab, India

Mukul Kumar Department of Food Technology and Nutrition, Lovely Professional University, Phagwara, Punjab, India

Sanjeev Kumar Department of Horticulture, School of Agriculture, Lovely Professional University, Phagwara, Punjab, India

Vikas Kumar Department of Food Science and Technology, Punjab Agricultural University, Ludhiana, Punjab, India

Mamta Department of Food Science, Nutrition and Technology, CSK Himachal Pradesh Krishi Vishvavidalaya, Palampur, Himachal Pradesh, India

Aniket More Department of Food Technology & Nutrition, School of Agriculture, Lovely Professional University, Phagwara, Punjab, India

Sonia Morya Department of Food Technology & Nutrition, School of Agriculture, Lovely Professional University, Phagwara, Punjab, India

Vikas Nanda Department of Food Engineering and Technology, Sant Longowal Institute of Engineering and Technology, Sangrur, Punjab, India

Nikhil Dnyaneshwar Patil Department of Food Technology and Nutrition, Lovely Professional University, Phagwara, Punjab, India

Amrita Poonia Department of Dairy Science and Food Technology, Institute of Agricultural Sciences, Banaras Hindu University, Varanasi, Uttar Pradesh, India

Prasad Rasane Department of Food Technology and Nutrition, School of Agriculture, Lovely Professional University, Phagwara, Punjab, India

Nadimintri Chandana Sri Sai Department of Dairy Science and Food Technology, Institute of Agricultural Sciences, Banaras Hindu University, Varanasi, Uttar Pradesh, India

Joginder Singh Department of Botany, Nagaland University, HQRS: Lumami, Zunheboto, Nagaland, India

Jyoti Singh Department of Food Technology and Nutrition, School of Agriculture, Lovely Professional University, Phagwara, Punjab, India

D. Sowdhanya Department of Food Technology and Nutrition, School of Agriculture, Lovely Professional University, Phagwara, Punjab, India

Shweta Suri Amity Institute of Food Technology, Amity University, Noida, Uttar Pradesh, India

Riaz Ullah Department of Pharmacognosy, College of Pharmacy, King Saud University, Riyadh, Saudi Arabia

Vaishnavi Department of Food Science, Nutrition and Technology, CSK Himachal Pradesh Krishi Vishvavidalaya, Palampur, Himachal Pradesh, India

Hitesh Verma Biofern Life Sciences Private Limited, Kolar, Karnataka, India

Introduction to Nutri-Cereals

1

Jyoti Singh, Prasad Rasane, Vikas Nanda, Sezai Ercisli, and Hitesh Verma

Abstract

The changes in dietary habits, population growth, and the unsustainable exploitation of natural resources exist a deficiency in the capacity to provide adequate and nutritious sustenance for all individuals within the present circumstances. Concurrently, the rapid depletion of natural plant resources underscores the urgency to investigate and develop alternative solutions. Numerous underutilized crops, in addition to the staples wheat and rice, are consumed today and have the potential to completely replace the staples. They possess higher amount of essential micronutrients such as minerals and B-complex vitamins, thereby presenting important role in addressing food security and nutritional well-being within developing countries. Given the growing recognition of millets' significance, the current chapter focuses on the potential contribution that millets might make to promote dietary diversity and balanced diets. It also offers recommendations for using millets to address food and nutrition security issues worldwide. Utilizing millets in conjunction with other major food crops to create food substitutes is a growing area for the food industry. Millets can also enhance immunity and health, which will aid in the fight against childhood and adolescent malnutrition.

J. Singh (✉) · P. Rasane
Department of Food Technology and Nutrition, School of Agriculture, Lovely Professional University, Phagwara, Punjab, India

V. Nanda
Department of Food Engineering and Technology, Sant Longowal Institute of Engineering and Technology, Sangrur, Punjab, India

S. Ercisli
Department of Horticulture, Faculty of Agriculture, Ataturk University, Erzurum, Turkey

H. Verma
Biofern Life Sciences Private Limited, Kolar, Karnataka, India

J. Singh et al. (eds.), *Cereals and Nutraceuticals*,
https://doi.org/10.1007/978-981-97-2542-7_1

Through the process of bio-fortification of staple crops, it can be effectively employed to mitigate the prevalence of micronutrient deficiencies. To add proteins and the aforementioned micronutrients, millets might be incorporated into bread, pastries, cakes, biscuits, and cookies. This could turn contemporary foods into functional foods.

Keywords

Nutri-cereals · Food security · Underutilized crops · Nutrition · Production · Health

1.1 Introduction

Growing concern over the escalating global population and shortage of water resources has fueled a heightened interest in cultivating drought-resistant crops that promote agricultural sustainability and nutritional profiles. In many regions of the world, research projects that aim to boost the output and value of these grains are receiving more funding and interest (Chandra et al. 2016). Alternative crops that are sustainable may be required to meet food demand and to enhance the farmers' financial situation.

Millets grains considered as "Nutri-cereals" or "Dryland cereals" are diminutive grains recognized for their cultivation in the elevated regions of central China around 8000 years prior (Louhar et al. 2020). Their inherent biological and techno-functional properties suggested for their consideration as a fundamental dietary food component (Saleh et al. 2013). It is noteworthy that millets were among the earliest grains consumed during the early stages of human civilization millennia ago, as illustrated by the historical practice in Northern China of crafting noodles from millet flour approximately 4000 years ago (Sharma et al. 2021). Millets are the seven small-seeded warm-season cereal grains, presenting promising prospects as sources of sustenance, animal feed, and fodder across diverse regions. These varieties include finger millet (*Eleusine coracana*), proso millet (*Panicum miliaceum*), pearl millet (*Pennisetum glaucum*), foxtail millet (*Setaria italica*), and barnyard millet (*Echinochloa utilis*) (Saleh et al. 2013; Shahidi and Chandrasekara 2013). Recognized for their sustainability and favorable environmental impact, they have been known as "the cereals of the future" (Sharma and Gujral 2019).

According to the World Health Organization (WHO) and the World Bank (2019), malnutrition is defined as a condition characterized by deficiencies, excesses, or irregularities in an individual's consumption of food and energy. This condition manifests in various adverse effects on newborns, including compromised immune function, susceptibility to chronic diseases, impaired survival rates, and inadequate organ development. As per the latest report from the United Nations Children's Emergency Fund (UNICEF) in 2019, India exhibits a newborn mortality rate of 8.8% and a 69% prevalence of malnutrition among children under the age of five (UNICEF 2019).

1.3.7 Little Millet

Little millet (Gajrao) is rich in protein and approximately 37–38% dietary fiber present (Rao et al. 2018). It is used in versatile food items such as snacks, infant food, and various processed foods. Additionally, it contains apigenin, an antioxidant known for its potential in combating cancer, diabetes, celiac disease, cardiovascular disease, and high cholesterol levels (Thakur and Tiwari 2019). Individuals with wheat intolerance can also find it beneficial. Despite its small size, little millet offers significant nutritional value comparable to other grains. It serves as an excellent source of essential minerals such as calcium, iron, zinc, and potassium, alongside B-vitamins. Moreover, it provides the body with essential fats that facilitate weight loss. Another advantage lies in its high fiber content, rendering it an ideal alternative to rice in dishes like pongal or kheer (Reddy 2017). Germinating tiny millets serve as excellent sources of α-amylase, offering superior purity and specific yield (Usha et al. 2011). Approximately 80% of the phenolic (caffeic, ferulic, and sinapic acids) and flavonoid (kaempferol and luteolin) compounds in little millets are found in the soluble fraction (Pradeep and Sreerama 2018).

1.3.8 Consumption Pattern and Consumer Acceptance

There is little evidence that millets have any impact on food products' sensory acceptability (Kumar et al. 2018). According to some studies, there is literature on both the decreased acceptance and the enhanced acceptability of the items when millets are added. High sensory acceptability was found for cookies made with pearl millet by Florence et al. (2014). Cookies were prepared with sorghum and the attained a sensory acceptability score of 7.1 out of a possible 9.0 points (Okpala et al. 2013). The blending of fermented sorghum flour, germinated pigeon pea flour, and cocoyam flour in the proportions of 66.6:16.7:16.7, the acceptability rating was 7.2. Conversely, diminished acceptability was observed with an increased proportion of sorghum content, as evidenced in a study focusing on extruded products derived from sorghum flour, maize flour, whey protein isolate, and defatted soy flour (Devi et al. 2013). Research suggests that the overall acceptability of such products tends to escalate when millets are amalgamated with other cereals, pulses, or legumes (Deshpande and Poshadri 2011). However, millets face limitations as a standalone food source due to the presence of anti-nutritional elements like phytic acid, tannins, and phenols, alongside considerations regarding their sensory attributes (Devi et al. 2013). Investigations have indicated notably high levels of phytic acid in biscuits containing pearl millet (Okpala et al. 2013). Similarly, studies conducted by Mbithi-Mwikya et al. (2000) have yielded comparable findings in composite mixes comprising unprocessed finger millet, kidney beans, peanuts, and mango puree.

Millets flour can be utilized as an ingredient in food preparations since it contains particular bioactive components as well as other crucial nutrient elements (protein, fiber, and minerals). Millets flour is less suitable than wheat flour due to its gluten-free nature. Chapatti made of millet flour could be prepared in advance by adding hot water, which causes the starch in the grain to partially gelatinize. Thus, the gelatinization process imparts the essential particle binding and aids in rolling the dough to form thin sheets. A hot plate (tawa) is used for baking the flattened sheet of dough (Salar et al. 2017; Siroha et al. 2016). Millets flour can be added to multigrain flour since it contains antioxidants and a number of crucial elements. The final preparation's texture, flavor, and color may all be enhanced by the addition of millet. Products made from millets flour that have been fortified have better sensory qualities and help diabetic patients control their blood glucose levels (Kang et al. 2008; Kulkarni et al. 2018). The pericarp area of the product (papad) made from finger millet grain retains gelatinized starch, giving the food its dark color.

Frying, however, can be used to enhance color. Parboiling could be used to improve the milling of millets (Verma and Patel 2013). Flakes of millet are used to make savory, sweet, and crispy fried foods. Using the extrusion method, vermicelli (noodles), a value-added product made from millet flour, could be produced. People's evolving eating and lifestyle patterns lead to a desire for nutrient-dense food products with a variety of sensory attributes. The expanding expectations of consumers could be satisfied by various items made from millets (sweets, crispy goods, flakes, khichdi, and noodles) (Kumar et al. 2018).

1.3.9 Cultivation Pattern Worldwide

According to USDA Foreign Agricultural Service (2022), India ranked first with 40% of total world millet production (12,500 thousand million tonnes), followed by Niger (11% of world production, 3400 thousand MT), China (9% of world production, 2700 thousand MT), Nigeria (6% of world production, 2000 thousand MT), Mali (6% of world production, 1800 thousand MT), Sudan (5% of world production, 1500 thousand MT), Ethiopia (4% of world production, 1100 thousand MT), Burkina Faso (3% of world production, 100 thousand MT), Senegal (3% of world production, 1000 thousand MT) and Baguirmi (2% of world production, 700 thousand MT) as shown in Fig. 1.1.

The majority of the world's millet is produced in Asia and Africa, which account for approximately 96.69% of the total amount, estimated at 27,572,961 tonnes. The United States (1.18%) and the European Union (1.81%) follow with respective production of approximately 335,698 tonnes and 514,248 tonnes. The subcontinent of Australia only makes up 36,112 tonnes, or 0.127%, of the entire amount generated worldwide. About half of the world's millet is produced by pearl millet, which produces 15 million tonnes. Foxtail millet, proso millet, and finger millet, which produce five million tonnes, four million tonnes, and more than three million tonnes, follow closely behind (USDA 2016–17). Pearl millet is an important crop for food security in many of the world's drought-prone areas.

The majority of millet produced for domestic and commercial use is grown in marginal areas by small-scale farmers. Most millet species have limited or inconsistent statistical documentation. The separate documentation style for millet species

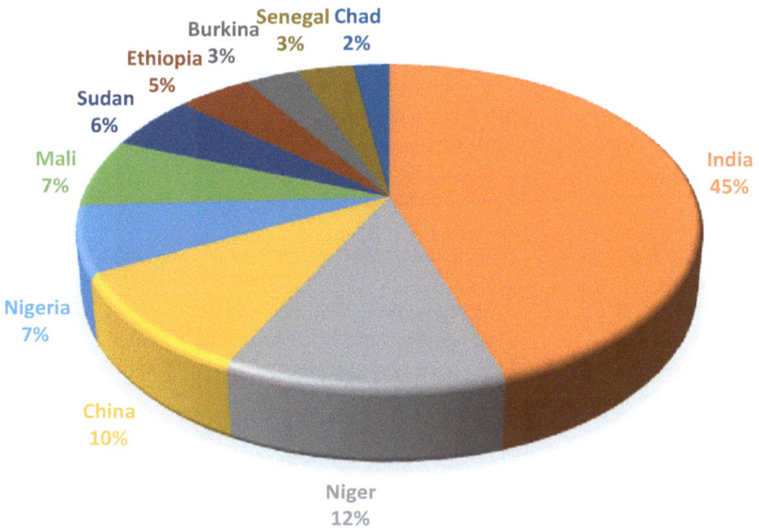

Fig. 1.1 World production of millets. (Source by USDA 2022)

is followed by just a few national and international organizations. Some organizations combine the statistics for millets, sorghum, and cereals, whereas others classify them as "coarse grains." Despite the fact that many of the statistics are merely approximations, studies based on these data should be used with caution. Similar to total global production, millet is cultivated on an estimated 30,697,888 ha in Africa and Asia, with the European Union coming in second with 344,533 ha and America coming in third with 166,260 ha (Chandra et al. 2021).

The analysis of millet production statistics by nation revealed that India, which produces about 40.62% of the world's total millet (estimated at around 11,560,000 t), is far ahead of Niger (13.32%) and China (7.02%), both of which produce about 1,996,856 t. African nations, which comprise eight of the top 10 millet-producing nations, account for 39.30% of the world's total production, estimated to be 11,185,551 t. India has been the world's top millet grower over the past few decades. However, millet output in African nations has significantly expanded recently. According to data analysis for millet cultivation area, the USDA estimates that Nigeria (6,998,776 ha), Niger (6,094,000 ha), and India (9,094,000 ha) are the next-largest producers (2,212,439 ha).

1.3.10 Health Benefits of Nutri-Cereals

The potential health benefits associated with the consumption of millets are showed in Table 1.1, These studies suggest that incorporating millets into one's diet may lead to a reduced risk of heart disease, protection against diabetes, enhanced digestive function, decreased cancer risk, detoxification of the body, strengthened

Table 1.1 Health benefits possessed by nutri-cereals

Name of nutri-cereals	Health-promoting property	Mechanism of action	References
All types of nutri-cereals	Anti-cancerous effect	Antioxidants, b-glucans, phytosterols, and lignins present in millets reduce growth of cancerous lesions	Saini et al. (2021)
Finger millet	Anti-tumorigenic effect	By generating short-chain fatty acids	Chandra et al. (2016)
		Alpha-amylase trypsin inhibitor shows anti-proliferative effect	
All types of nutri-cereals	Cardiovascular disease prevention	Fiber, namely b-glucans, policosanols, phytosterols lowers cholesterol level	Kaur et al. (2020)
All types of nutri-cereals	Anti-diabetic effect	Complex carbohydrates present in millet break down slowly and allow blood sugar to rise slowly	Kumar and Kumar (2015)
		Increased protein and fiber content slows down the rate of digestion of starch	
		Polyphenols prevent the α-amylase and α-glycosidase activity	
All types of nutri-cereals	Immune boosting effect	Arabinoxylans, inulin, and xylooligosaccharides promote growth of gut flora	Prashanth and Muralikrishna (2016), Banerjee et al. (2017), Palaniappan et al. (2017)
Foxtail millet		Bioactive peptides like prolamines provide anti-inflammatory effect by reducing pro-inflammatory cytokines	
	Management of celiac disease	High fiber and no gluten help in eradicating constipation, bloating, cramps, and excessive gas problems	Saleh et al. (2013)

immunity and respiratory functions, as well as defense against various degenerative conditions such as metabolic syndrome and Parkinson's disease (Chandrasekara and Shahidi 2012). The important nutrient and compounds like resistant starch, oligosaccharides, lipids, and antioxidants including phenolic acids, avenanthramides, flavonoids, lignans, and phytosterols are believed to contribute to these health benefits associated with millets (Edge et al. 2005).

Millets have been shown in pharmacological investigations to be possible functional components and to exhibit nutraceutical benefits, mostly due to their antioxidant capacity. Millets contains various bioactive compounds that diminish the risk of several degenerative chronic ailments, encompassing cardiovascular diseases, diabetes, diverse malignancies, hypertension, myocardial infarctions, and tumors

(Thakur and Tiwari 2019). Primarily scrutinized for their hypoglycemic properties, prebiotic attributes, anti-inflammatory traits, and capacity to mitigate cancer and cardiovascular afflictions, millets have garnered considerable research attention.

Millets are the good bio-fortification agents, boasting a notable iron content pivotal for combating anemia. Functioning as a cost-effective functional food, millets possess a significant calcium concentration, thereby mitigating the susceptibility to calcium deficiency among vulnerable demographics such as young children, pregnant women, and lactating mothers (Saini et al. 2021).

1.3.10.1 Anti-cancerous Effect

Millets possess different types of bioactive compounds such as phytosterols, β-glucans, antioxidants, and lignins, which contribute to protection against various types of cancers, including those affecting the breast, prostate, and colorectal regions. Specifically, dietary sitosterol (SIT) present in millets offers protection against chemically induced colon cancer, as evidenced by a notable reduction in tumor formation observed in rats administered with SIT (Awad et al. 1996). Moreover, lignins exert anti-cancer effects by producing short-chain fatty acids such as propionate, acetate, and butyrate, thereby promoting the proliferation of beneficial bifidobacteria. Butyrate, in particular, plays a pivotal role in impeding the progression of cancerous lesions and inhibiting the growth of malignant tumors, thereby exhibiting a secondary chemopreventive effect (Chen et al. 1993). Chronic myeloid leukemia (CML) is characterized by the uncontrolled proliferation of myeloid cells within the bone marrow. Finger millet, as a nutri-cereal, offers promising potential in treating numerous infectious disorders. This efficacy is attributed to the presence of RBI (ragi bifunctional inhibition), a compound found in finger millet seed extract, which acts as a bifunctional agent by simultaneously inhibiting trypsin and alpha-amylase. Consequently, finger millet seed extract demonstrates anti-proliferative effects on CML k562 cells (Chandra et al. 2016).

1.3.10.2 Cardiovascular Diseases

Millets possess high levels of antioxidant properties that play a crucial role in managing dyslipidemia, thereby reducing the risk of coronary heart disease. The presence of apigenin and flavonoids contributes to these antioxidant effects, while the inclusion of dietary fiber, specifically β-glucans, policosanols, and phytosterols, exhibits anti-cholesterolemic properties (Kaur et al. 2012).

Sireesha et al. (2011) demonstrated that aqueous extract from foxtail millet exhibited anti-hyperglycemic and anti-lipidemic effects in streptozotocin-induced diabetic mice. In this study, diabetic rats were administered a dose of 300 mg of aqueous extract per kilogram (kg) of body weight, resulting in a significant reduction (by 70%) in blood glucose levels after 6 h. Moreover, the treated diabetic rats exhibited decreased levels of triglycerides, total LDL (low-density lipoproteins), VLDL (very low-density lipoproteins), and HDL (high-density lipoproteins) cholesterol compared to diabetic rats that did not receive treatment, indicating the hypolipidemic effects of the aqueous extract. Furthermore, Lee et al. (2010) investigated the impact of millet consumption on lipid levels and C-reactive protein concentration. Their findings revealed that hyperlipidemic rats fed foxtail millet experienced reduced levels of triglycerides.

1.3.10.3 Anti-diabetic Effect

The millets are showing beneficial effects on diabetes mellitus. This positive impact is primarily attributed to their high fiber content, which influences carbohydrate digestion and absorption. Soluble dietary fiber, found abundantly in millets, slows down carbohydrate digestion and absorption, potentially reducing the need for insulin. Conversely, dietary insoluble fiber accelerates intestinal transit time, leading to reduced carbohydrate absorption. Millets are rich in complex carbohydrates that break down gradually into simple sugars, resulting in a gradual increase in blood sugar levels (Kumar and Kumar 2015).

Moreover, they exhibit hypoglycemic effect due to their elevated protein and fiber content, which slows down starch digestion and regulates postprandial hyperglycemia. Geetha et al. (2020) found that a diet incorporating millets significantly lowered the glycemic index (below 55) and glycemic load (below 20) compared to other cereals, thereby reducing blood glucose levels in diabetic individuals. Additionally, Vedamanickam et al. (2020) observed that millet consumption positively influenced lipid profiles, systolic and diastolic values, and blood sugar levels in diabetic subjects. The anti-diabetic properties of millets are attributed to natural polyphenols that inhibit the activity of enzymes such as α-amylase and α-glycosidase, thereby delaying carbohydrate digestion and affecting glucose absorption in the intestinal tract (Taslimi and Gulcin 2017). Furthermore, research by Choi et al. (2005) highlighted the significant impact of dietary protein from Korean foxtail millet on improving insulin sensitivity and lipid metabolism, as evidenced by a notable decrease in insulin levels in rats fed with foxtail millet.

1.3.10.4 Immune Boosting Effect

Prebiotic are the substances that play a pivotal role in enhancing health through the modulation of intestinal microbiota, augmentation of immunological functions, and inhibition of detrimental microorganism proliferation. Chronic inflammation has been identified as a significant factor contributing to tumorigenesis (Sharma et al. 2021). Various prebiotic compounds (arabinoxylans, inulin, and xylooligosaccharides) are abundant in the bran and seed coats of diverse millet varieties (Prashanth and Muralikrishna 2016; Banerjee et al. 2017; Palaniappan et al. 2017). Arya and Shakya (2020) explored the prebiotic potential of multi-millet functional beverages and demonstrated that millets possess the capability to foster gut flora growth, as indicated by prebiotic activity scores. Additionally, Palaniappan et al. (2017) revealed that xylooligosaccharides derived from finger millet exhibit potent prebiotic effects on Lactobacillus plantarum and suppress the growth of four pathogenic microorganisms by generating short-chain fatty acids. Phenolic compounds present in millets have been shown to possess anti-inflammatory properties in both in vitro and in vivo studies (Shi et al. 2017). Bioactive peptides from foxtail millet prolamins have been found to attenuate inflammation in murine macrophages, outperforming similar compounds and reducing the production of pro-inflammatory cytokines (Ji et al. 2020). The anti-inflammatory effects of millet grain bioactive peptides, hydrolyzed under controlled in vitro gastrointestinal conditions (Jakubczyk et al. 2019). Furthermore, Hu et al. (2020) demonstrated the anti-inflammatory effects of foxtail millet peptides on RAW264.7 cells, evidenced by the suppression of nitric oxide production and pro-inflammatory cytokines such as interleukin-6 and tumor necrosis factor alpha.

1.3.10.5 Celiac Disease

Celiac disease, an immune-mediated enteropathy triggered by gluten consumption in genetically susceptible individuals, presents a significant health concern (Sarita and Singh 2016). For those with celiac disease and gluten sensitivity, the avoidance of gluten is paramount due to its adverse effects. Millets emerge as a favorable alternative given their gluten-free nature, contrasting with the gluten content found in common cereal grains such as wheat (Saleh et al. 2013). The adoption of millets in the diet not only addresses the dietary restrictions imposed by celiac disease but also offers potential health benefits by regulating the digestive process. This regulation contributes to enhanced nutrient absorption and a reduced risk of severe gastrointestinal disorders such as gastric ulcers or colon cancer. Additionally, the high fiber content in millets aids in alleviating common gastrointestinal discomforts including constipation, excessive gas, bloating, and cramps.

The management of celiac disease often necessitates a significant dietary shift away from gluten-containing cereals. Substituting these grains with gluten-free alternatives like millets, rice, corn, sorghum, amaranth, buckwheat, quinoa, and wild rice can offer relief and nutritional diversity (Ambati and Sucharitha 2019). Millets, particularly notable for their gluten-free composition, hold promise for inclusion in various food products and beverages, thus contributing to the expanding market demand for gluten-free options.

1.3.11 Nutritional Importance of Nutri-Cereals

The recent research findings have delved into enhancing the suitability of foxtail millet, proso millet, and barnyard millet for infant consumption by understanding the molecular characteristics of the waxy starch present in these grains. Comparative genomics, leveraging the genetic similarity of finger millet to cereals, has facilitated the identification of quantitative trait loci and genes pertinent to protein quality. Notably, the transgenic zinc transporter expression has led to the development of high-zinc grains, while transcriptomics has unveiled numerous calcium sensor genes implicated in calcium absorption, translocation, and accumulation in finger millet. To mitigate phytate, anti-nutrient techniques such as RNA interference and genome editing utilizing zinc finger nucleases (ZFNs), transcription activator-like effector nucleases (TALENs), and clustered regularly interspaced short palindromic repeats (CRISPR) are imperative (Singh et al. 2019).

The effects of independent variables including soaking time, temperature and germination time on antioxidant activity and bioactive compounds of foxtail millet using central composite rotatable design were studied. The findings revealed a significant increase in the overall phenolic and flavonoid contents of foxtail millet with prolonged independents variables, followed by antioxidant activity. Optimal combinations of germination variables for producing germinated foxtail millet flour with heightened antioxidant activity (90.5%), total phenol content (45.67 mg GAE/100 g), and total flavonoid content (30.52–43.96 mg RU/g) were identified at a soaking time of 15.84 min and a germination temperature of 25 °C (Sharma et al. 2015). Compared to ungerminated foxtail millet flour, the optimized germinated version exhibited increased yields of sodium (69.45 mg/kg), calcium (25.62 mg/kg), magnesium (107.16 mg/kg), dietary fiber (27.42 g/kg), and protein (14.32 g/100 g)/kg.

In regions like India, China, and Bangladesh, foxtail millet (*Setaria italica*), known as tenai, kakun, and navane, is predominantly cultivated as a rain-fed crop. Reports suggest that foxtail millet offers diverse nutritional and therapeutic advantages and is recommended in Ayurveda and Unani practices (Adekunle 2012). Its non-glutinous and non-acidic nature makes it easily digestible, setting it apart from grains like buckwheat and quinoa (Prashant et al. 2005). Due to their high lysine content, millets hold promise as functional dietary ingredients and supplementary protein sources for various cereals (Mohamed et al. 2009). The reduction of antinutrients like polyphenols and saponins during germination likely enhances mineral bioavailability. Furthermore, germinated foxtail millet has been found to exhibit significant micronutrient concentrations after 3 days, suggesting its potential as a functional meal rich in antioxidant phenolics and flavonoids, beneficial for mitigating oxidative stress-induced non-communicable diseases (Silva et al. 2013).

A recent investigation reported the natural antioxidants present in edible flours of little millets. Total carotenoid content in millet cultivars ranged from 78 to 366 μg/100 g, with finger and proso millets exhibiting higher total tocopherol content compared to foxtail and tiny millets (Asharani et al. 2010). Despite being more commonly used as bird feed in economically developed nations, foxtail millet's physicochemical attributes and health-functional properties, along with processing technologies to enhance these qualities, are under investigation. By employing appropriate processing techniques, anti-nutrients like phytic acid can be minimized, rendering them nearly undetectable. Additionally, millets are noted for their antioxidant properties, low glycemic index, and hypolipidemic characteristics.

With its ability to thrive in challenging terrains and requiring minimal agricultural inputs, foxtail millet has garnered increased research attention and economic interest. It holds promise in improving nutritional and functional food security, especially in the West, by enhancing taste and flavor while retaining its biochemical protective properties (Yano et al. 2017; Wu et al. 2014). Millets can be incorporated into baked goods such as bread, cakes, cookies, and biscuits to augment protein and micronutrient content. Notably, the recipe for chapatti, a popular Indian bread, often incorporates ragi, a common term encompassing all millets. In summary, millets offer substantial potential as functional foods worldwide. Further research is needed to reduce phytate concentration and enhance antioxidant capacity by modulating phenolic and flavonoid levels, crucial for the prevention of non-communicable diseases (NCDs).

1.4 Food Security Issues

Food security is defined as a condition wherein individuals possess both physical and financial means to access nutritious food, thereby fulfilling their dietary requirements (World Food Summit 1996). Hatab et al. (2019) suggest that challenges regarding food security are often simplified to the mere availability of agricultural commodities such as livestock. However, it is argued that these challenges are more complex than merely bolstering supply levels. Numerous elements, including urbanization and accessibility, are among the difficulties mentioned by Hatab et al. (2019). The elements that contribute to food insecurity are also described as institutional failures and the structures and procedures that govern economies and societies (Pangaribowo et al. 2013).

According to Abegaz's (2017) study on the state of food security in Ethiopia, a primary determinant is the initial rain shock induced by climate change. A casual action to prevent food insecurity is not always useful, according to Fraval et al. (2020). Therefore, for interventions to be effective, proxies of interventions and a fuller understanding of the different possible paths are essential. To overcome these intricate obstacles, multisectoral strategies and planning are needed. According to Pangaribowo et al. (2013), combining the important socioeconomic and environmental characteristics of a particular entity with the indicators of food insecurity is the best way to address the issues of food and nutrition security.

Despite the difficulties facing food security in underdeveloped nations, the importance of locally grown and underutilized grains like millet cannot be understated. Conventional grain harvests are insufficient to address these issues (Muthamilarasan et al. 2021). The expansion of small grains like millet could worsen poverty among the underprivileged population, especially in light of recent events like the extraordinary COVID-19 epidemic (Padulosi et al. 2009). Millets are the crops most likely to withstand extreme weather and contribute to the stability of food security in the context of climatic change.

Minor millets (finger, kodo, foxtail, proso, little, and barnyard) have the genetic adaptation to grow well on a range of soils, with varying rainfall patterns, photoperiods, and marginal areas (Padulosi et al. 2009). Because of these qualities, millets can eventually lead to food security in areas with harsh climates by taking the place of staples like wheat and rice. Millet is still considered a neglected agro-biodiversity, despite its potential to improve the agricultural system and food security for the poor in Sub-Saharan Africa (Gari 2020).

1.5 Future Challenges

Millets undergo essential processing stages encompassing primary procedures such as wetting, dehulling, and milling, as well as secondary techniques including fermentation, malting, extrusion, flaking, popping, and roasting. Given their significance as a household staple, both traditional and industrial processing methods must be considered, catering to diverse enterprises ranging from small to large scales (Obilana and Manyasa 2002; Hamad 2012). Due to the diminutive size of millet grains, dehulling proves to be less advantageous and leads to nutrient loss. Grinding millets can be accomplished manually at the household level or through machinery in cottage, small-to-medium scale, and large-scale industrial settings.

Africa has a long heritage of producing malt from sorghum and millet, which is used to make beverages with lactic acid and alcohol fermentation as well as baby food (Adekunle 2012). There are three basic steps in the traditional malting methods used in many developing nations: soaking, germination, and drying. Each operation's time and circumstances are highly variable, which leads to highly variable malt and derives product quality. Traditional African beers like Burukutu and Pito differ from those in the West in several ways, including the fact that they lack hops and are frequently sour and low in carbonation. These beers are both traditional and industrially produced (Anukam and Reid 2009; Amadou et al. 2011).

The development of cookies and confectionery, beverages, weaning foods, and beer are some of the newer industrial raw uses for millets (Laminu et al. 2011; Anukam and Reid 2009). Commonly found in the market are essential commodities such as grits, flour, and various grain-based products including millet, sorghum, and corn meals. Blends of sorghum, maize, and wheat are employed in crafting tender biscuits and cookies, while there has been a notable surge in research concerning cakes and bread alternatives to wheat, yielding promising outcomes in recent times (Akeredolu et al. 2005; Laminu et al. 2011).

Although the industry holds immense potential, its growth has been restrained by the current limited in industrial malting capacity. To mitigate the high cost of imported barley, a considerable proportion of beer varieties marketed in developing countries incorporate indigenous grains such as millet. The number of problems encountered by these sectors often leads to a decline in product quality and adversely affects overall utilization rates.

In the sectors of non-alcoholic beverages and weaning foods, several challenges hinder the broader adoption of millet within developing countries. These challenges encompass concerns such as the preservation quality of storage grains, nutrients degradation post-processing, the prohibitive expense associated with imported machinery, and the limited diversity of cultivars (Akeredolu et al. 2005; Laminu et al. 2011; Adekunle 2012). The soft, readily digestible foods must constantly be introduced to complement the infants early feeding during weaning process.

Conversely, it has been shown that weaning food blends including fermented pearl millet and roasted cowpea in 70:30 and 60:40 ratios exhibit reduced phytic acid content and increased in vitro protein digestibility. In a clinical experiment, it was showed that a process weaning diet consisting of pearl millet-conophor nut flour encouraged development (Akeredolu et al. 2005; Laminu et al. 2011).

1.6 Conclusion

The staple food for millions of poor individuals in Asia and Africa is still millets. Millets are a beneficial complement to a balanced diet since, like many other cereals, they are satisfying and have a high energy content from carbohydrates. If millets are paired with other protein sources, they can compensate for the deficiency of some amino acids, such lysine. Millets may be used to make a wide range of processed meals and food products that have industrial significance because of their accessibility and price. It contain high concentration of bioactive components and potent antioxidants, millet has been demonstrated to offer a multitude of health advantages, including anti-diabetic, anti-aging, antimicrobial, and chronic disease protectant properties. Various health-related problems and chronic disorders may benefit from millet and millet-based fortified functional products. Given the size of the global food business, there is an opportunity to value cereal grain-based functional and processed food items. Expanding the range of uses for millet grains would depend critically on the successful enhancement of these characteristics. Future trends should concentrate on the consumption of millet in industrialized nations, which could contribute to its industrial revolution.

References

Abegaz KH (2017) Determinants of food security: evidence from Ethiopian Rural Household Survey (ERHS) using pooled cross-sectional study. Agric Food Secur 6:70. https://doi.org/10.1186/s40066-017-0153-1

Adekunle AA (2012) Agricultural innovation in sub-Saharan Africa: experiences from multiple-stakeholder approaches. Forum for Agricultural Research in Africa, Ghana

Akeredolu IA, Addo AA, Akeredolu OA (2005) Clinical evaluation of pearl millet Technol conophor weaning mix as supplementary food for Nigerian children. Braz Arch Biol 48(4):531–536

Amadou I, Gbadamosi OS, Guo-Wei L (2011) Millet-based traditional processed foods and beverages—a review. Cereal Food World 56(3):115–121

Ambati K, Sucharitha KV (2019) Millets-review on nutritional profiles and health benefits. Int J Recent Sci Res 10(7):33943–33948

Anukam KC, Reid G (2009) African traditional fermented foods and probiotics. J Med Food 12(6):1177–1184

Arya SS, Shakya NK (2020) High fiber, low glycaemic index (GI) pre-biotic multigrain functional beverage from barnyard, foxtail and kodo millet. LWT 135:109991

Asharani VT, Jayadeep A, Malleshi NG (2010) Natural anti-oxidants in edible flours of selected small millets. Int J Food Prop 13:4150

Awad AB, Chen YC, Fink CS, Hennessey T (1996) Beta sitosterol inhibits HT-29 human colon cancer cell growth and alters membrane lipids. Anticancer Res 16:2797–27804

Awika JM (2011) Major cereal grains production and use around the world. In: Advances in cereal science: implications to food processing and health promotion. American Chemical Society, Washington, DC, pp 1–13

Banerjee D, Chowdhury R, Bhattacharya P (2017) Optimization of extraction process of inulin from Indian millets (jowar, bajra and ragi)—characterization and cost analysis. J Food Sci Technol 54:4302–4314

Bhattacharjee R, Khairwal IS, Bramel PJ, Reddy KN (2007) Establishment of a pearl millet [*Pennisetum glaucum* (L.) R. Br.] core collection based on geographical distribution and quantitative traits. Euphytica 155:35–45

Bukhari MA, Ayub M, Ahmad R, Mubeen K, Waqas R (2011) Impact of different harvesting intervals on growth, forage yield and quality of three pearl millet (*Pennisetum americanum* L.) Cultivars. Int J Agr Vet Med Sci 5(3):307–315. https://doi.org/10.5455/ijavms.20

Chandra D, Chandra S, Sharma AK (2016) Review of finger millet (Eleusine coracana (L.) Gaertn): a power house of health benefiting nutrients. Food Sci Human Wellness 5:149–155

Chandra AK, Chandora R, Sood S, Malhotra N (2021) Global production, demand, and supply. In: Millets and pseudo cereals. Woodhead Publishing, Duxford, pp 7–18

Chandrasekara A, Shahidi F (2012) Bioaccessibility and anti-oxidant potential of millet grain phenolics as affected by simulated in vitro digestion and microbial fermentation. J Funct Foods 4:226–237

Chen F, Cole P, Mi ZB, Xing LY (1993) Corn and wheat-flour consumption and mortality from esophageal cancer in Shanxi, China. Int J Cancer 53:902–906

Chhikara N, Abdulahi B, Munezero C, Kaur R, Singh G, Panghal A (2019) Exploring the nutritional and phytochemical potential of sorghum in food processing for food security. Nutr Food Sci 49:318–322

Choi YY, Osada K, Ito Y, Nagasawa T, Choi MR, Nishizawa N (2005) Effects of dietary protein of Korean foxtail millet on plasma adiponectin, HDL-cholesterol, and insulin levels in genetically type 2 diabetic mice. Biosci Biotechnol Biochem 69(1):31–37

Dayakar RB, Bhaskarachary K, Arlene Christina GD, Sudha Devi G, Tonapi A (2017) Nutritional and health benefits of millets. ICAR Indian Institute of Millets Research (IIMR), Hyderabad

Deshpande HW, Poshadri A (2011) Physical and sensory characteristics of extruded snacks prepared from foxtail millet based composite flours. Int Food Res J 18(2):751–756

Deshpande SS, Mohapatra D, Tripathi MK, Sadvatha RH (2015) Kodo millet-nutritional value and utilization in Indian foods. Cent Inst Agric Eng J Grain Process Storage 2:16–23

Devi NL, Shobha S, Tang X, Shaur SA, Dogan H, Alavi S (2013) Development of protein-rich sorghum-based expanded snacks using extrusion technology. Int J Food Prop 16(2):263–276

Edge MS, Jones JM, Marquart L (2005) A new life for whole grains. J Am Diet Assoc 105(12):1856–1860

Farrell W (2011) Plant guide for billion-dollar grass (*Echinochloa frumentacea*). USDA-Natural Resources Conservation Service Plant Materials Program

Florence SP, Urooj A, Asha MR, Rajiv J (2014) Sensory, physical and nutritional qualities of cookies prepared from pearl millet (*Pennisetum typhoideum*). J Food Process Tech 5(10):1

Fraval S, Yameogo V, Ayantunde A et al (2020) Food security in rural Burkina Faso: the importance of consumption of own-farm sourced food versus purchased food. Agric Food Secur 9:2. https://doi.org/10.1186/s40066-020-0255-z

Gari JA (2020) Review of the African millet diversity. http://www.fao.org/fileadmin/templates/esw/esw_new/documents/Links/publications_other/6_millets.pdf. Accessed 23 Nov 2020

Geetha K, Yankanchi GM, Hulamani S, Hiremath N (2020) Glycemic index of millet based food mix and its effect on pre diabetic subjects. J Food Sci Technol 57:2732–2738

Gupta A, Sood S, Agrawal PK, Bhatt JC (2013) Under-utilized food crops of Himalayan region: utilization and prospective. In: Newer approaches to biotechnology. Narendra Publishing House, Delhi, pp 101–120

Habiyaremye C, Matanguihan JB, D'Alpoim Guedes J, Ganjyal GM, Whiteman MR, Kidwell KK, Murphy KM (2017) Proso millet (*Panicum miliaceum* L.) and its potential for cultivation in the pacific northwest, U.S.: a review. Front Plant Sci 7:1–17. https://doi.org/10.3389/fpls.2016.01961

Hamad SH (2012) The microbial quality of processed date fruits collected from a factory in Al-Hofuf City, Kingdom of Saudi Arabia. Emir J Food Agric 24(2):105–112

Hannaway DB, Larson C (2004) Forage fact sheet: pearl millet (Pennisetum americanum). Oregon State University, Corvallis

Hatab AA, Cavinato MER, Lagerkvist CJ (2019) Urbanization, livestock systems and food security in developing countries: a systematic review of the literature. Food Secur 11(2):279–299

Hu S, Yuan J, Gao J et al (2020) Anti-oxidant and anti-inflammatory potential of peptides derived from the in vitro gastrointestinal digestion of germinated and heat-treated foxtail millet (Setaria italica) proteins. J Agric Food Chem 68:9415–9426

Jakubczyk A, Szymanowska U, Karas M, Złotek U, Kowalczyk D (2019) Potential anti-inflammatory and lipase inhibitory peptides generated by in vitro gastrointestinal hydrolysis of heat treated millet grains. CyTA-J Food 17:324–333

Ji Z, Mao J, Chen S, Mao J (2020) Anti-oxidant and anti-inflammatory activity of peptides from foxtail millet (Setaria italica) prolamins in HaCaT cells and RAW264. 7 murine macrophages. Food Bioscience 36:100636

Kalinova J (2007) Nutritionally important components of Proso millets (*Panicum miliceum* L.). Food 1(1):91–100

Kang RK, Jain R, Mridula D (2008) Impact of indigenous fiber rich premix supplementation on blood glucose levels in diabetics. Am J Food Tech 3(1):50–55

Kannan SM, Thooyavathy RA, Kariyapa RT, Subramanian K, Vijayalakshmi K (2013) Seed production techniques for cereals and millets. In: Vijayalakshmi K (ed) Seed node of the revitalizing rainfed agriculture network Centre for Indian knowledge systems (CIICS). PM Digital Products, Chennai, pp 1–39

Karuppasamy P (2015) Overview on millets. Trends Biosci 8(13):3269–3273

Kaur H, Sharma S (2020) An overviewof Barnyard Millet (Echinochloafrumentacea). J Pharmacogn Phytochem 9:819–822

Kaur KD, Jha A, Sabikhi L, Singh AK (2012) Significance of coarse cereals in health and nutrition: a review. J Food Sci Technol 51(8):1429–1441

Krishnamurthy L, Upadhyaya HD, Gowda CLL, Kashiwagi J, Purushothaman R, Singh S, Vadez V (2014) Large variation for salinity tolerance in the core collection of foxtail millet (*Setaria italica* L.) germplasm. Crop Pasture Sci 65(4):353–361. https://doi.org/10.1071/CP13282

Kulkarni DB, Sakhale BK, Giri NA (2018) A potential review on millet grain processing. Int J Nutr Sci 3(1):1018

Kumar A, Kumar S (2015) Promoting minor millets for food and nutritional security. In: III International symposium on underutilized plant species 1241, August 2015. pp 571–576

Kumar A, Tomer V, Kaur A, Kumar V, Gupta K (2018) Millets: a solution to agrarian and nutritional challenges. Agric Food Secur 7(1):1–15

Laminu HH, Modu S, Numan AI (2011) Production, in vitro protein digestibility, phytate content and acceptability of weaning foods prepared from pearl millet (*Pennisetum typhoideum*) and cowpea (*Vigna unguiculata*). Int J Nutr Metabol 3(9):109–113

Lee SH, Chung IM, Cha YS, Park Y (2010) Millet consumption decreased serum concentration of triglyceride and C-reactive protein but not oxidative status in hyperlipidemic rats. Nutr Res 30(4):290–296

Lin HC, Sheu SY, Sheen LY, Sheu PW, Chiang W, Kuo TF (2020) The gastroprotective effect of the foxtail millet and adlay processing product against stress-induced gastric mucosal lesions in rats. J Tradit Complement Med 10(4):336–344

Louhar G, Bana RS, Kumar V, Kumar H (2020) Nutrient management technologies of millets for higher productivity and nutritional security. Indian J Agric Sci 90:2243–2250

Malik S (2015) Pearl millet-nutritional value and medicinal uses. Int J Adv Res Innov Ideas in Edu 1(3):414–418

Mbithi-Mwikya S, Ooghe W, Van Camp J, Nagundi D, Huyghebaert A (2000) Amino acid profile after sprouting, autoclaving and lactic acid fermentation of finger millet (*Elusine coracana*) and kidney beans (*Phaseolus vulgaris* L.). J Agric Food Chem 48:3081–3085

Mohamed TK, Zhu K, Issoufou A, Fatmata T, Zhou H (2009) Functionality, in vitro digestibility and physico-chemical properties of two varieties of defatted foxtail millet protein concentrates. Int J Mol Sci 10:522438

Muthamilarasan M, Prasad M (2021) Small millets for enduring food security amidst pandemics. Trends Plant Sci 26(1):33–40. https://doi.org/10.1016/j.tplants.2020.08.008

Obilana AB, Manyasa E (2002) Millets. In: Belton PS, Taylor JRN (eds) Pseudo cereals and less common cereals: grain properties and utilization potential. Springer, New York, pp 177–217

Okpala L, Okoli E, Udensi E (2013) Physico-chemical and sensory properties of cookies made from blends of germinated pigeon pea, fermented sorghum, and cocoyam flours. Food Sci Nutr 1(1):8–14

Padulosi S, Mal B, Ravi SB, Gowda J, Gowda KTK, Shanthakumar G, Yenagi N, Dutta M (2009) Food security and climate change: role of plant genetic resources of minor millets. Indian J Plant Genet Resour 22(1):1–16

Palaniappan A, Balasubramaniam VG, Antony U (2017) Pre-biotic potential of xylooligosaccharides derived from finger millet seed coat. Food Biotechnol 31:264–280

Pangaribowo, EH, Gerber, N, Torero, M (2013). Food and nutrition security indicators: a review. ZEF Working Paper No. 108. https://ssrn.com/abstract=2237992 or https://doi.org/10.2139/ssrn.2237992

Pant SR, Irigoyen S, Doust AN, Scholthof KB, Mandadi KK (2016) Setaria: a food crop and translational research model for C4 grasses. Front Plant Sci 7(1885):1–5. https://doi.org/10.3389/fpls.2016.01885

Pradeep PM, Sreerama YN (2018) Phenolic anti-oxidants of foxtail and little millet cultivars and their inhibitory effects on α-amylase and α-glucosidase activities. Food Chem 247:46–55

Prashant SH, Namakkal SR, Chandra TS (2005) Effect of the anti-oxidant properties of millet species on oxidative stress and glycemic status in alloxan-induced rats. Nutr Res 2005(25):110920

Prashanth MR, Muralikrishna G (2016) Evaluation of prebiotic properties of purified arabinoxylans isolated from finger millet (Eleusine coracana, v. Indaf 15) bran. Trends Carbohydr Res 8:50–58

Rao BD, Ananthan R, Hariprasanna K, Bhatt V, Rajeswari K, Sharma S, Tonapi VA (2018) Nutritional and health benefits of nutri cereals. Nutrihub TBI, ICAR_Indian Institute of Millets Research (IIMR), Hyderabad

Reddy OSK (2017) Smart millet and human health. Green Universe Environmental Services Society, Hyderabad

Saini S, Saxena S, Samtiya M, Puniya M, Dhewa T (2021) Potential of underutilized millets as nutri-cereal: an overview. J Food Sci Technol 58(12):4465–4477

Salar RK, Purewal SS, Sandhu KS (2017) Fermented pearl millet (Pennisetum glaucum) with in vitro DNA damage protection activity, bioactive compounds and anti-oxidant potential. Food Res Int 100:204–210

Saleh AS, Zhang Q, Chen J, Shen Q (2013) Millet grains: nutritional quality, processing, and potential health benefits. Compr Rev Food Sci Food Saf 12:281–295

Sarita ES, Singh E (2016) Potential of millets: nutrients composition and health benefits. J Sci Innov Res 5(2):46–50

Satish L, Rathinapriya P, Rency SA, Ceasar SA, Prathibha M, Pandian S et al (2016) Effect of salinity stress on finger millet (*Eleusine coracana* (L.) Gaertn): histochemical and morphological analysis of coleoptile and coleorhizae. Flora-Morphol Distribut Funct Ecol Plants 222:111–120. https://doi.org/10.1016/j.flora.2016.04.006

Shahidi F, Chandrasekara A (2013) Millet grain phenolics and their role in disease risk reduction and health promotion: a review. J Funct Foods 5:570–581

Sharma B, Gujral HS (2019) Influence of nutritional and antinutritional components on dough rheology and in vitro protein and starch digestibility of minor millets. Food Chem 299:125115

Sharma S, Saxena DC, Riar CS (2015) Anti-oxidant activity, total phenolics, flavonoids and antinutritional characteristics of germinated foxtail millet (Setaria italica). Cogent Food Agric 1:1081728. https://doi.org/10.1080/23311932.2015.1081728

Sharma R, Sharma S, Dar BN, Singh B (2021) Millets as potential nutri-cereals: a review of nutrient composition, phytochemical profile and techno-functionality. Int J Food Sci Technol 56(8):3703–3718

Shi J, Shan S, Li H, Song G, Li Z (2017) Anti-inflammatory effects of millet bran derived-bound polyphenols in LPS-induced HT-29 cell via ROS/miR-149/Akt/NF-jB signaling pathway. Oncotarget 8:74582

Silva LR, Pereira MJ, Azevedo J (2013) Glycine max (L.) Merr., Vigna radiata L. and Medicago sativa L. sprouts: a natural source of bioactive compounds. Food Res Int 50:16775. https://doi.org/10.1016/j.foodres.2012.10.025

Singh RB, Khan S, Chauhan AK, Singh M, Jaglan P, Yadav P et al (2019) Millets as functional food, a gift from Asia to Western World. In: The role of functional food security in global health. Academic Press, London, pp 457–468. https://doi.org/10.1016/b978-0-12-813148-0.00027-x

Sireesha Y, Kasetti RB, Nabi SA, Swapna S, Apparao C (2011) Antihyperglycemic and hypolipidemic activities of *Setaria italica* seeds in STZ diabetic rats. Pathophysiology 18:159–164

Siroha AK, Sandhu KS, Kaur M (2016) Physicochemical, functional and anti-oxidant properties of flour from pearl millet varieties grown in India. J Food Meas Charact 10(2):311–318

Taslimi P, Gulcin I (2017) Antidiabetic potential: invitro inhibition effects of some natural phenolic compounds on a-glycosidase and a-amylase enzymes. J Biochem Mol Toxicol 31:e21956

Thakur M, Tiwari P (2019) Millets: the untapped and underutilized nutritious functional foods. Plant Arch 19:875–883

UNICEF (2019) For every child, reimagine. UNICEF Annual Report 2019. UNICEF, New York. https://www.unicef.org/sites/default/files/2020-06/UNICEF-annual-report-2019_1.pdf

United States Department of Agriculture (2016–17) Foreign Agricultural Service

USDA, Foreign Agricultural Service (2022) Millet 2022 world production. https://ipad.fas.usda.gov/cropexplorer/cropview/commodityView.aspx?cropid=0459100#. Assessed 10 July 2022

Usha B, Krishna Veni G, Muni Kumar D, Hemalatha KPJ (2011) Partial characterization of α-amylase from germinating little millets (Panicum sumatrense). J Phytology 3(1):1–8

Vedamanickam R, Anandan P, Bupesh G, Vasanth S (2020) Study of millet and non-millet diet on diabetics and associated metabolic syndrome. Biomedicine 40:55–58

Verma V, Patel S (2013) Value added products from nutri-cereals: finger millet (Eleusine coracana). Emir J Food Agric 25:169–176

WHO, United Nations Children's Fund, World Bank (2019) Levels and trends in child malnutrition. https://www.who.int/nutgrowthdb/jme-2019-key-findings.pdf?ua=1

World Food Summit (1996) The world free from hunger. World Food Summit, Rome

Wu Y, Lin Q, Cui T, Xiao H (2014) Structural and physical properties of starches isolated from six varieties of millet grown in China. Int J Food Prop 17:234460

Yano A, Takakusagi M, Oikawa K, Nakajo S, Sugawara T (2017) Xanthophyll levels in foxtail millet grains according to variety and harvesting time. Plant Prod Sci 20(1):13643

Zhang YY, Lu Y, Zhang B, He L, Liu LL, Wang XC et al (2016) The relationship between the gene SiLCYB related to carotenoid synthesis and the colored formation of foxtail millet. Mol Plant Breed 14:1341–1351

Barley

2

Prasad Rasane, Jyoti Singh, Sawinder Kaur,
Mahendra Gunjal, Vikas Kumar, and Amine Assouguem

Abstract

Barley grains stand as the fourth most significant grain crop around the globe, and it is widely used for the food manufacturing, animal husbandry, and brewing sectors. Barley's distinctive good chemical composition and associated health advantages have attracted growing interest from agricultural and food scientists. It contains a wide range of bioactive phytochemicals with promising potential for promoting health. However, the complete scope of its beneficial nutritional properties remains insufficiently investigated due to its limited use in different types of food products. Thus, there is a need within the food processing industry to innovate and develop distinct, nutritious food items incorporating barley that cater to consumer preferences. It has attracted more attention as a potentially beneficial dietary choice for individuals. It is an excellent source of dietary fiber, especially the kind of functional foods that are noticeably enhanced with beta-glucan, a beneficial component that has several health benefits. Apart from these, highland barley showed unique properties including anti-inflammatory, anticancer, antidiabetic, antibacterial, and anti-obesity qualities, in addition to controlling hyperglycemia and hyperlipidemia. The chapter emphasizes the significant potential of barley to be used as a nutritious and useful food, and it may be instructive for improving how barley is used in food processing.

P. Rasane (✉) · J. Singh · S. Kaur · M. Gunjal
Department of Food Technology and Nutrition, School of Agriculture, Lovely Professional University, Phagwara, Punjab, India

V. Kumar
Department of Food Science and Technology, Punjab Agricultural University, Ludhiana, Punjab, India

A. Assouguem
Department of Plant Protection and Environment, National School of Agriculture, Meknes, Morocco

J. Singh et al. (eds.), *Cereals and Nutraceuticals*,
https://doi.org/10.1007/978-981-97-2542-7_2

Keywords

Barley · Dietary fiber · Health benefits · Processing · Production · Antinutrients

2.1 Introduction

Barley grain ranks as the fourth most significant grain crop globally and has been utilized since antiquity, with ancestral communities initially gathering and utilizing wild barley around 10,000 years ago, and it became a domesticated crop about 7000 years ago (Geng et al. 2022; Dai et al. 2012; Haas et al. 2019). Compared to other cereal crops (rice, maize, and wheat), barley grains are more resistant to salt, drought, and desert climates, which makes it possible to cultivate them in a variety of environmental conditions around the globe. Although barley is used for many things, such as feed and brewing, it is still a staple in some places, including Tibet in China (Sakellariou and Mylona 2020).

The hull-free barley is more valuable in the food processing industry than hulled barley because it is higher in protein, carbs, and β-glucan (Pejcz et al. 2017; Sterna et al. 2017). Referred to locally as "qingke," highland barley cultivated in the Qinghai–Tibet Plateau exhibits distinct growth patterns and chemical compositions, boasting a wealth of bioactive compounds carbohydrates, minerals, vitamins, polyphenols, flavonoids, and β-glucan (Obadi et al. 2021). This distinctive nutritional profile has sparked considerable attention regarding the potential of highland barley as a nutrient-rich dietary option.

Moreover, the exploration of barley grain as a valuable dietary resource has attracted significant interest due to advancements in our understanding of the biochemical makeup of barley grains and their impact on human well-being. The advantageous effects associated with barley consumption are largely linked to its elevated levels of tocols and dietary fiber, such as glucan. A high grain glucan content that is usually close to 10 times that of wheat sets barley apart (Geng et al. 2021). It is well known that eating a diet high in glucan content will bolster the human immune system and protecting against a variety of diseases like type 2 diabetes, hypertension, stroke, and cardiovascular diseases (Maheshwari et al. 2019; Tosh and Bordenave 2020).

Researchers working on barley and similar other cereal crops face the ongoing task of developing varieties rich in beneficial chemical compounds. They aim to produce crops suitable for incorporation into healthy food products items. The improving barley quality leads to the development of better, healthier food from barley. To meet market needs and increase relevant quality, it is crucial to understand the chemical and functional components of barley grains, as well as their genetic regulation and health benefits.

2.2 Morphology of Barley Grain

The plant belongs to the Poaceae family, which is distinguished by the production of fruits with seeds called a caryopsis. Anatomically, the grain is made up of endosperm, embryo, and husk (the tissue surrounding the endosperm, which makes up about 7–12% of the grain size) (Sharma and Kotari 2017). A single cultivar (cv) of barley cannot adapt to all environmental conditions, even though it can be grown under various climatic conditions. As a result, each nation has its breeding program to create a proper cultivar. Barley seeds possess distinguishing characteristics that facilitate classification based on various factors including their growth morphology, the presence or absence of hulls, starch composition, color characteristics of their aleurone layer, their height during growth, and an arrangement of seeds on the stalk, which may occur in either two or six rows (Farag et al. 2022). The barley kernel (grain) is usually 7–12 mm long and spindle-shaped. The cross-sectional structure of the mature barley kernel comprises the rachilla, caryopsis, and hull components.

2.3 Origin and Background

Although the precise origin of barley remains unclear, it is thought to have originated in Egypt, Ethiopia, the Near East, or Tibet. However, compelling evidence suggests the likelihood of multiple sites of origin, potentially beginning in the Iberian Peninsula and extending into North Africa. Nonetheless, it is established that barley was among the earliest grains to undergo domestication. Before 10,000 BC, cultivation of barley was evident in the Middle East, though subsequent adoption in China and India is probable. By 1500–850 BC, barley cultivation had spread to the Korean Peninsula, alongside millet, wheat, and legumes. The development of barley with six rows is believed to have occurred after 6000 BC.

In the ancient Egypt era (3200 BC–30 BC), beer alcoholic beverages (brewed from malt) and barley bread made up a person's entire diet. Barley (*Hordeum vulgare*) has been a fundamental dietary staple food for humans since prehistoric times. Strong evidence of the cultivation's extremely long history comes from its employment in ancient Hindu and Greek religious rituals. There has been some debate on where the origin of cultivated barley is. There has been significant disagreement over the true origin of cultivated barley. The barley grain remnants are found at historic locations in the Middle East's Fertile Crescent. The crop was domesticated from the wild plant *H. spontaneum* around 10,000 years ago.

The comprehensive dataset obtained from various sources, including 367 barley variations collected from the Middle Eastern International Center for Agricultural Research in Dry Areas (ICARDA) in Aleppo, Syria, and comparable national institutions in diverse regions, facilitated archeological interpretations. Consequently, researchers concluded that the Fertile Crescent serves as the origin of barley cultivation. This interpretation corroborates the dominant and widely accepted hypothesis regarding the origins of cultivated barley (Vasan et al. 2014).

Barley emerged as one of the first crops cultivated within the geographical expanse of the Fertile Crescent, denoting a region in Western Asia characterized by favorable water resources, including proximity to the Nile River within northeastern Africa (Badr et al. 2000). It appeared at the same duration as einkorn and emmer wheat. *H. vulgare* ssp. *spontaneum* is found in regions ranging from Tibet in the east to North Africa and Crete in the west. According to some researchers, the previous records of wild barley go back to the Epipaleolithic at Ohalo II, which is located close to the southernmost point of the Sea of Galilee. An estimate of the age of the remains was 8500 BCE. According to scholars claim that Mesopotamia, particularly the Jarmo region of modern-day Iraq, is where the oldest evidence comes from.

2.4 Production and Cultivation Practices

In the late twentieth century, the global barley grain was cultivated approximately 55 million hectares of land producing around 140 million tonnes of barley annually. This crop demonstrates notable versatility and remarkable adaptability across various environmental conditions. The cultivation of barley cultivation occurs in regions unsuitable for thriving cereals (maize and rice), extending into arctic or subarctic zones, with some varieties even reaching near the subtropical regions. The majority of regions with a Mediterranean climate include *Hordeum* species. Additionally, the genus is found in regions with both an oceanic and a continental climate (Robinson et al. 2007). The comparison with other types of small grains, barley grains exhibit comparatively greater resistance to arid temperatures.

Barley ranks among the top four most significant cereals globally based on its production size and cultivated acreage. The world typically harvests over 140 million tonnes (Mt) of barley from a cultivated area spanning less than 50 million hectares, per year (Mha) (FAO 2018). Moreover, barley demonstrates remarkable versatility among the cereals that have developed the fastest genetically to be highly suited to a variety of worldwide climates (Garstang et al. 2011; Ingvordsen et al. 2015). This plant grows in climates ranging from the arctic and subarctic zones to the subtropical region, outside of those where maize, wheat, and rice are native. Its great genetic diversity aids in its capacity to adapt to a variety of environmental conditions. Research is being done to find novel features that may help to enhance the crop's sustainability. The genetic elements that enable synchronizing the plant's vegetative cycle with its environment are primarily responsible for this trait. This advancement facilitated the development of late winter cultivar varieties optimized for harnessing the agricultural capacity of temperate zones and early spring varieties suited for regions characterized by prolonged cold spells and abbreviated spring-summer seasons. The species has been able to adapt to conditions like those in North Africa and the Middle East due to its strong drought resilience.

The growing of barley takes place in a wide range of climates, from wet zones of Europe to parts of South America to desert zones of Africa and Asia (Kling et al. 2004; Tricase et al. 2018a, b). The yield of barley has fluctuated over time, going from 1.39 (in 1960) to 2.99 t/ha, depending on climatic factors, soil qualities, agricultural techniques (such as irrigation), as well as a variety of crops and technical advancements (in 2017). Zimbabwe, New Zealand, Chile, and Switzerland have the highest yields with respective yields of 7.5, 7.05, 6.50, and 6.17 t/ha (USDA 2018). At present, barley grain is employed as feed in industry and serves as a primary dietary component in developing nations. However, malt and beer constitute its most significant application from an economic standpoint. In Ethiopia, as well as the Andean and Himalayan regions, it serves as a significant subsistence crop. Instead, it is utilized in Canada as swine feed (Tricase et al. 2018b).

According to FAO data, today's barley crop is used for 70% of animal feed, 21% for the malting, brewing, and distilling sectors, and 6% for human consumption. A small amount of barley grain is also utilized to make biofuels due to a growing interest in renewable energy (Griffey et al. 2010).

One of the crops that may handle some abiotic stress is barley (*Hordeum vulgare*). Farmers who are motivated by market demands often grow wheat in terrain better suited for barley, although such areas are simple to interchange for barley. While there exists flexibility in reallocating the cultivation land between barley and wheat, the overall allocation of land toward barley cultivation is determined by a multitude of factors beyond mere crop rotation. These factors encompass consumer preferences and market price (Abebaw 2021).

2.5 Chemical Composition

In recent years, there has been a growing interest in utilizing barley more regularly as human food due to its good nutritional value. Recently, there has been a growing interest in utilising barley more regularly as human food due to its good nutritional value and beneficial chemical components such as iron, zinc, and calcium (6, 3.3, and 50 mg/100 g raw material). Apart from these whole grains, barley also has significant amounts of dietary fiber (14.8 g/100 g raw material) (Langridge 2018). According to findings from the European Food Safety Authority (EFSA), the prescribed daily allowance of different types of nutrients for adults is contingent upon gender distinctions. For example, the advised thresholds encompass zinc (8–11 mg), iron (8–18 mg), and calcium (750 mg) (EFSA 2017). Barley exhibits significantly higher protein content (10–20%) compared to corn (9.5%), rice bran (10–16%), and wheat (9.5%). Barley protein is a perfect ingredient for dietary supplements since it also includes important amino acids and has desired functional characteristics (such as flexibility, water-holding capacity, and emulsifying ability) (Wang et al. 2010; Huang 2020). According to studies, barley has a high quantity of dietary fiber, particularly β-glucan (34.41 g/kg), followed by oats (23.5 g/kg), rye (15.6 g/kg), and wheat (5.6 g/kg) on a dry weight basis (Sullivan et al. 2013; Šimic et al. 2017; Menkovska et al. 2017). Two of the many important health benefits of regular beta-glucan consumption are the control of blood glucose and cholesterol levels (Tong et al. 2015; Wang et al. 2017; Velikonja et al. 2019).

Barley grain is distinguished as a significant level of dietary source of antioxidants due to its elevated content of vitamin E isomers (tocotrienols and tocopherols) compared to alternative cereal grains (Panfili et al. 2003). Antioxidants, including vitamin E, play a pivotal role in human nutrition by conferring considerable defense against free radicals, and barley grain particularly stands out among grain crops for encompassing all eight isomers of vitamin E (Granda et al. 2018). The concentrations of vitamin E within barley grains manifest variability contingent upon genotype, offering potential avenues for formulating breeding strategies aimed at augmenting barley's utility for functional food applications (Do et al. 2015; Martínez et al. 2018).

Barley flour has become more popular recently because of its wide range of bioactive compounds. As a result, it has been used in many value-added food products, including bread, cereal for breakfast, and other bakery goods (Mariotti et al. 2014; Collar and Angioloni 2014; Blandino et al. 2015; Narwal et al. 2017). The sprouted barley flour contains a good amount of nutritional and functional components, and it preserved 87% of the initial β-glucan content and exhibited elevated concentrations of ascorbic acid, riboflavin, and phenolic compounds compared to ungerminated barley grains (Rico et al. 2020). The expanding nutrient content and the potential health advantages of barley grains and their by-products might lead to new commercial opportunities for food makers and farmers by creating innovative functional food products.

2.6 Bioactive and Functional Components

In order to cure or prevent illnesses associated with oxidative stress, antioxidants have gained popularity recently (Hurtado-Fernández et al. 2010). The phytoestrogens, vitamin E, and phenolic compounds are the bioactive compounds found in barley that have strong biological activity and can boost the nutritional value of products. When comparison with other cereals grains, barley grains possess the higher γ-tocotrienol, one of the vitamin E isomers (Zielinski et al. 2001). They are a crucial dietary component for both individuals maintaining good health and those coping with metabolic disorders, owing to their rich content of diverse bioactive compound profiles. It is important to note that while cereal crops serve as primary sustenance for humans, their antioxidant properties remain inadequately investigated (Shvachko et al. 2021).

Barley grains contain tocols (tocopherols and tocotrienols), and it is a higher percentage than among all grains (Panfili et al. 2008). Research findings indicate that when both types are ingested, α-tocotrienol exhibits superior antioxidant activity compared to α-tocopherol, although its absorption is less efficient following oral consumption (Do et al. 2015). It contains phenolic compounds, both in bound and free forms. Barley extract has a high concentration of flavanols (catechin, procyanidins, and prodelphinidins) in the free form and phenolic acids (coumaric, sinapic, and ferulic acids) in the bound form present (Abdel-Aal et al. 2012; Rao et al. 2018).

Typically, there are two types of barley: hulled and hull-less (naked). Hulled barley has to be processed in order to change mainly its nutritional and phytochemical makeup before it can be eaten by people (Liu et al. 2009). There is an uneven distribution of the bioactive components in barley grain (endosperm, bran, and germ fractions). β-glucan is mostly present in the inner layers of the grain, whereas tocotrienols and phenolic compounds are concentrated in the outer layers present (Sullivan et al. 2010; Gangopadhyay et al. 2018). The germ of the grain has a very high concentration of tocopherols. Consequently, barley fractionation using abrasion or milling can produce fractions with significant levels of advantageous bioactive (Irakli et al. 2020).

2.7 Antinutrients and Toxic Compounds

Phytic acid and inositol are the forms of phosphorus (P) that are present in various types of cereal grains. It has the ability to bind minerals (iron and zinc), and it is considered as anti-nutrient because it affects the solubility, digestibility, and absorption of these elements. Increased phytase enzyme causes phytate to be hydrolyzed into myo-inositol and inorganic phosphate, which improves nutrition utilization (Farag et al. 2022). A novel variety of barley grains with lower phytic acid content and higher phytase activity was observed (Dai et al. 2011). The QTL assays for phytase and phytase protein indicate that barley's chromosome 5 H has significant quantitative trait loci (QTL) for phytase activity.

Antinutrients may also arise when grains are processed. The roasting of barley to enhance its sensory qualities has been associated with the synthesis of acrylamide chemicals, which are potentially harmful to health and should be avoided at low quantities. According to Schlörmann et al. (2019), varying levels of acrylamide were found in different regions of the barley plant, including the kernels (322 g/kg), thin flakes (586 g/kg), and thick flakes (804 g/kg). The antinutrients found in barley are referred to as trichothecenes and the toxic fungal by-products identified in various barley products contaminated *Fusarium* and *Stachybotrys* with molds. When *fusarium* species of mold infect barley grains, tetracyclic sesquiterpenoid mycotoxins known as trichothecenes are produced. It is important to keep an eye on their concentration because they pose health hazards. The sample sizes of 59 barley and barley products were thoroughly examined to look for different trichothecenes. Deoxynivalenol (DON) emerged as the predominant type A trichothecene in the samples examined, alongside T-2 and HT-2 toxins, as well as type B trichothecenes. Significantly, malting barley showed high concentrations of type A trichothecenes contamination, while winter barley showed high levels of type B trichothecenes contamination. Conversely, dehulled, naked, and pearl barley cultivars demonstrated the lowest levels of mycotoxins, suggesting that processing techniques, such as malting, impact the presence of mycotoxins in barley. Moreover, various mycotoxins, including fumonisin B1, deoxynivalenol (DON), and zearalenone, were detected in both barley and malted barley cultivated in South Africa. In addition, aflatoxin B1 and ochratoxin A were present in barley malt alcohol (Maenetje and Dutton 2007).

2.8 Health Benefits

A barley grain presents several functional bioactive substances (β-glucan, tocols, and resistant starch) that have numerous health positive health effects. Barley β-glucan exhibits promise in enhancing gastrointestinal well-being and mitigating the level of blood glucose alongside cholesterol levels. Tocols have been observed to diminish serum cholesterol levels, whereas resistant starch demonstrates the capacity to promote intestinal health and decrease blood glucose levels.

2.9 Hypoglycemic Effect

Diabetes can be treated with a variety of dietary adjustments, pharmaceuticals, weight-loss strategies, and lifestyle modifications. Various epidemiological research studies have shown that eating barley is linked to a lower risk of diabetes (Adolphe and Fitzpatrick 2014; Benkeblia 2014).

The grains contain a high level of nutritional and fiber content and are considered to be the main contributors to barley's benefits; however, they haven't been found yet. Oxidative stress is widely believed to be a major pathophysiological pathway that eventually results in insulin resistance, −cell dysfunction, poor glucose tolerance, and the establishment of diabetes. Despite the dearth of in vivo studies on the effects of barley, phytochemical compounds on diabetes were studied. By using the phytochemicals found in barley, oxidative stress, a major contributing element to the pathophysiology of diabetes can be reduced (Evans et al. 2003; Mertens et al. 2003; Matsuda and Shimomura 2013). Consequently, the mitigation of oxidative stress through the phytochemicals holds promise in halting the initiation and advancement of both obesity and diabetes (Okarter and Liu 2010).

The barley grains are the good source of phytochemicals (phenolic acids, flavonoids, phytosterols, and tocols), and they are useful in preventing diabetes and obesity. These compounds show the strongest antioxidant properties. However, it is believed that the main compound responsible for cereals' antioxidant advantages particularly for barley is total phenolic acids (Gamel and Abdel-Aal 2012; Adom and Liu 2002). The presence of phenolic acids has been linked to the prevention of chronic illness because they contain unsaturated carboxylic groups and have demonstrated anti-oxidative properties that are on par with or higher than those of catechins (Siddhuraju et al. 2007). Zielinski et al. (2001) reported that in their research, methanolic extracts (80%) derived from whole barley grains exhibited a substantial correlation coefficient (0.96) between the collective phenolic constituents and their anti-oxidative capacities.

In another study, Goupy et al. (1999) investigated the phytochemical components of barley's anti-oxidative properties. The barley grain contains flavonoids, phenolic acids, and tocopherols (α, β, γ), all of which have potent anti-oxidative properties. These compounds exhibit individual antiradical capacities of roughly 4.76, 0.34, and 0.89, respectively, as evaluated by their interaction with 2,2-diphenyl-1-picrylhydrazyl (DPPH). Furthermore, the anti-oxidative properties of various barley flavonoids tested ranged from 16.4% to 65.5%, while that of phenolic acids varied between 12% and 31.2%. The bioactive compound tocopherol showed an anti-oxidative range of 49.1% to 55.2%, determined by its ability to suppress the co-oxidation of β-carotene in a linoleate model system. The antioxidant properties of barley grains result in the inhibition of lipoxygenase activity. The anti-oxidative potential of flavonoids (24.8%), phenolic acids (65.4%), and tocopherols (2.9%) was observed.

The various studies have been carried out to study the effect of isolated β-glucan and whole grain barley in altering diabetes indicators in individuals with diabetes, obesity or overweight, moderately diabetic, and healthy individuals (Chillo et al. 2011; Higa et al. 2019; Liu et al. 2015; Tosh 2013). The majority of these studies additionally enhanced or included grain barley and β-glucan to a variety of meals, such as bread, porridge, tortillas, and pasta, in order to evaluate their impact on glycemic management. Reducing the post-prandial glycemic response is widely supported by the results of these clinical trials and experimental studies (Idehen et al. 2020).

2.10 Anticancer Activity

According to epidemiological research, diet and lifestyle have a role in 20–80% of cancer-related mortality in humans (Stewart and Wild 2015). Dietary components, particularly those that counteract reactive oxygen species, can strengthen the immune system and protect DNA damage, hence lowering the chance of developing cancer (Seifried et al. 2007). It contains various types of bioactive compounds and their derivatives that show anti-oxidative and immunomodulatory properties that have been related to cancer prevention. The majority of research on barley's ability to chemoprevent carcinogenesis has been conducted in vitro, with an emphasis on the disease's moderation and the benefits of barley fiber (β-glucan). Furthermore, phytochemicals present in germinated barley flour (GBF), especially phenolic acids, contribute to positive health benefits (Floridi et al. 2003). According to Kanauchi et al. (2008), the GBF influences the pathophysiology of colon cancer in its early stages and prevents the transformation of hyper-proliferative epithelia. Compared to the control diet, the quantity of GBF accelerated the production of acetate and raised the amount of luminal short-chain fatty acids, mainly butyrate, which is formed in the colon. The colonic mucosa of the F344 rats in the GBF group had aberrant crypt foci expression, -catechin formations, and decreased succinate synthesis.

McIntosh et al. (1993) examined the impact of several diets containing high-fiber sources of foods derived from wheat and barley. This study had shown that consuming insoluble fiber from barley was more effective than consuming soluble fiber-enriched commercial bran from barley and oat in reducing the incidence, load, mass index, and development of dimethyl-induced tumors. The lowest tumor frequency (70%), tumor mass index (1.20), and tumor burden per 10 rats were found in spent barley grain and its by-products obtained from the brewing industry. Some research study indicates that polyphenols may have more health benefits than previously thought (Fraga et al. 2010). There are health benefits such as as enhanced endothelium function, cellular signaling, and vital intestinal defense against undigested polyphenols associated with fiber. Barley may be a good dietary source of antioxidant compounds with antiradical and anti-proliferative activities due to its high content of phenolic compounds (Zhao and Moghadasian 2008). The FA inhibits tumor growth as shown in several studies (Kumar and Pruthi 2014; Murakami et al. 2002). It has also been suggested that it can stop big bowel cancer from developing (Kawabata et al. 2000).bIt is still uncertain if the protective effect of phytochemicals stems from their antioxidant activity or some other mechanism, despite a correlation existing between phenolic compound consumption and a decreased risk of acquiring numerous types of chronic diseases (Idehen et al. 2017).

2.11 Cardiovascular Disease Prevention

In the US, CVD is one of the main causes of death. A total of 229.6 fatalities per 100,000 Americans were linked to CVD in 2011, and one in seven deaths in the US occurred from coronary heart disease alone in 2013 (Mozaffarian et al. 2016). Elevated blood cholesterol levels are considered to have a major effect on the occurrence of cardiovascular diseases. Furthermore, research studies have shown that free radicals and other oxidant compounds are primarily responsible for the bulk of brain disorders linked to CVDs, such as stroke and Alzheimer's disease (Uchida 2000; Brennan and Hazen 2003; Daffu et al. 2013; Stampfer 2006). It has been shown that natural antioxidant compounds (polyphenols) protect against the oxidative processes that these substances trigger, potentially improving cardiovascular health (Ghafoor 2015; Mahmoudi et al. 2015; Lahouar et al. 2014; Frei 2012).

The various strategies are employed to reduce the risk of cardiovascular diseases (CVDs), encompassing measures such as restricting dietary cholesterol consumption and employing pharmacological agents known as statins to lower blood cholesterol levels. The presence of phytosterols in barley grains underscores their significance in promoting cardiovascular well-being. The consumption of barley grains may help manage CVD without the need for medication or special diets. It is a good source of phytonutrients such as phytosterols it can form micelles in the intestinal lumen by competing with cholesterol, which inhibits the absorption of cholesterol and increases secretion and cholesterol regulation. A recent study showed that at doses of 15 mL/day, barley oils have sufficient level of total phytosterols content (0.18–1.44 g/15 g oil) to considerably lower low-density lipoprotein (LDL) cholesterol level (Moreau et al. 2007).

Polyphenols demonstrate a capacity to scavenge free radicals, playing a crucial role in preventing cardiovascular diseases (CVDs). Extracted from black highland barley, these polyphenols exhibit potent antioxidant activity. Administering black highland polyphenol extract to mice at a dosage of 0.6 g/g relative to body weight led to a decrease in their atherosclerosis index, total cholesterol, and LDL cholesterol levels. Furthermore, there was a notable increase of 17.80% in high-density lipoprotein cholesterol levels (Shen et al. 2016). The barley sprout extract affects AMP-activated protein kinase, which is a cellular sensor of energy metabolism and regulates cholesterol metabolism. Mice given barley extract (19.65 mg/g of body weight) exhibited a 24% reduction in intracellular total cholesterol levels and an 18% reduction in free cholesterol levels (Lee et al. 2015).

2.12 Anti-microbial Activity

Various natural plant components, such as dietary spices, have demonstrated promising antibacterial effects in vitro against a variety of foodborne pathogens and spoilage organisms (Mostafa et al. 2018). Highland barley bran extracts containing β-glucan and subjected to chemical modification through carboxymethylation effectively combat *Staphylococcus aureus* (Song et al. 2020). Several studies have indicated that chemical modification of polysaccharides results in unique biological properties (Bezerra et al. 2017; Song et al. 2020). The aqueous extract derived from purple highland barley bran contains elevated levels of anthocyanins, which help inhibit bacterial growth, particularly against *Pseudomonas aeruginosa* PAO1 and *Salmonella enterica* ATCC10398 (Zhang et al. 2020).

Scientific developments in the highland barley active peptides were noted in the current review. The greatest antibacterial efficacy is observed in HB proteins which have undergone extraction and trypsin hydrolysis for a duration of 4 h. While showing minimal or no impact on Gram-negative bacteria, this peptide demonstrates inhibitory effects on Gram-positive bacteria. The barleycin is antimicrobial peptide and holds promising potential for applications in medicine and as a food preservative (Pei et al. 2018). These results suggest that WHB is more effective against bacteria than refined barley. The antibacterial properties of HB peptides have been validated by several in vitro studies.

2.13 Anti-obesity and Anti-dyslipidemic Effect

In various regions worldwide, there exists a clear correlation between obesity and chronic health conditions like diabetes, cardiovascular diseases, and hypertension. The World Health Organization (WHO) recognized obesity as a prevalent chronic condition in its 2004 report (Jin et al. 2017). According to WHO, reducing the intake of fatty and sugary foods while increasing the consumption of whole grains, fruits, and vegetables is recommended to lower the risk of obesity and its associated health-related issues. Recent research indicates that regularly including whole grains in one's diet can significantly reduce the chances of obesity (Koo et al. 2018; Xia et al. 2017). Therefore, it seems prudent to integrate whole grains into your diet as a strategy for managing weight.

2.14 Improve Intestinal Function

Amylose and amylopectin constitute the primary forms of starch, forming semi-crystalline structures characterized by both crystalline and amorphous regions. The crystalline regions within starch granules arise from the linear segments of the amylopectin chain. The digestibility of starch granules is predominantly influenced by their semi-crystalline structure (Perla et al. 2017). Starch, categorized into three distinct types—rapidly digestible, slowly digestible, and resistant starch—based on how quickly it breaks down in the small intestine, has attracted considerable research interest, especially regarding slowly digestible and resistant starch, owing to their advantageous health effects. The gradual breakdown of slowly digestible starch in the small intestine leads to a delayed and consistent release of glucose into the bloodstream, thereby explaining its low glycemic index (Bello-Perez et al. 2020). Instead of being digested in the small intestine, resistant starch makes it to the big intestine (Perla et al. 2017). In the large intestine, resistant starch and β-glucan are fermented to produce short-chain fatty acids (SCFA), particularly butyrate and propionate. These fatty acids have an impact on the intestine by giving epithelial cells, which create a healthy colonic mucosa, a source of energy (Demartino and Cockburn 2020). The fermentation of β-glucan and resistant starch in the large intestine produces short-chain fatty acids (SCFAs), particularly butyrate and propionate. These SCFAs have an effect on the gastrointestinal tract by giving epithelial cells, which are in charge of preserving a healthy colonic mucosa, a crucial source of energy (Demartino and Cockburn 2020).

2.15 Processing

To maximize barley's potential, many researches were conducted to determine the impact of various processing methods on the barley's nutritional content as shown in Fig. 2.1 effect of different processing methods on the barley's nutritional content.

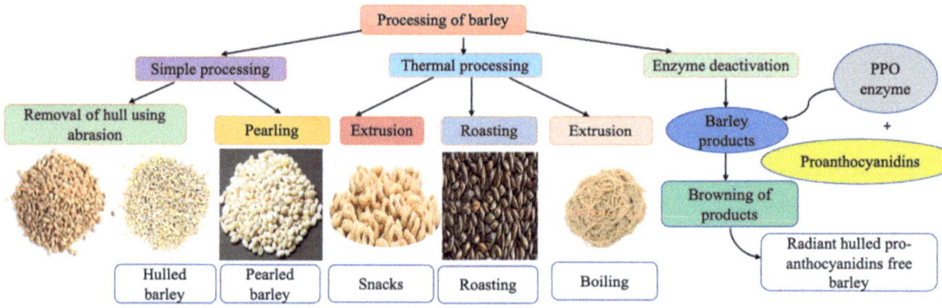

Fig. 2.1 Processing methods for barley

2.15.1 Simple Processing

The most well-known method of processing barley is to remove the hull using abrasion (grains are exposed to a spinning abrasive surface). In all genotypes of barley, protein, ash, total polyphenol content, and polyphenol oxidase (PPO), activity was dramatically reduced by abrasion. Pearling, another well-known processing method involves further removing the leftover hull from dehulled or genetically hull-less grains. The husk and outer layers of the kernel are removed during pearling, the earliest type of processing, to create a consistent grain shape. The amount of total fiber decreased as a result of pearling, whereas the amount of beta-glucan in the kernel rose.

2.15.2 High-Temperature Processing Methods

2.15.2.1 Extrusion Cooking Processing

Extrusion cooking is a high-speed processing technique characterized by the application of high temperature and pressure over a short period. This method is widely used in the production of various processed food items, such as infant foods, snacks, ready-to-eat breakfast cereals, and pet foods. Research conducted by Vasanthan et al. (2002) has demonstrated its influence on the fiber composition of barley. It has been shown to have an impact on the fiber content of barley. The implementation of extrusion technique resulted in an increase in the fiber content of barley flour by converting insoluble dietary fiber (IDF) into soluble dietary fiber (SDF), generating more SDF, and triggering retrogradation of amylose via trans-glycosidation (Vasanthan et al. 2002). Furthermore, it was shown that crushing IDF finely increased its swelling capacity and antioxidant content, possibly improving the health-promoting qualities of barley dietary fiber (Du et al. 2014). This extrusion process affects the starch, enhances dietary fiber solubility, and significantly raises antioxidant and total phenolic content. It was shown that the extrusion process increased free and soluble-conjugated phenolic content by 200–300% in terms of certain phenolic compounds. The most prevalent were vanillic, syringic, and p-coumaric acids, in that order (Wang et al. 2019).

2.15.3 Roasting and Other Thermal Processing

Roasting barley at low or moderate temperatures can enhance the sensory qualities of barley-based products without affecting their protein, fat, carbohydrate, and beta-glucan content (Schlörmann et al. 2019). Maillard reaction products, such as those found in coffee, are said to have improved barely perceptible characteristics. Various thermal processing methods, including micronization, roasting, and conditioning, have been shown to effectively reduce microbial contamination due to their thermal impact, thus potentially boosting the health benefits associated with β-glucan in different barley-based products (Boyd et al. 2017). This highlights the potential of heat treatment to enhance the overall value of barley across various aspects.

The chemopreventive effect on colon cells and other biological consequences of roasting barley flakes at 160–180 °C for 20 min remained unchanged (Schlörmann et al. 2021). Schlörmann et al. (2019) reported that toasting significantly decreased viscosity by 1.9, 2.4, and 2.7 times for kernels, thin flakes, and thick flakes, respectively. Barley would probably be affected by such a drop in viscosity since it is able to absorb nutrients. The exact method of operation is still unclear, but thermal heat treatment of highland barley was found to promote the growth of good bacteria, lower the glycemic index, and reduce insulin resistance in rats fed a high-fat diet (Zheng et al. 2020). It is interesting to note that traditional roasting and boiling process enhanced the phenolic profiles and antioxidant activity of barley (Wang et al. 2019), which may be related to the formation of melanoidins with possible antioxidant benefits (Carvalho et al. 2014; De Paula et al. 2017; Tuersuntuoheti et al. 2020).

Wang et al. (2019) reported that microwave heat treatment decreased total phenolic compounds and flavonoid concentration while having a significant impact on polyphenol oxidase (PPO) activity and non-enzymatic browning. In a recent study, the availability of phenolic chemicals in whole and bran barley was examined about the impacts of steam explosions. Ferulic acid and p-coumaric acid, which constitute the main phenolic acids, were found to be 9 times, 59 times, and 48 times greater in barley bran than in untreated one, respectively. In comparison to water-cooked samples, steam explosion demonstrated a rise in free and soluble-conjugated ferulic acids of 119.8% and 193.0% in whole barley extracts (Wang et al. 2019). Fermentation with lactic acid bacteria also increased to free phenolic components, with ferulic acid showing the greatest percentile increase of 82%. When barley is fermented with *Lactobacillus reuteri*, symbiotic products may be produced that contain increased levels of bioactive substances such c-aminobutyric acid (GABA), 1,3-pro-panediol, and histamine (Wang et al. 2019).

2.16 Enzyme Deactivation

The dark coloring pigmentation observed in food items derived from barley presents a challenge in terms of consumer acceptance and limits the use of barley in various food preparations. Consequently, the deactivation of enzymes becomes imperative in many food products. Dough incorporating barley flour undergoes discoloration due to the presence of phenolic compounds. Despite their relatively low concentrations, the activity of polyphenol oxidase (PPO) in barley leads to increased oxidation of catechin compared to proanthocyanidin (Farag et al. 2022). When barley products are heated, the phenolic compounds, catechin and proanthocyanidin, trigger a browning reaction. Microwave baking has emerged as the optimal method for deactivating PPO enzymes, preserving the nutritional value and functional properties of highland barley protein. Moreover, studies have explored the effects of steaming, frying, and microwave baking on enzyme deactivation.

2.17 Value Addition

Barley, among the earliest and most adaptable of cereal crops, holds a prominent status in human sustenance throughout the annals of history, as corroborated by extensive historical documentation and archeological findings spanning millennia of human evolution. Barley is now being used in an expanding number of products, such as flour, breakfast cereals, functional foods, and beverages, as a result of consumers' growing awareness of the possible nutritional benefits of this grain (Sullivan et al. 2013). The different products formulated from barley are shown in Table 2.1. Before there were any written records, barley powder was used to make flour and other items. The wheat-based flour gradually replaced barley as the main raw component in flour products and barley flour used for the various products such as noodles, bread, steamed buns, biscuits, and tortillas. Noodles made from wheat flour have been proven to become darker and more heterogeneous on the surface when barley bran is added (Izydorczyk et al. 2005). However, because of the quicker cooking times and higher nutritional value, it does provide consumers with convenience.

Various studies have proven that microwave processing and addition of additives improve the quality attributes as weel as the shelf life of the noodles (Tuersuntuoheti et al. 2019). The study also stated that adding 20% barley powder to noodles affects their quality and antioxidant capabilities (Piya et al. 2013). The outcome showed that barley powder noodles had lower hardness and viscosity than noodles made with wheat powder, but they had better antioxidant properties. Furthermore, research study has shown that the sensory attributes of instant noodles are satisfactory with the incorporation of barley powder constituting less than 20%, while the optimal

Table 2.1 Applications of barley on value-added food products

Type of food products	Formulations	Results and findings	References
Barley noodles	Bran of barley into wheat flour	Increase in heterogeneity, decreased cooking time	Izydorczyk et al. (2005)
	Microwave-treated barley noodles	Increased shelf life	Tuersuntuoheti et al. (2019)
	20% of barley powder addition to noodles	Low hardness and viscosity but high antioxidant properties	Piya et al. (2013)
Bread		Increase the amount of dietary fiber	Holtekjølen et al. (2008)
Tortillas	100% barley flour	Low amylose and high glucan content increase consumer acceptability	Ames and Rhymer (2006)
Barley tea		Improve blood fluidity and skin temperature	Ashigai et al. (2018)
Barley porridge		Improve blood glucose levels	Thondre et al. (2012)
Barley wine		Health nourishing effect	Zhang et al. (2013, 2017)

balance in terms of sensory parameters such as color, flavor, and taste is achieved at a 10% incorporation level (Hou and Jimenez 2013).

Barley contains a significant amount of dietary fiber, making barley loaves a healthier food option (Holtekjlen et al. 2008). Consumer acceptance of tortillas made entirely from barley flour (barley cultivars with high glucan content and low amylose concentration) has been demonstrated (Ames and Rhymer 2006). The products prepared from barley have excellent nutritional quality and will appeal to customers who value their health in the future, especially if their unfavorable impacts on appearance and texture can be mitigated. A traditional refreshing beverage called barley tea is popular in China, Japan, Korea, and other nations. Barley teas, also called "Oriental coffee," possess the pleasant aroma. The Chinese medical dictionary Materia Medica has long listed the therapeutic properties of barley tea, stating that it may help with digestion, empty stomach, and eliminate fat. Previous research has suggested that barley tea may increase cutaneous blood flow, improve blood fluidity, and elevate skin temperature (Ashigai et al. 2018). Furthermore, Gas Chromatography-Olfactometry (GC-O) analysis and analytical sensory assessment have been employed to discern the constituents contributing to the aroma and flavor attributes of barley tea (Goto et al. 2019). In another study by Ma and Jia (2014), a method encompassing baking, grinding, and enzymatic hydrolysis of barley was used to formulate a barley beverage emulsion.

The β-glucan can be used as a thickening agent in place of pectin, gum arabic, and alginates in culinary products, including sauces, ice creams, salad dressings, and beverages (Mishra 2020). For such purposes in beverages, barley β-glucan is appropriate since it allows for a soft mouthfeel. The use of 6 g/day of β-glucan in drinks over a 12-week period significantly lowers fasting insulin and the insulin-resistance index in an attempt to lessen the post-prandial response to insulin (Wu et al. 2020). One of the primary reasons for the lower prevalence of cardiac ailments and colon cancer in Tibet, contrary to initial expectations, is attributed to the cultural practice of consuming Tsangpa mixed with barley flour, regarded as a therapeutic medicine (Tashi et al. 2013). A polyphenol found in barley called protocatechualdehyde inhibits HDAC2-mediated cyclin D1 to have anticancer effects on human colorectal cancer cells (Jeong and Lee 2013). Salrok has the highest lunasin content in barley grain (99.0 g/g), while Nonghwa has the lowest (12.7 g/g), and may be significantly involved in the prevention of cancer in populations that consume barley (Zeng et al. 2013). Twenty amino acids necessary for the synthesis of energy, cell growth, and regeneration are present in barley grass (Lahouar et al. 2017; Obadi et al. 2021).

More than 10 millennia have passed since the beginning of barley wine's history (Gous and Fox 2017). Wine fermented with highland barley (HB) as the main raw ingredient is a unique beverage from the Qinghai–Tibet Plateau, which is well known in China for its unique flavor. The HB wine is one of the first known alcoholic drinks in Western China, with a modest alcohol percentage (5–10% vol). The Qingke wine is manufactured from three main ingredients: glacier water from the Himalayas (above 3000 m), HB, a special fermentation starter derived from a pea and HB combination (Qian et al. 2019). Notably, modern techniques have enabled the industrial-scale production of highland barley wine, marking a significant milestone in its journey.

2.18 Future Challenges and Opportunities

The phytochemical composition of barley kernels is distinct and consists of a variety of compounds such as β-glucan, polyphenolic compounds (flavonoids and phenolic acids), polysaccharides (arabinoxylan) and phytosterols (sitosterol and campesterol) as well as tocols (β-tocotrienol, α-tocotrienol, β-tocopherol, and α-tocopherol). Additionally, they serve as a significant reservoir of dietary fiber, vitamins E and B, and an array of minerals including copper, zinc, iron, selenium, and magnesium. Therefore, barley can be used as a supplement to several main cereals, such as wheat and rice, to provide well-balanced nutrition. The unique chemical components of highland in vitro and in vivo results as well as research on chronic diseases have shown that it has significant positive effects on human health. Anti-inflammatory, anticancer, antidiabetic, antibacterial, anti-obesity, antioxidant, anti-fatigue, anti-aging, and cholesterol-lowering activities are among these positive effects.

The different methods used to extract nutrient compounds and the goods they yield enhancement should be thoroughly studied. The product development methods must be applied in order to produce new highland barley food products that make health claims. Further study is needed to fully explore the potential of HB sprouts as a valuable health food that can also be utilized in the production of unique culinary products. Testing and clinical research have not focused much on barley's nutritional significance, and more research is needed to comprehend how nutrient components alter throughout the digestion process.

2.19 Conclusion

Barley stands out among cereals crops due to its considerable bioactive compounds and offering numerous health benefits. The beta-glucan content is noteworthy, as it contributes to mitigating the risk of colon cancer and aids in the regulation of blood glucose levels, cholesterol, and visceral adiposity. Additionally, barley grains contain a notable amount of protein and mineral source, particularly in underdeveloped and rural regions, given their resilience to adverse environmental conditions. Because it may be utilized to precisely meet the needs of a diet that is high in fiber and low in calories, barley has been identified as a desired healthy food. It is generally known that barley grain contains constituent characteristics, nutritional components, and health-promoting components. More research studies should be needed to denouement novel food items that are well-liked by consumers and markets, enhance barley's quality to make it more fit for food processing, and make greater use of these attributes while processing barley.

References

Abdel-Aal ESM, Choo TM, Dhillon S, Rabalski I (2012) Free and bound phenolic acids and total phenolics in black, blue, and yellow barley and their contribution to free radical scavenging capacity. Cereal Chem 89:198–204
Abebaw G (2021) Review on structure, functional and nutritional composition of barley (Hordeum vulgare). J Nutr Food Process 4(2):1–8. https://doi.org/10.31579/2637-8914/046
Adolphe J, Fitzpatrick K (2014) Barley and diabetes. Hemoglobin 1:6
Adom KK, Liu RH (2002) Antioxidant activity of grains. J Agric Food Chem 50:6182–6187
Ames N, Rhymer C (2006) Utilization of diver hulless barley properties to maximize food product quality. Cereal Food World 51(1):23–28. https://doi.org/10.1094/CFW-51-0023
Ashigai H, Taniguchi Y, Matsukura Y, Ikeshima E, Nakashima K, Mizutani M, Yajima H (2018) Roasted barley extract affects blood flow in the rat tail and increases cutaneous blood flow and skin temperature in humans. J Agric Food Chem 66(5):1251–1257. https://doi.org/10.1021/acs.jafc.7b04930
Badr A, Sch R, Rabey HE, Effgen S, Ibrahim HH, Pozzi C et al (2000) On the origin and domestication history of barley (Hordeum vulgare). Mol Biol Evol 17(4):499–510
Bello-Perez LA, Flores-Silva PC, Agama-Acevedo E et al (2020) Starch digestibility: past, present, and future. J Sci Food Agric 100(14):5009–5016
Benkeblia N (ed) (2014) Polysaccharides: natural fibers in food and nutrition. CRC Press, London
Bezerra LS, Magnani M, Castro-Gomez RJH, Cavalcante HC, da Silva TAF, Vieira RLP et al (2017) Modulation of vascular function and anti-aggregation effect induced by (1→ 3)(1→ 6)-β-d-glucan of Saccharomyces cerevisiae and its carboxymethylated derivative in rats. Pharmacol Rep 69(3):448–455. https://doi.org/10.1016/j.pharep.2017.01.002
Blandino M, Locatelli M, Gazzola A, Coïsson JD, Giacosa S, Travaglia F, Bordiga M, Reyneri A, Rolle L, Arlorio M (2015) Hull-less barley pearling fractions: nutritional properties and their effect on the functional and technological quality in bread-making. J Cereal Sci 65:48–56
Boyd L, Holley R, Storsley J, Ames N (2017) Effect of heat treatments on microbial load and associated changes to β-glucan physicochemical properties in whole grain barley. Cereal Chem 94(2):333–340
Brennan ML, Hazen SL (2003) Emerging role of myeloperoxidase and oxidant stress markers in cardiovascular risk assessment. Curr Opin Lipidol 14:353–359
Carvalho DO, Correia E, Lopes L, Guido LF (2014) Further insights into the role of melanoidins on the antioxidant potential of barley malt. Food Chem 160:127–133. https://doi.org/10.1016/j.foodchem.2014.03.074
Chillo S, Ranawana DV, Pratt M, Henry CJK (2011) Glycemic response and glycemic index of semolina spaghetti enriched with barley β-glucan. Nutrition 27(6):653–658
Collar C, Angioloni A (2014) Nutritional and functional performance of high β-glucan barley flours in bread making: mixed breads versus wheat breads. Eur Food Res Technol 238:459–469
Daffu G, del Pozo CH, O'Shea KM, Ananthakrishnan R, Ramasamy R, Schmidt AM (2013) Radical roles for RAGE in the pathogenesis of oxidative stress in cardiovascular diseases and beyond. Int J Mol Sci 14:19891–19910
Dai F, Qiu L, Ye L, Wu D, Zhou M, Zhang G (2011) Identification of a phytase gene in barley (*Hordeum vulgare* L.). PLoS One 6(4):e18829. https://doi.org/10.1371/journal.pone.0018829
Dai F, Nevo E, Wu DZ et al (2012) Tibet is one of the centers of domestication of cultivated barley. Proc Natl Acad Sci U S A 109(42):16969–16973
De Paula R, Rabalski I, Messia MC, Abdel-Aal E-SM, Marcon E (2017) Effect of processing on phenolic acids composition and radical scavenging capacity of barley pasta. Food Res Int 102:136–143. https://doi.org/10.1016/j.foodres.2017.09.088
Demartino P, Cockburn DW (2020) Resistant starch: impact on the gut microbiome and health. Curr Opin Biotechnol 61:66–71
Do TDT, Cozzolino D, Muhlhausler B, Box A, Able AJ (2015) Antioxidant capacity and vitamin E in barley: effect of genotype and storage. Food Chem 187:65–74

Du B, Zhu F, Xu B (2014) Physicochemical and antioxidant properties of dietary fibers from Qingke (hull-less barley) flour as affected by ultrafine grinding. Bioact Carbohydr Diet Fibre 4(2):170–175. https://doi.org/10.1016/j.bcdf.2014.09.003

EFSA (2017) Dietary reference values for nutrients summary report. European Food Safety Authority, Parma

Evans JL, Goldfine ID, Maddux BA, Grodsky GM (2003) Are oxidative stress−activated signaling pathways mediators of insulin resistance and β-cell dysfunction? Diabetes 52:1–8

FAO-FAOSTAT (Food and Agriculture Organization of the United Nations) (2018) Crops [Internet]. www.fao.org/faostat/en/#data/QC. Accessed 20 July 2022

Farag MA, Xiao J, Abdallah HM (2022) Nutritional value of barley cereal and better opportunities for its processing as a value-added food: a comprehensive review. Crit Rev Food Sci Nutr 62(4):1092–1104

Floridi S, Montanari L, Marconi O, Fantozzi P (2003) Determination of free phenolic acids in wort and beer by coulometric array detection. J Agric Food Chem 51:1548–1554

Fraga CG, Galleano M, Verstraeten SV, Oteiza PI (2010) Basic biochemical mechanisms behind the health benefits of polyphenols. Mol Asp Med 31:435–445

Frei B (2012) Natural antioxidants in human health and disease. Academic Press, London

Gamel T, Abdel-Aal ESM (2012) Phenolic acids and antioxidant properties of barley wholegrain and pearling fractions. Agric Food Sci 21:118–131

Gangopadhyay N, Harrison SB, Brunton NP, Hidalgo-Ruiz JL, Gallagher E, Rai DK (2018) Brans of the roller-milled barley fractions rich in polyphenols and health-promoting lipophilic molecules. J Cereal Sci 83:213–221

Garstang JR, Spink JH, Suleimenov M, Schillinger WF, McKenzie RH, Tanake DL, Ceccarelli S, Grando S, Paynter BH, Fettell NA (2011) Cultural practices: focus on major barley-producing regions. In: Ullrich SE (ed) Barley: production, improvement, and uses. Wiley-Blackwell, Oxford, pp 221–281. https://doi.org/10.1002/9780470958636.ch1

Geng L, Li MD, Xie SG et al (2021) Identification of genetic loci and candidate genes related to β-glucan content in barley grain by genome-wide association study in international barley core selected collection. Mol Breed 41(1):6

Geng L, Li M, Zhang G, Ye L (2022) Barley: a potential cereal for producing healthy and functional foods. Food Qual Saf 6:fyac012

Ghafoor K (2015) Optimized extraction of phenolic compounds from barley (*Hordeum vulgare* L.) seed and their radical scavenging properties. J Food Process Preserv 39:793–799

Goto Y, Mio K, Matsuoka T, Izue M, Ayuzawa N, Tokitomo Y (2019) Study on the flavors of barley tea using the flavor wheel and gas chromatography-olfactometry. Nippon Shokuhin Kagaku Kogaku Kaishi 66(9):351–359. https://doi.org/10.3136/nskkk.66.351

Goupy P, Hugues M, Boivin P, Amiot MJ (1999) Antioxidant composition and activity of barley (*Hordeum vulgare*) and malt extracts and of isolated phenolic compounds. J Sci Food Agric 79:1625–1634

Gous PW, Fox GP (2017) Amylopectin synthesis and hydrolysis–understanding isoamylase and limit dextrinase and their impact on starch structure on barley (Hordeum vulgare) quality. Trends Food Sci Technol 62:23–32. https://doi.org/10.1016/j.tifs.2016.11.013

Granda L, Rosero A, BeneŠová K, Pluhácˇková H, Neuwirthová J, Cerkal R (2018) Content of selected vitamins and antioxidants in colored and nonpigmented varieties of quinoa, barley, and wheat grains. J Food Sci 83:2439–2447

Griffey C, Brooks W, Kurantz M, Thomason W, Taylor F, Obert D, Moreau R, Flores R, Sohn M, Hicks K (2010) Grain composition of Virginia winter barley and implications for use in feed, food and biofuels production. J Cereal Sci 51:41–49. https://doi.org/10.1016/j.jcs.2009.09.004

Haas M, Schreiber M, Mascher M (2019) Domestication and crop evolution of wheat and barley: genes, genomics, and future directions. J Integr Plant Biol 61(3):204–225

Higa M, Fuse Y, Miyashita N, Fujitani A, Yamashita K, Ichijo T, Aoe S, Hirose T (2019) Effect of high β-glucan barley on post-prandial blood glucose levels in subjects with normal glucose tolerance: assessment by meal tolerance test and continuous glucose monitoring system. Clin Nutr Res 8(1):55–63

Holtekjølen AK, Baevre AB, Rødbotten M, Berg H, Knutsen SH (2008) Antioxidant properties and sensory profiles of breads containing barley flour. Food Chem 110(2):414–421. https://doi.org/10.1016/j.foodchem.2008.02.054

Hou G, Jimenez V (2013) Development of barley-fortified instant noodles. AACC Int 2:53–54. https://doi.org/10.1094/CPLEX-2013-1001-26B

Huang H (2020) Content analysis of vitamins, dietary fibers and amino acids in a wide collection of barley (*Hordeum vulgare* L.) from Tibet, China. Bioinformation 16:314–322

Hurtado-Fernández E, Gómez-Romero M, Carrasco-Pancorbo A, Fernández-Gutiérrez A (2010) Application and potential of capillary electroseparation methods to determine antioxidant phenolic compounds from plant food material. J Pharm Biomed Anal 53:1130–1160

Idehen E, Tang Y, Sang S (2017) Bioactive phytochemicals in barley. J Food Drug Anal 25(1):148–161

Idehen E, Wang W, Sang S (2020) Health benefits of barley for diabetes. J Food Bioact 12:e12246

Ingvordsen CH, Backes G, Lyngkjær MF, Peltonen-Sainio P, Jensen JD, Jalli M, Jahoore A, Rasmussen M, Mikkelsen TN, Stockmarr A, Jørgensen RB (2015) Significant decrease in yield under future climate conditions: stability and production of 138 spring barley accessions. Eur J Agron 63:105–113

Irakli M, Lazaridou A, Mylonas I, Biliaderis CG (2020) Bioactive components and antioxidant activity distribution in pearling fractions of different Greek barley cultivars. Foods 9(6):783

Izydorczyk MS, Lagassé SL, Hatcher DW, Dexter JE, Rossnagel BG (2005) The enrichment of Asian noodles with fiber-rich fractions derived from roller milling of hull-less barley. J Sci Food Agric 85(12):2094–2104. https://doi.org/10.1002/jsfa.2242

Jeong JB, Lee S-H (2013) Protocatechualdehyde possesses anticancer activity through downregulating cyclin D1 and HDAC2 in human colorectal cancer cells. Biochem Biophys Res Commun 430(1):381–386. https://doi.org/10.1016/j.bbrc.2012.11.018

Jin Q, Yu H, Wang X, Li K, Li P (2017) Effect of the molecular weight of water-soluble chitosan on its fat-/cholesterol-binding capacities and inhibitory activities to pancreatic lipase. PeerJ 5:e3279. https://doi.org/10.7717/peerj.3279

Kanauchi O, Oshima T, Andoh A, Shioya M, Mitsuyama K (2008) Germinated barley foodstuff ameliorates inflammation in mice with colitis through modulation of mucosal immune system. Scand J Gastroenterol 43:1346–1352

Kawabata K, Yamamoto T, Hara A, Shimizu M, Yamada Y, Matsunaga K, Tanaka T, Mori H (2000) Modifying effects of ferulic acid on azoxymethane-induced colon carcinogenesis in F344 rats. Cancer Lett 157:15–21

Kling JG, Hayes PM, Barley USE (2004) Genetics and breeding. In: Wrigley C, Corke H, Walker C (eds) Encyclopedia of grain science. Elsevier, Oxford, pp 27–37

Koo HC, Poh BK, Abd Talib R (2018) The GReat-Child™ trial: a quasi-experimental intervention on whole grains with healthy balanced diet to manage childhood obesity in Kuala Lumpur. Malaysia Nutr 10(2):156

Kumar N, Pruthi V (2014) Potential applications of ferulic acid from natural sources. Biotechnol Rep 4:86–93

Lahouar L, El Arem A, Ghrairi F, Chahdoura H, Salem HB, El Felah M, Achour L (2014) Phytochemical content and antioxidant properties of diverse varieties of whole barley (*Hordeum vulgare* L.) grown in Tunisia. Food Chem 145:578–583

Lahouar L, Ghrairi F, El Arem A, Medimagh S, El Felah M, Salem HB, Achour L (2017) Biochemical composition and nutritional evaluation of barley Rihane (Hordeum vulgare L.). Afr J Tradit Complement Altern Med 14(1):310–317

Langridge P (2018) Economic and academic importance of barley. In: The barley genome. Springer, Berlin

Lee JH, Lee SY, Kim B, Seo WD, Jia Y, Wu C et al (2015) Barley sprout extract containing policosanols and polyphenols regulate AMPK, SREBP2 and ACAT2 activity and cholesterol and glucose metabolism in vitro and in vivo. Food Res Int 72:174–183

Liu K, Barrows FT, Obert D (2009) Dry fractionation methods to produce barley meals varying in protein, beta-glucan, and starch contents. J Food Sci 74:487–499

Liu L, Wang X, Li Y, Sun C (2015) Post-prandial differences in the amino acid and biogenic amines profiles of impaired fasting glucose individuals after intake of highland barley. Nutrients 7(7):5556–5571

Ma XJ, Jia KH (2014) 大麦料乳状液的定性 [Research on stabilization of barley beverage emulsion]. Food Ferment Ind 40(7):235–240. https://doi.org/10.13995/j.cnki.11-1802/ts.2014.07.044

Maenetje PW, Dutton MF (2007) The incidence of fungi and mycotoxins in South African barley and barley products. J Environ Sci Health B 42(2):229–236. https://doi.org/10.1080/03601230601125644

Maheshwari G, Sowrirajan S, Joseph B (2019) β-Glucan, a dietary fiber in effective prevention of lifestyle diseases—an insight. Bioact Carbohydr Diet Fibre 19:100187

Mahmoudi T, Oveisi MR, Jannat B, Behzad M, Hajimahmoodi M, Sadeghi N (2015) Antioxidant activity of Iranian barley grain cultivars and their malts. Afr J Food Sci 9:534–539

Mariotti M, Garofalo C, Aquilanti L, Osimani A, Fongaro L, Tavoletti S, Hager A-S, Clementi F (2014) Barley flour exploitation in sourdough bread-making: a technological, nutritional and sensory evaluation. LWT Food Sci Technol 59:973–980

Martínez M, Motilva MJ, López de las Hazas MC, Romero MP, Vaculova K, Ludwig IA (2018) Phytochemical composition and β-glucan content of barley genotypes from two different geographic origins for human health food production. Food Chem 245:61–70

Matsuda M, Shimomura I (2013) Increased oxidative stress in obesity: implications for metabolic syndrome, diabetes, hypertension, dyslipidemia, atherosclerosis, and cancer. Obes Res Clin Pract 7:e330–e341

McIntosh G, Jorgensen L, Royle P (1993) The potential of an insoluble dietary fiber-rich source from barley to protect from DMH-induced intestinal tumors in rats. Nutr Cancer 19:213–221

Menkovska M, Damjanovski D, Levkov V, Gjorgovska N, Knezevic D, Nikolova N, Stanoev V (2017) Content of B-glucan in cereals grown by organic and conventional farming. Banat J Biotechnol 8:39–47

Mertens A, Verhamme P, Bielicki JK, Phillips MC, Quarck R, Verreth W, Stengel D, Ninio E, Navab M, Mackness B, Mackness M (2003) Increased low-density lipoprotein oxidation and impaired high-density lipoprotein antioxidant defense are associated with increased macrophage homing and atherosclerosis in dyslipidemic obese mice LCAT gene transfer decreases atherosclerosis. Circulation 107:1640–1646

Mishra N (2020) Cereal β Glucan as a functional ingredient. In: Innovations in food technology. Springer, Singapore, pp 109–122. https://doi.org/10.1007/978-98

Moreau RA, Flores RA, Hicks KB (2007) Composition of functional lipids in hulled and hulless barley in fractions obtained by scarification and in barley oil. Cereal Chem 84(1):1–5

Mostafa AA, Al-Askar AA, Almaary KS, Dawoud TM, Sholkamy EN, Bakri MM (2018) Antimicrobial activity of some plant extracts against bacterial strains causing food poisoning diseases. Saudi J Biol Sci 25(2):361–366. https://doi.org/10.1016/j.sjbs.2017.02.004

Mozaffarian D, Benjamin EJ, Go AS, Arnett DK, Blaha MJ, Cusman M, Da SR, de Ferranti S, Després JP, Fullerton HJ, Howard VJ (2016) Executive summary: heart disease and stroke statistics—2016 update a report from the American Heart Association. Circulation 133:447–454

Murakami A, Nakamura Y, Koshimizu K, Takahashi D, Matsumoto K, Hagihara K, Taniguchi H, Nomura E, Hosoda A, Tsuno T, Maruta Y (2002) FA15, a hydrophobic derivative of ferulic acid, suppresses inflammatory responses and skin tumor promotion: comparison with ferulic acid. Cancer Lett 180:121–129

Narwal S, Kumar D, Sheoran S, Verma RPS, Gupta RK (2017) Hulless barley as a promising source to improve the nutritional quality of wheat products. J Food Sci Technol 54:2638–2644

Obadi M, Sun J, Xu B (2021) Highland barley: chemical composition, bioactive compounds, health effects, and applications. Food Res Int 140:110065

Okarter N, Liu RH (2010) Health benefits of whole grain phytochemicals. Crit Rev Food Sci Nutr 50:193–208

Panfili G, Fratianni A, Irano M (2003) Normal phase high-performance liquid chromatography method for the determination of tocopherols and tocotrienols in cereals. J Agric Food Chem 51:3940–3944

Panfili G, Fratianni A, Criscio TD, Marconi E (2008) Tocol and b-glucan levels in barley cultivars and in pearling by-products. Food Chem 107:84–91

Pei J, Feng Z, Ren T, Jin W, Li X, Chen D et al (2018) Selectively screen the antibacterial peptide from the hydrolysates of highland barley. Eng Life Sci 18(1):48–54. https://doi.org/10.1002/elsc.201700118

Pejcz E, Czaja A, Wojciechowicz-Budzisz A et al (2017) The potential of naked barley sourdough to improve the quality and dietary fibre content of barley enriched wheat bread. J Cereal Sci 77:97–101

Perla AMR, Pamela CFI, Luis ABE (2017) Starch structure influences its digestibility: a review. J Food Sci 82(9):2016–2023

Piya MK, McTernan PG, Kumar SJJE (2013) Adipokine inflammation and insulin resistance: the role of glucose, lipids and endotoxin. J Endocrinol 216(1):T1–T15. https://doi.org/10.1530/JOE-12-0498

Qian YL, An Y, Chen S, Qian MC (2019) Characterization of Qingke liquor aroma from Tibet. J Agric Food Chem 67(50):13870–13881. https://doi.org/10.1021/acs.jafc.9b05849

Rao S, Santhakumara AB, Chinkwo KA, Blanchard CL (2018) Q-TOF LC/MS identification and UHPLC-online ABTS antioxidant activity guided mapping of barley polyphenols. Food Chem 266:323–328

Rico D, Peñas E, del Carmen García M, Martínez-Villaluenga C, Rai DK, Birsan RI, Frias J, Martín-Diana AB (2020) Sprouted barley flour as a nutritious and functional ingredient. Foods 9:296

Robinson L, Juttner J, Milligan A, Lahnstein J, Eglinton JK, Evans DE (2007) The identification of a barley haze active protein that influences beer haze stability: cloning and characterisation of the barley SE protein as a barley trypsin inhibitor of the chloroform/methanol type. J Cereal Sci 45:348

Sakellariou M, Mylona PV (2020) New uses for traditional crops: the case of barley biofortification. Agronomy 10(12):1964

Schlörmann W, Zetzmann S, Wiege B, Haase N, Greiling A, Lorkowski S, Dawczynski C, Glei M (2019) Impact of different roasting conditions on chemical composition, sensory quality and physicochemical properties of waxy-barley products. Food Funct 10(9):5436–5445. https://doi.org/10.1039/c9fo01429b

Schlörmann W, Atanasov J, Lorkowski S, Dawczynski C, Glei M (2021) Study on chemopreventive effects of raw and roasted b-glucan-rich waxy winter barley using an in vitro human colon digestion model. Food Funct 12:10615

Seifried HE, Anderson DE, Fisher EI, Milner JA (2007) A review of the interaction among dietary antioxidants and reactive oxygen species. J Nutr Biochem 18:567–579

Sharma P, Kotari S (2017) Barley: impact of processing on physi-cochemical and thermal properties—a review. Food Rev Intl 33(4):359–381. https://doi.org/10.1080/87559129.2016.1175009

Shen Y, Zhang H, Cheng L, Wang L, Qian H, Qi X (2016) In vitro and in vivo antioxidant activity of polyphenols extracted from black highland barley. Food Chem 194:1003–1012

Shvachko NA, Loskutov IG, Semilet TV, Popov VS, Kovaleva ON, Konarev AV (2021) Bioactive components in oat and barley grain as a promising breeding trend for functional food production. Molecules 26(8):2260

Siddhuraju P, Manian S (2007) The antioxidant activity and free radical-scavenging capacity of dietary phenolic extracts from horse gram (Macrotyloma uniflorum (Lam.) Verdc.) seeds. Food Chem 105(3):950–958

Šimic G, Horvat D, Dvojkovic K, Abicic I, Vuletic MV, Tucak M, Lalic A (2017) Evaluation of total phenolic content and antioxidant activity of malting and hulless barley grain and malt extracts. Czech J Food Sci 35:73–78

Song J, Chen H, Wei Y, Liu J (2020) Synthesis of carboxymethylated β-glucan from naked barley bran and its antibacterial activity and mechanism against Staphylococcus aureus. Carbohydr Polym 242:116418. https://doi.org/10.1016/j.carbpol.2020.116418

Stampfer M (2006) Cardiovascular disease and Alzheimer's disease: common links. J Intern Med 260:211–223

Sterna V, Zute S, Jansone I et al (2017) Chemical composition of covered and naked spring barley varieties and their potential for food production. Pol J Food Nutr Sci 67(2):151–158

Stewart B, Wild CP (2015) World cancer report 2014. WHO, Lyon

Sullivan P, O'Flaherty J, Brunton N, Gee VL, Arendt E, Gallagher E (2010) Chemical composition and microstructure of milled barley fractions. Eur Food Res Technol 230:579–595

Sullivan P, Arendt E, Gallagher E (2013) The increasing use of barley and barley byproducts in the production of healthier baked goods. Trends Food Sci Technol 29(2):124–134. https://doi.org/10.1016/j.tifs.2012.10.005

Tashi N, Yawei T, Xingquan Z (2013) Food preparation from hulless barley in Tibet. In: Advance in barley sciences. Springer, Dordrecht, pp 151–158. https://doi.org/10.1007/978-94-007-4682-4_13

Thondre PS, Wang K, Rosenthal AJ, Henry CJ (2012) Glycaemic response to barley porridge varying in dietary fibre content. Br J Nutr 107(5):719–724

Tong L-T, Zhong K, Liu L, Zhou X, Qiu J, Zhou S (2015) Effects of dietary hull-less barley β-glucan on the cholesterol metabolism of hypercholesterolemic hamsters. Food Chem 169:344–349

Tosh S (2013) Review of human studies investigating the post-prandial blood-glucose lowering ability of oat and barley food products. Eur J Clin Nutr 67(4):310–317

Tosh SM, Bordenave N (2020) Emerging science on benefits of whole grain oat and barley and their soluble dietary fibers for heart health, glycemic response, and gut microbiota. Nutr Rev 78:13–20

Tricase C, Lamonaca E, Ingrao C, Bacenetti J, Giudice AL (2018a) A comparative life cycle assessment between organic and conventional barley cultivation for sustainable agriculture pathways. J Clean Prod 172:3747–3759. https://doi.org/10.1016/j.jclepro.2017.07.008

Tricase C, Amicarelli V, Lamonaca E, Rana RL (2018b) Economic analysis of the barley market and related uses. Grasses Food Feed 10:25–46

Tuersuntuoheti T, Wang Z, Zheng Y, Wang S, Wang Z, Wu Y, Liang S, Li X, Zhang M (2019) Study on the shelf life and quality characteristics of highland barley fresh noodles as affected by microwave treatment and food preservatives. Food Sci Nutr 7(9):2958–2967. https://doi.org/10.1002/fsn3.1151

Tuersuntuoheti T, Wang Z, Duan M, Asimi S, Ren X, Wang Z, Zheng Y, Wu Y, Liang S, Zhang M (2020) Noodle processing, storage time and cooking affect the antioxidant activities and phenolic compounds content of Qingke barley noodles. Int J Food Sci Technol 55(7):2730–2739. https://doi.org/10.1111/ijfs.14526

Uchida K (2000) Role of reactive aldehyde in cardiovascular diseases. Free Radic Biol Med 28:1685–1696

USDA/FAS (United States Department of Agriculture, Foreign Agricultural Systems) (2018) Market and Trade Data/PSD Online/Custom Query [Internet]. https://apps.fas.usda.gov/psdonline/app/index.html#/app/advQuery. Accessed 20 July 2022

Vasan A, Mani M, Boora P (2014) Barley foods and health: opportunities ahead. In: Proceedings of the 2014 international conference on intelligent agriculture (IPCBEE), vol 63. IACST Press, Singapore, pp 88–93

Vasanthan T, Gaosong J, Yeung J, Li J (2002) Dietary fiber profile of barley flour as affected by extrusion cooking. Food Chem 77(1):35–40. https://doi.org/10.1016/S0308-8146(01)00318-1

Velikonja A, LipoglavŠek L, Zorec M, Orel R, AvguŠtin G (2019) Alterations in gut microbiota composition and metabolic parameters after dietary intervention with barley beta glucans in patients with high risk for metabolic syndrome development. Anaerobe 55:67–77

Wang C, Tian Z, Chen L, Temelli F, Liu H, Wang Y (2010) Functionality of barley proteins extracted and fractionated by alkaline and alcohol methods. Cereal Chem 87:597–606

Wang Y, Harding SV, Thandapilly SJ, Tosh SM, Jones PJH, Ames NP (2017) Barley β-glucan reduces blood cholesterol levels via interrupting bile acid metabolism. Br J Nutr 118:822–829

Wang J, Sun B, Cao R (2019) Bioactive factors and processing technology for cereal foods. Springer, Singapore

Wu W, Qiu J, Wang A, Li Z (2020) Impact of whole cereals and processing on type 2 diabetes mellitus: a review. Crit Rev Food Sci Nutr 60(9):1447–1474. https://doi.org/10.1080/10408398.2019.1574708

Xia X, Li G, Ding Y, Ren T, Zheng J, Kan J (2017) Effect of whole grain Qingke (Tibetan Hordeum vulgare L. Zangqing 320) on the serum lipid levels and intestinal microbiota of rats under high-fat diet. J Agric Food Chem 65(13):2686–2693. https://doi.org/10.1021/acs.jafc.6b05641

Zeng Y-W, Yang J-Z, Pu X-Y, Du J, Yang T, Yang S-M, Zhu W-H (2013) Strategies of functional food for cancer prevention in human beings. Asian Pac J Cancer Prev 14(3):1585–1592. https://doi.org/10.7314/APJCP.2013.14.3.1592

Zhang KZ, Deng K, Luo HB, Zhou J, Wu ZY, Zhang WX (2013) Antioxidant properties and phenolic profiles of four Chinese Za wines produced from hull-less barley or maize. J Inst Brew 119(3):182–190. https://doi.org/10.1002/jib.81

Zhang K, Pan Y, Zou W, Zhou L, Wu Z, Zhang W (2017) Nutritive assessment of amino acids for three Chinese Zajius produced from hull-less barley. J Inst Brew 123(4):587–593. https://doi.org/10.1002/jib.464

Zhang Y, Lin Y, Huang L, Tekliye M, Rasheed HA, Dong M (2020) Composition, antioxidant, and anti-biofilm activity of anthocyanin-rich aqueous extract from purple highland barley bran. LWT-Food Sci Technol 125:109181. https://doi.org/10.1016/j.lwt.2020.109181

Zhao Z, Moghadasian MH (2008) Chemistry, natural sources, dietary intake and pharmacokinetic properties of ferulic acid: a review. Food Chem 109:691–702

Zheng B, Zhong S, Tang Y, Chen L (2020) Understanding the nutritional functions of thermally-processed whole grain highland barley in vitro and in vivo. Food Chem 310:125979. https://doi.org/10.1016/j.foodchem.2019.125979

Zielinski H, Ciska E, Kozlowska H (2001) The cereal grains: focus on vitamin E. Czech Food Sci 19:182–188

Maize

3

Amanjyoti, Jyoti Singh, D. Sowdhanya, Prasad Rasane, Joginder Singh, Sezai Ercisli, Hitesh Verma, and Riaz Ullah

Abstract

Maize (*Zea mays* subsp. *mays*) an ancient tetraploid is inherited genes from its wild predecessor, teosinte (*Zea mays* ssp. *parviglumis*) and is a few of the major cultivated crop. Maize is symbolic and culturally significant in many civilizations, frequently being used in customs and ceremonies. Because of the genetic diversity seen in maize, numerous varieties that can flourish in a range of climates and environments have been developed. They also increase nutrient contents, and maize plants can improve the fertility of the soil. For many people worldwide, maize is a staple grain that makes up a sizable amount of their daily caloric intake. Maize contains several antioxidants such as lutein, zeaxanthin, and carotenoids, which are required for maintenance of eye health and reduce the oxidative stress. Moreover, maize is a good source of vitamins and minerals which are important for immune system maintenance, bone health, and neuron function. The production of meat, eggs, and dairy products is supported by the use of maize in animal feed. It is used as a raw material in many different indus-

Amanjyoti · J. Singh (✉) · D. Sowdhanya · P. Rasane
Department of Food Technology and Nutrition, School of Agriculture, Lovely Professional University, Phagwara, Punjab, India

J. Singh
Department of Botany, Nagaland University, HQRS: Lumami, Zunheboto, Nagaland, India

S. Ercisli
Department of Horticulture, Faculty of Agriculture, Ataturk University, Erzurum, Turkey

H. Verma
Biofern Life Sciences Private Limited, Kolar, Karnataka, India

R. Ullah
Medicinal Aromatic and Poisonous Plants Research Center College of Pharmacy King Saud University, Riyadh, Saudi Arabia

J. Singh et al. (eds.), *Cereals and Nutraceuticals*,
https://doi.org/10.1007/978-981-97-2542-7_3

tries, such as the manufacturing of biofuels, corn syrup, and corn oil. With its ability to be processed into many forms like cornmeal, cornflakes and corn flour, maize is a highly adaptable ingredient in culinary arts. It is also been used in several industry for the manufacturing of paper, biopolymers, biofuels from the waste produced by its process for the effective and ecofriendly utilization of the maize by-products. This chapter overall focuses on the consumption, production pattern, nutrient profile, pharmacological benefits and its application in the industry.

Keywords

Maize · Functional · Nutritional · Application · Food industry · Antioxidants

3.1 Origin and Background

Maize (*Zea mays* subsp. *mays*) is an ancient tetraploid crop which has a gene flow from the wild ancestor crop, teosinte (*Zea mays* ssp. *Parviglumis*) (Huang et al. 2021). The progenitor of maize is configured with a dense structure called "fruitcase." The teosinte glume architecture1 (tga1) plays a chief role in maize domestication in converting inedible capsuled grain (teosinte) to edible food grain. The amino acid substitution of Lys6-to-Asn6 directs TGA1 in controlling the morphological growth of hardened fruitcase structure which results in the development of grain out of the enclosure (Wang et al. 2015). The factors such as biotic interactions, human culture, human health, colonization, competition between genes, and transformation of gene allele from wild crop cause occurrence of domestication. Due to their agronomic performance and highly adaptive nature under various environments, the maize plant was denoted as human generated "monstrosity" (Stitzer and Ross-Ibarra 2018). An earlier domestication was observed in the Balsas region of southwest Mexico about 9000 years ago. The domesticated modern maize comprises multiple gene flow from teosinte found in various geographical regions (Ramos-Madrigal et al. 2016). After the domestication, maize crop has propagated to America after the discovery of Columbus. From America, it got distributed toward the southwestern USA about 4500 years ago and to the coastal South America around 6700 years ago (Wang et al. 2017). Based on the genetic, historic, and geographical information, the spread of maize landraces got initiated from Mexico, and diverged to Mexican lowlands and highlands to America Northern Flints. The routeway of maize distribution passed over Caribbean Islands to Spain, Northern South America, and Southern South America. It even got routed down toward Italy and ends up in the European Northern Flint (Brandenburg et al. 2017). There is historical evidence related to the consumption and distribution of maize widely across the world. The maize crop caught people's eye when it was used to prepare bread instead of cassava flour to extend its shelf life. The maize bread was called "magic pan" in Spanish by Gonzalo Fernández de Oviedo and it was one of the reasons for the spread of maize landraces from the Caribbean Islands to Spain followed by the Mediterranean area by the sixteenth century and got terminated in Europea (Revilla et al. 2022).

According to Bonavia (2013), the word "maize" is believed to have originated from the ancient word "mahiz" from the extinct Arawakan language, Taino, which was spoken by the native inhabitants of pre-Hispanic America. Maize was the most treasured crop among the Mayan, Aztec, and Olmec civilizations of ancient Mexico, according to archeological evidence (Bonavia 2013). The history of maize cultivation and origin in Mexico has contributed significantly to the region's ongoing cultural identity. Numerous artifacts from Central American indigenous societies illustrate the significance of maize through symbolism and incorporation into jewelry, works of art, and other handicrafts. In religious murals and creation tales all over Latin America, maize is depicted (Tanumihardjo et al. 2020).

Throughout the Americas, maize naturally traveled on trade routes and ingratiated itself into local civilizations (Bedoya et al. 2017). As one of the "Three Sisters"—maize, squash, and beans—it developed into a significant food crop for Native Americans (Mt. Pleasant 2016). Throughout South America, maize spreads and is now a common ingredient in many regional foods, including tamales and arepas in Colombia and Venezuela, indio viejo and guirilas in Nicaragua, humitas in Bolivia and Chile, and locro in Argentina. Maize likely arrived in Africa during the 1500s thanks to the Portuguese sailors' mid-trading route across the Atlantic. Maize was also brought to Asia by European traders (Ranum et al. 2014).

3.2 Production and Cultivation Practices

At the global scale, *Zea mays* (maize) ranks as the third most consequential crop, based on harvested land area. It exhibits medium to optimal climatic suitability across diverse regions, encompassing the eastern United States, Mexico, Brazil, Argentina, several other South American nations, Central and Southern Africa, substantial portions of Europe, Southern India, Eastern China, and the coastlines of Australia. Notably, over 60% of the African continent possesses varying degrees of climatic suitability for maize cultivation, while this proportion rises to 77% in South America. Europe emerges as the region with the most extensive climatic suitability for maize farming, covering nearly 82% of its landmass. Conversely, projections indicate a contraction in the areas conducive to maize cultivation in Africa, with notable shifts from medium to marginal climatic suitability in Angola, Zambia, Mozambique, and the Congo (Ramirez-Cabral et al. 2017).

India's diverse agro-climates support adaptable maize cultivation. Maize production in India requires 500 to 600 mm of annual rainfall, dependent on variety duration. Dry-season maize areas are expanding due to higher productivity compared to wet-season areas, providing quality fodder for livestock in both dry and green forms (Mageto et al. 2017). Eastern India experiences notable rainfall variability and a pronounced incidence of water shortages, leading to the manifestation of Water Deficit Stress (WDS) affecting crop cultivation during both the wet and dry seasons. Meteorological data revealed a heightened occurrence of WDS, particularly between the months of August and September during the wet season and subsequent to the initial fortnight of February in the dry season (Sah et al. 2020).

Irrespective of geographical location, the adoption of reduced tillage with subsurface application of fertilizers resulted in a significant reduction in CO_2 emissions when contrasted with the conventional agricultural system. Specifically, this reduction amounted to 35% in the Żelazna locale and a substantial 52% decrease in emissions in the Czesławice area. These findings underscore the environmental benefits associated with the implementation of reduced tillage practices in maize cultivation in mitigating soil-derived CO_2 emissions (Rutkowska et al. 2018).

During the fallow season, two cover crops, hairy vetch and barley, with different C/N ratios, were grown in different green manure treatments. Their aboveground biomass was integrated into the soil before transplanting maize seedlings. Maize was grown without the use of chemical fertilization in the green manure treatments, while NPK fertilizers were used in a separate treatment. The use of plastic film mulching significantly increased grain productivity by 8% to 33% compared to no-mulching conditions (Cuello et al. 2015).

The anticipated rise in transporting high-value cereal crop straw for animal feed and bedding necessitates the increased use of locally available vegetative materials such as *Sesbania, Brassica, Jatropha,* and *Leucaena* for ex situ mulching. Notably, these mulching materials have exhibited significant positive effects on maize grain yields, with *Sesbania* leading at 18.3%, followed by *Brassica* at 17.2%, *Jatropha* at 13.7%, and *Leucaena* at 9.2%. Moreover, system productivity gains, ranging from 10.7% to 13.1%, have been observed with the use of *Sesbania*, *Jatropha,* and *Brassica* mulching compared to no-mulch conditions (Jat et al. 2015).

The biolog EcoPlates method is a widely accepted tool renowned for its sensitivity in discerning viable microorganisms and providing a distinctive microbial community composition fingerprint. Notably, within the surface layer, microbial communities exhibit a more favorable profile in no-tillage systems with crop residue mulch when compared to plough-tillage systems lacking residue mulch (Gałązka et al. 2017).

In the realm of agriculture, nitrogen (N) application techniques, such as soil versus foliar methods, significantly influence crop productivity. Increased N fertilizer application, up to 300 kg/ha (equivalent to 4.8 g N/pot), enhances crop growth rate and total dry matter production. The crucial agronomic parameters, including grain per cob, 1000-grain weight, and final grain yield, exhibit superior performance under soil N application, particularly at 100% field capacity with 300 kg N/ha, as compared to foliar N application methods (Hammad et al. 2018).

In the context of maize cultivation, achieving potential yields of up to 7.0 Mg/ha is feasible in Hyderabad when employing single-cross/double-cross hybrids like DHM-11, DHM-117, and implementing balanced fertilization, along with soil and moisture conservation practices. Additionally, in Arjia, maize grain yields as high as 8 Mg/ha have been reported through the adoption of High-Yielding Varieties (HYVs) such as HM10, PM3, P3501, and the implementation of Resource Management Practices (RMPs) (Rao et al. 2015).

It was observed that the integrated approach of reduced tillage coupled with maize-cowpea intercropping, followed by the retention of mustard residue, demonstrated heightened system productivity and enhanced net benefits. Intercropping, as a pivotal component, markedly amplifies total system productivity, attributable not only to cowpea's role as an additional concurrent crop with maize but also due to its elevated market value, which is 1.5 to 2.0 times that of maize. Significantly, the adoption of minimum tillage practices in conjunction with maize-cowpea intercropping yields the highest recorded soil quality index, substantiating its profound influence on soil health and overall quality (Pradhan et al. 2018).

Augmenting plant density represents an efficacious strategy to amplify maize yield. Nevertheless, lodging emerges as a significant impediment hindering the realization of future maize yield increments, particularly in the context of high-density planting. Elevated maize plant density elicits beneficial effects on biomass and yield, primarily by accelerating the rate of internode elongation while diminishing the duration of internode thickening. This phenomenon results in elongated internodes with reduced diameters, ultimately contributing to enhanced maize productivity (Jun et al. 2017).

The implementation of the Ridge-Furrow Planting with Fertilizer Management (RFPFM) technique resulted in a notable increase in maize yield in the year 2013 and a substantial increase in the year 2014, when considering both cultivars in combination. The elevated grain yield observed in the context of RFPFM was consistently associated with greater kernel count per ear and increased grain weight, rather than a change in the harvest index (Li et al. 2017).

The rapid emergence of the Rice Maize System (RMS) in South Asia is primarily driven by its potential to increase farm profitability. RMS offers cost-saving opportunities by shifting from conventional tillage to reduced or zero-till systems. This transition leads to improved soil conditions, resulting in increased crop productivity, particularly in maize, and higher profitability when compared to traditional rice and maize cultivation practices (Singh et al. 2016).

In maize cultivation, there has been a consistent rise in average plant density observed in countries such as the United States, Canada, Brazil, China, and other significant corn-producing nations. For instance, Minnesota recorded an increase of approximately 19,000 plants per hectare between 1930 and 1970, with plant density contributing significantly, at 21%, to the overall yield improvement. Similarly, Kansas reported a continual upward trend in plant density, registering an annual increase of 597 plants per hectare between 1939 and 2009 (Assefa et al. 2018).

Global maize production averages 1127 million metric tons annually as per 2016–18, OECD/FAO, 2019. It is adaptable to various agro-ecologies with varying temperatures, altitudes, latitudes, land, and soil types, resulting in differing yields per hectare. North America is the primary maize producer, with Asia, particularly East Asia, following closely (Grote et al. 2021).

The emerging trend in precision maize planters involves increased automation and intelligent technologies for improved planting precision and user-friendliness. Variable planters can now adjust seeding populations based on varying soil fertility, optimizing yield potential and seed conservation. In hilly and mountainous terrain, maize planters should be lightweight, compact, and suitable for low-power tractors due to small farms and rough terrain. Medium-sized farms require precision and efficiency, necessitating middle-sized planters with high-speed, precise seeding capabilities. In double cropping regions with wheat residue, anti-blocking mechanisms are crucial components of maize planters (Li et al. 2016).

3.3 Consumption Pattern

Maize grain, recognized as Zea mays, ranks as the third most widely consumed cereal for human sustenance, following rice and wheat, encompassing both direct pathways and various forms of processing, either in their raw or treated states. The global human consumption of maize stands at 18.5 kg/capita annually, constituting 11% of the average annual cereal consumption of 175 kg/capita worldwide during the period of 2014–2018, excluding beverage-derived cereals, as reported by the Food and Agriculture Organization (FAO Stat 2021). The utilization of maize as a primary food source is notably more prevalent in low- and lower-middle-income countries (L/LM-ICs), where it accounts for 43% of total maize production earmarked for human consumption. Africa, in particular, showcases a robust affinity for maize in its dietary landscape, with maize representing 54% of the total cereal consumption in the continent. This predilection is especially pronounced in eastern and southern Africa, where maize comprises 66% of the cereal-based diet (Erenstein et al. 2022).

Maize finds its place on the tables of 161 countries across the globe, with consumption levels surpassing 50 kg/capita annually in 22 nations. This concentration of maize consumption is predominantly observed in eastern and southern Africa, where nine countries exhibit such high levels, and Latin America, with seven countries demonstrating similarly substantial maize consumption trends (FAO Stat 2021).

The most significant advancements in production were achieved in rice, wheat, and maize, which are commonly referred to as the Green Revolution crops. In the period spanning from 1970 to 2010, maize production experienced a remarkable threefold increase, rising from 282 million metric tons. Maize exhibited the most substantial production surplus in relation to consumption, amounting to approximately 730 million metric tons in 2010, constituting 57% of the overall surplus (Pingali 2015).

In 2014, Mexico's white corn production accounted for 18% of its total output, but a shift in public policy has led to Northern Mexico, particularly Sinaloa, becoming the main supplier of white maize for the flour and tortilla industry, contributing significantly to domestic output and commercialization. Yellow and hybrid maize were influenced by international commodity prices, whereas markets for specialty colored maize varieties are growing locally and regionally (Appendini and Quijada 2016).

Maize and sorghum are cereals predominantly consumed by lower-income populations, with approximately 13% of the poor's food expenditures allocated to maize, compared to 2.6% for the wealthy. Maize is the primary calorie contributor, constituting nearly 20% of the average adult equivalent's calorie intake. In the cereal category, maize and sorghum offer the most economical calorie prices (Worku et al. 2017).

The daily consumption of maize flour per person varied among sites and age-groups. Among children under 5 years old, consumption ranged from 0.1 to 0.25 kg/day in Nandi and Makueni, respectively. Maize and sorghum were typically consumed in the form of ugali, porridge, and muthokoi (traditionally dehulled maize cooked with beans or pigeon peas). These cereals were also used for weaning children, prepared as uji, githeri, and ugali (Kang'Ethe et al. 2017).

On average, South Africans predominantly consume maize, wheat, vegetables, milk, potatoes, and sugar. A study found that the most commonly consumed food items among South African adults were maize, while children aged 1–5 and 6–9 primarily consumed maize porridge. There is innovation in value-added breakfast cereals, and lower-income consumers opt for maize as a more cost-effective carbohydrate source (Ronquest-Ross et al. 2015).

Quality protein maize (QPM) hybrids possess superior nutritional qualities and are employed in the production of high-quality food. QPM-based foods enhance protein quality in the diets of cereal-consuming infants. While white maize is the conventional food choice, yellow genotypes are utilized in areas such as Brazil, China, and Argentina. In Africa, there is growing interest in utilizing yellow corn for biofortification with zinc and provitamin A (Serna-Saldivar and Carrillo 2019).

In China's commercial feed, maize and soybean meal are dominant components, growing in significance. The proportion of maize in energy feed will increase from 83.0% in 2010 to 84.9% in 2020 and is projected to reach 90.4% by 2030. Maize usage for feed will rise from 118,100 thousand tons in 2010 to 153,400 thousand tons in 2020, with a further increase to 173,800 thousand tons by 2030 (Dong et al. 2015).

Utilization of whole crop maize silage for cattle feed and biogas production has significantly grown. In Denmark, maize cultivation for forage has increased by 90% in the past decade. Whole crop maize silage is notable for its high starch content. Cows fed maize silage hybrids with improved aNDFom digestibility which may experience elevated milk protein concentration, potentially due to increased rumen microbial protein synthesis (Krämer-Schmid et al. 2016).

There is a widely accepted consensus that maize became a significant dietary component after AD 800–1000. This shift is substantiated by the abundance of maize fragments in archeological sites post-AD 1000 and dietary isotope records. Maize consumption consistently rose, particularly by the thirteenth century, in line with the archaeobotanical evidence for the same period (Emerson et al. 2020).

Maize is now a predominant component, constituting over 50% of local diets in Southern and Eastern Africa, particularly among smallholder farmers. Across the continent, decorticated and degermed maize flour is a staple, often boiled into a stiff porridge, known as "mealie pap" in South Africa and Lesotho, "sadza" in Zimbabwe, "ugali" in Kenya and Tanzania, and "nshima" in Zambia and Malawi. In Ethiopia, maize gained significance during drought periods and continues to expand due to improved varieties, increased farmer access, and rising market demand (Tanumihardjo et al. 2020).

Maize starch, due to its high fermentability, serves as an adjuvant in high-gravity beer production. Pre-gelatinized maize or corn flakes can shorten mashing times significantly. In bioethanol production, maize excels in starch content, crop yield, and ethanol yield. Its usage includes animal feed (64%), human food (16%), industrial starch and beverages (19%), and seeds (1%). Maize holds a prominent industrial crop status, with 83% of its products allocated to the starch and feed industry (Dabija et al. 2021).

In Mexico, there are over 600 maize-based food products and dishes, with approximately half utilizing nixtamalization—a process involving cooking maize in lime. This includes staple items like tortillas, chips, snacks, tamales, porridges, and cookies. Across Latin America, maize is a key ingredient in various dishes, such as beverages, snacks, and flatbreads. A notable example is arepas, a thick, round flatbread popular in Panama, Venezuela, and Colombia, made from precooked, decorticated, and degermed maize kernels (Palacios-Rojas et al. 2020).

3.4 Chemical Composition

Maize is one of the nutritionally rich cereal grains which is consumed globally. It serves as a major food resource for its energy and protein content, meeting the dietary requirement of the human population (Enyisi et al. 2014). The chemical composition of maize is balanced with sufficient amounts of protein, carbohydrates, fiber, fat, oils, vitamins, and minerals as shown in Table 3.1. The protein content is

Table 3.1 Chemical composition of maize

Nutritional composition	Quantity	References
Moisture	11.6–20%	Enyisi et al. (2014), Hwang et al. (2016)
Total ash	1.34–4.82%	Yadav and Supriya (2014), Eleazu et al. (2020)
Energy	358–365 Kcal	Gwirtz and Garcia-Casal (2014), Deribe and Kassa (2020)
Protein	6.1–13.53%	Kataria (2014), Vaswani et al. (2016), Swapna et al. (2020)
Fat	2.86–3.84%	Dall'Asta et al. (2015), Oladeji et al. (2016)
Fiber	1.20–1.85%	Okonkwo and Agharandu (2017), Rodriguez-Salinas et al. (2020)
Total carbohydrate	71.88–72.89 g	Rouf Shah et al. (2016), Oluba and Oredokun-Lache (2018)
Total sugars	0.64–8.78 g	Ndukwe et al. (2015), Oas and Adams (2022)
Starch	63.72–72.17%	Ndukwe et al. (2015), Prasanthi et al. (2017)
Vitamins	**Quantity**	**References**
Vitamin A	11 μg	Nuss and Tanumihardjo (2010)
Riboflavin (Vitamin B1)	0.06–0.20 mg	Suri and Tanumihardjo (2016), Jeon et al. (2021)
Thiamine (Vitamin B2)	0.35–0.42 mg	Murdia et al. (2016), Bathla et al. (2019)
Niacin (Vitamin B3)	1.13–3.63 mg	Oas and Adams (2022)
Pantothenic acid (Vitamin B5)	0.42 mg	Gwirtz and Garcia-Casal (2014)
Vitamin B6	0.62 mg	Gwirtz and Garcia-Casal (2014)
Vitamin E	1.80%	Mankambou et al. (2021)
Folate	19 μg	Nuss and Tanumihardjo (2010)

quantified in terms of Grain Crude Protein Concentration (GCPC) (Oladapo et al. 2017). It is consumed as a staple food in many countries worldwide and is entitled as a "poor man's nutricereal" considering its nutritional composition, stability in production, availability, cost-efficient, and fulfills the satiety of low socioeconomic people (Yadav and Supriya 2014).

3.5 Macronutrients Composition

3.5.1 Moisture Content

The moisture content is one of the crucial factors in determining the shelf life of the product. Since maize falls under the cereal grains category, the moisture content ranges between 11% and 20%. The post-harvest handling techniques like threshing, cleaning, and drying reduce the moisture content of the fresh commodity suitable to store in dry conditions for a longer period (Oluba and Oredokun-Lache 2018).

3.5.2 Total Carbohydrates

Zea mays, a cereal grain commonly referred to as maize or maize, exhibits a substantial carbohydrate composition. Carbohydrates form the predominant constituent within the chemical framework of maize. It is a high calorific value food containing around 70% of carbohydrates. The carbohydrate molecules are more concentrated toward the endosperm region which acts as a storage chamber in the grains. The endosperm covers 81% of the structural composition of maize (Gwirtz and Garcia-Casal 2014). The principal carbohydrate found in maize is starch, which contributes significantly to its energy content. Starch serves as the primary storage polysaccharide, facilitating the accumulation of glucose molecules in plant tissues and serving as an easily accessible energy source upon consumption (Yu and Moon 2021). Additionally, maize contains minor quantities of monosaccharides (glucose, fructose, and sucrose) and dietary fiber, both soluble and insoluble in nature. Monosaccharides confer the characteristic sweet taste to fresh maize, while dietary fiber imparts invaluable health benefits. The abundant carbohydrate content of *Zea mays* renders it a substantial source of energy, making it an indispensable staple in human and animal diets alike (Nazari et al. 2020).

3.5.3 Protein

Maize exhibits a modest proportion in terms of protein content. On average, the protein content of maize amounts to approximately 9% of its total weight. Nonetheless, it is noteworthy that the protein quality of maize is relatively suboptimal when compared to other grains and legumes. This can be attributed to the deficient presence of certain indispensable amino acids, notably lysine, and tryptophan, which are crucial for human nutritional requirements (Zhang et al. 2016). To augment the protein nutritive value of maize, it is frequently consumed in conjunction with complementary protein-rich foods or subjected to processing techniques that synergistically improve its amino acid profile. Furthermore, advancements in agricultural practices and genetic breeding strategies have been undertaken to enhance maize varieties, emphasizing elevated protein content and enhanced amino acid composition (Tien Lea et al. 2016).

3.5.4 Fat

The lipophilic composition of maize is subject to inherent variability dependent on varietal genotypic distinctions and environmental stimuli. On average, maize typically contains approximately 4–6% lipids per unit weight (Mukri et al. 2022). The predominant lipids found in maize predominantly comprise unsaturated fatty acids, notably linoleic acid and oleic acid, which have been recognized for their beneficial impact on human physiological well-being. However, maize also encompasses trace amounts of saturated fatty acids (Saini et al. 2020). It is noteworthy that the lipid content of maize is relatively modest in comparison to other oleaginous crops such as soybean or sunflower. Nevertheless, maize oil, derived from the germ of maize kernels, is conventionally employed in gastronomic preparations, utilized as an ingredient in food formulations, and integrated into animal feed formulations. The presence of lipids in maize contributes to its energy density and imparts sensory attributes and nutritional profiles to a diverse array of maize-derived food products (Sanjeev et al. 2014).

3.5.5 Dietary Fiber

The dietary fiber content of Zea mays (maize) has been observed to have an average value of approximately 2 g/100 g of raw maize. However, this value is subject to variability based on the specific maize variety and processing methods employed (Krimmer et al. 2019). It is important to note that the fiber content of maize may be altered by cooking, canning, or other forms of processing. Nevertheless, it is generally acknowledged that the bran and germ, which are components of the maize kernel rich in dietary fiber, are predominantly found in whole maize kernels (Hussain et al. 2021). Consequently, the consumption of whole maize, such as fresh or frozen maize on the cob, is expected to yield a higher dietary fiber content compared to refined maize products where the bran and germ may have been removed through refining processes. In summary, the inclusion of *Zea mays* in a well-rounded diet comprising diverse fruits, vegetables, whole grains, and legumes can contribute to achieving an adequate intake of dietary fiber (Kumar et al. 2013).

3.6 Micronutrient Composition

3.6.1 Vitamins

The vitamin content of *Zea mays* is influenced by various factors including cultivar selection, agricultural practices, and environmental conditions. On average, 100 grams of raw maize is found to contain approximately 0.5 mg of vitamin B3 (niacin), 0.2 mg of vitamin B5 (pantothenic acid), and trace amounts of other B-vitamins such as B1 (thiamine), B2 (riboflavin), B6 (pyridoxine), and folate (Azam et al. 2022). Moreover, maize demonstrates modest quantities of vitamin C, vitamin E, and vitamin K. It is crucial to acknowledge that these values are subject to fluctuations influenced by factors such as genetic variation, crop maturity, and post-harvest handling. Furthermore, the processing methods employed, such as cooking techniques (e.g., boiling or steaming), can affect the bioavailability and retention of vitamins in maize (Suri and Tanumihardjo 2016). Therefore, for optimal vitamin assimilation from *Zea mays*, it is recommended to consume fresh or minimally processed maize as part of a diverse and nutritionally balanced diet incorporating other vitamin-rich food sources.

3.6.2 Minerals

Through a comprehensive analysis of the mineral composition of *Zea mays*, a remarkable plethora of essential minerals emerges, underscoring its potential as a valuable source of micronutrients. On average, 100 g of raw maize contains approximately 0.35 mg of iron, 28 mg of magnesium, 127 mg of phosphorus, and 270 mg of potassium. Furthermore, maize exhibits minor quantities of other minerals such as calcium, zinc, copper, and manganese (Bojtor et al. 2022). It is worth noting that these values are subject to variations influenced by factors including maize cultivar selection, agricultural practices, and soil characteristics (Chukwudi et al. 2021). Additionally, the mineral content of maize can be influenced by post-harvest processing and cooking methods. Incorporating *Zea mays* into the diet can contribute to the intake of these vital minerals, which are essential for numerous physiological processes within the human body (Goredema-Matongera et al. 2021).

3.6.3 Bioactive and Functional Components

Phenolic acid distribution was noted in various fractions of yellow maize, with bound phenolic acids found in the bran and endosperm of white maize. Purple and waxy maize contained flavonoid compounds, while yellow maize did not. Whole grain cereals also contain essential bioactive compounds like vitamins, minerals, trace elements, and carotenoids, with vitamin E, particularly α-tocopherol, serving as a potent free radical scavenger (Das and Singh 2015).

3.6.4 Polyphenols

Specialty maize samples consistently showed high phenolic levels in pericarps and germs, with sweet corn exhibiting the highest content in both free and bound forms. In contrast, popcorn had the lowest free phenolic content. The total phenolic content (free + bound) in pericarp was 35–56 times greater than that in the endosperm across maize genotypes (Das and Singh 2015). The presence of phenolic acids, typically within the range of 40–90 ppm, primarily accounts for the sour, bitter, and astringent taste characteristics observed in maize. The predominant phenolic acids identified in maize are primarily ferulic acids, primarily existing in a bound form (Meena et al. 2022). Phenolic compounds in various maize genotypes influence their distinct flavor and color profiles. Blue and purple maize grains exhibit significantly higher TPA concentrations compared to yellow and red varieties. This difference likely arises from a combination of genetic factors and the structural condensation of flavanols (catechin or epicatechin) and anthocyanins (Suriano et al. 2021).

Polyphenols identified in fresh, prepared, and processed corn samples include 4-OH benzoic acid, vanillic acid, syringic acid, o-coumaric acid, caffeic acid, p-coumaric acid, and ferulic acid. Corn grains are a significant reservoir of polyphenol antioxidants, particularly phenolic acids like ferulic, caffeic, and p-coumaric acids. These compounds serve as antioxidants and immune system regulators in humans (Prasanthi et al. 2017). Ferulic acid is prominently found in bound extracts and is a distinctive component in cereals like maize. Phenolic compounds identified and quantified in maize kernels include glycosides of p-coumaric acid, syringic acid, kaempferol, and naringenin (Félix-Medina et al. 2021).

3.6.5 Flavonoids

There exist over 7000 identified flavonoids, widely distributed across diverse plant tissues and organs, exhibiting species-specific variations. Within maize, flavonoids are detectable in leaves, silks, pollens, and seeds. Notably, four significant flavonoids, namely eriodictyol, luteolin, isoorientin, and maysin, are found in different parts of the maize plant (Zhang et al. 2018).

3.6.6 Anthocyanins

Colored maize varieties are distinguished by their abundant anthocyanin content, which imparts vibrant pigmentation like blue, purple, and red, in contrast to the subdued colors of white or yellow maize. The primary anthocyanins in colored maize, namely cyanidins, pelargonidins, and peonidins, are concentrated in the pericarp and aleurone layer of the maize endosperm (Suri and Tanumihardjo 2016). Anthocyanin content in unprocessed blue maize flour was measured at 26.2 mg CGE/100 g in DW. Variability in the anthocyanin content of raw blue maize was influenced by genetic diversity among maize genotypes, regional adaptation, environmental factors, climatic conditions, abiotic stresses, grain properties, and, notably, the relative proportions of kernel anatomical components, with the pericarp and aleurone layer being richer in anthocyanins (Chavarín-Martínez et al. 2019).

3.6.7 Carotenoids

In ground, raw orange maize, lutein and zeaxanthin constituted 63% of the total carotenoids, while provitamin A carotenoids, specifically β-carotene and β-cryptoxanthin, comprised the remaining 37% of the total carotenoid content (Beta and Hwang 2018). Yellow maize grains, particularly in the horny and floury endosperm, contain a substantial amount of carotenoid pigments. These pigments fall into two categories: carotenes, which are oxygen-free hydrocarbons, and xanthophylls like lutein and zeaxanthin, which are hydrocarbons containing oxygen (Rouf Shah et al. 2016). The pigmented genotypes displayed a statistically significant elevation in zeaxanthin levels while demonstrating reduced quantities of lutein. Among these genotypes, the nano variety exhibited the greatest concentration of β-cryptoxanthin and β-carotene, pivotal compounds that serve as essential precursors for the synthesis of vitamin A (Capocchi et al. 2017). The carotenoid profiles of five corn types (white, yellow, high carotenoid, blue, and red) were analyzed. Lutein content was most abundant in yellow corn (4.05 μg/g dry weight) and lowest in white corn (0.01 μg/g dry weight). Zeaxanthin content was highest in yellow corn (3.53 μg/g dry weight), while white corn exhibited the lowest zeaxanthin levels (0.06 μg/g dry weight), in accordance with expectations (Lopez-Martinez and Garcia 2015).

Carotenoid levels in the pericarp and germ fractions were relatively low. Lutein and zeaxanthin were the predominant carotenoids, constituting 18.5% to 51.9% and 28.6% to 53.3% of the total carotenoid content, respectively. Among these fractions, the endosperm of maize had the highest lutein content (5.2 mg/kg, DW), followed by the pericarp fraction (2.7 mg/kg, DW), and the lowest lutein content was observed in the germ fraction (1.1 mg/kg, DW) (Sun et al. 2022).

3.6.8 Bioactive Peptides

The maize prolamin fraction, primarily composed of α-zein, is the predominant component, constituting approximately 50% to 60% of the total protein content. Zein-derived peptides exhibit hydrophobic properties due to their non-polar amino acid composition, consisting mainly of leucine (19.3%), alanine (8.3%), and proline (9.0%) (Trinidad-Calderon et al. 2021). β- and δ-zeins exhibit solubility in aqueous alcohols when a reducing agent is present, while γ-zeins are soluble in both aqueous and alcoholic solvents in the presence of salt and reducing agents. However, the zein protein composition and concentration vary depending on corn genotypes, protein body distribution within the grain, and maturity (Singh et al. 2019).

3.6.9 Fatty Acids

In maize grains, the predominant fatty acids identified were linoleic acid (56.5%) and oleic acid (27.8%). Trace amounts of stearic, linolenic, arachidic, and eicosenoic acids were also observed. The nutritional quality index (NQI) is established based on diverse indices related to the composition of fatty acids. Maize can be regarded as a superior source of unsaturated fatty acids (UFA) since its UFA-to-saturated fatty acid (SFA) ratios surpass 1.6 (Hançerlioğulları et al. 2024). The total tocopherol content ranges from 767 to 1344 mg/kg of oil. Among all the tocopherols, γ-tocopherol (583–1048 mg/kg of oil) consistently emerged as the predominant derivative across all hybrids followed by α-tocopherol, δ-tocopherol, and β-tocopherol. α-tocopherol was notably associated with linoleic acid, which serves as the primary polyunsaturated fatty acid (PUFA) found in corn oil (Kaplan et al. 2017).

3.6.10 Health Implications

Maize exhibits a notable starch abundance and a substantial presence of antioxidant constituents, encompassing phenolic compounds (e.g., phenolic acids), carotenoids (e.g., β-carotene), and xanthophylls (e.g., lutein and zeaxanthin) (D'Amato et al. 2019). Pigmented maize serves as a significant source of anthocyanins and various phenolic compounds, primarily concentrated in the pericarp and monolayered aleurone. Anthocyanins play a crucial role in scavenging free radicals, potentially mitigating or treating chronic degenerative conditions such as atherosclerosis, aging, diabetes, hypertension, inflammation, and cancer as shown in Fig. 3.1. The anthocyanin content ranged from 646 to 1052 mg Cy3Glu/kg in maize (Urias-Lugo et al. 2015a). Blue maize with floury endosperm exhibits superior nutritional value compared to most dent variants. It boasts higher mineral and flavonoid content, particularly anthocyanins found in the pericarp and aleurone layer. These anthocyanins, present in colored kernels, have potential as functional ingredients due to their antioxidant properties (Urias-Peraldí et al. 2013).

Maize exhibits a low cholesterol and fat content. Its whole grain fiber content serves to mitigate the risk of cardiovascular diseases and diabetes, while also fortifying the immune system. Consequently, maize finds suitability in weight reduction regimens and cholesterol-lowering diets. Additionally, the nutritional advantages of maize are influenced by its vitamin A content, which functions as an antioxidant, thereby contributing to disease prevention, including cancer (Murdia et al. 2016). The total crude lipid content ranged from 3.4% to 5.15%. Among the ten fatty acids analyzed, linoleic acid was the most abundant, followed by oleic, palmitic, and stearic acids. Maize hybrids exhibited the high levels of PUFAs (51.4–58.9%) and low levels of SFAs (13.6–17.2%), making them suitable for PUFA-rich diets. Incorporating appropriate levels of MUFAs and PUFAs with enhanced LC-PUFAs is recommended for improved health benefits (Saini et al. 2020).

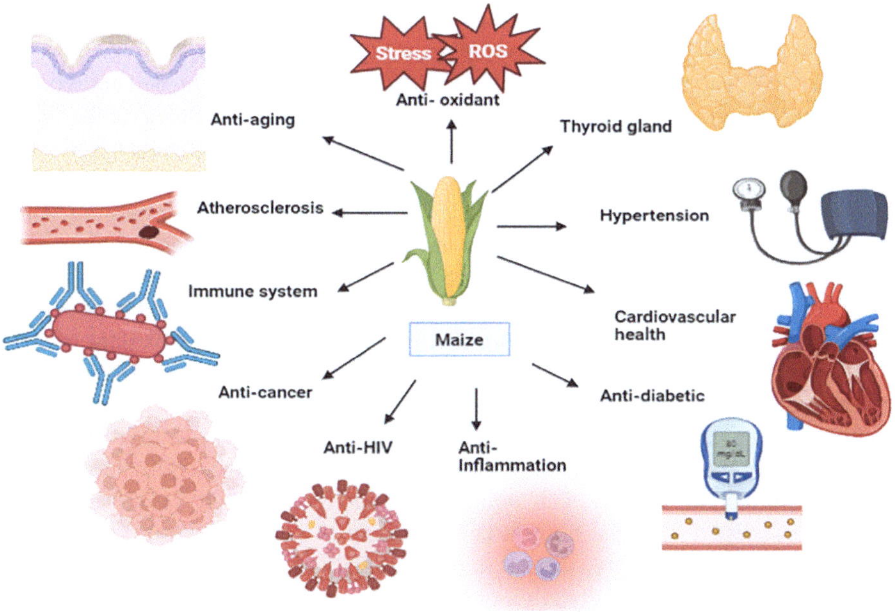

Fig. 3.1 Showing the pharmocological benefits of maize

It offers benefits for rheumatism by aiding limb and joint mobility. Vitamins K, C, and A, along with carotenes, enhance immunity and thyroid gland function. Maize oil, rich in vitamin E, supports antioxidant activity, bolstering the immune system. Additionally, maize contains Galanthus nivalis agglutinin (GNA), which exhibits anti-HIV activity, making it valuable in HIV disease management (Saeed and Saeed 2020). Purple corn pigments have been found to mitigate high-glucose-induced mesangial inflammation, expansion, and hyperplasia by inhibiting the inflammatory activity of interleukin-8 (IL-8). Both anthocyanin-rich water fractions (WF) and phenolic-rich ethyl acetate fractions (EAF) from purple corn exhibit dose-dependent anti-mutagenic activity against the food mutagen Trp-P-1. Purple corn's antioxidant properties offer protection against oxidative degradation and damage in biological systems (Lao et al. 2017).

Infants and young children who were provided with Quality Protein Maize (QPM) characterized by elevated levels of lysine and tryptophan exhibited enhanced growth in terms of both weight and height compared to those consuming conventional maize. Maize hybrids featuring increased levels of provitamin A (proA), lysine, and tryptophan represented a comprehensive nutritional resource, with the potential to address malnutrition more comprehensively (Hossain et al. 2023). Whole maize grains, including whole cornmeal, whole corn flours, and popcorn, demonstrated a notable reduction in aortic stiffness associated with obesity, potentially mitigating cardiovascular disease (CVD) risks. Resistant starch derived from corn displayed positive effects in lowering the risk of type 2 diabetes, with RS2 specifically enhancing insulin sensitivity in men, while women exhibited lower sensitivity to changes in circulating free fatty acids (Siyuan et al. 2018).

Maize lines enriched with flavonoids exhibit significantly heightened antioxidant activity, potentially promoting health. Extensive research has focused on the flavonoid biosynthesis pathway in maize, particularly the well-characterized sub-branches of anthocyanin and phlobaphene biosynthesis at genetic and molecular levels. Anthocyanin functions as antioxidants, effectively inhibiting pro-inflammatory cytokine signaling cascades and thereby mitigating inflammation (Wu et al. 2021). Maize kernels contain a greater quantity of phenolic compounds compared to rice, wheat, and oats. The health advantages associated with these phenolic compounds in maize kernels are ascribed to their potent antioxidant properties, which have the capacity to diminish the likelihood of oxidative stress-induced diseases. It is widely acknowledged that consistent consumption of maize kernels significantly contributes to the reduction in the risk of chronic ailments (Zhang et al. 2020). A maize variety with a purplish-black waxy appearance exhibited the most elevated antioxidant activity. This heightened antioxidant activity was found to be directly associated with the highest concentration of anthocyanins present in the maize kernels. Furthermore, the antioxidant activity of anthocyanins was significantly diminished upon methylation of the 4'-OH group (Zhu 2018).

Among various corn varieties, blue corn demonstrated the most pronounced anti-proliferative effects on cancer cell lines, specifically HepG2, H-460, MCF-7, and PC-3, as well as tumorigenic clones and lymphocytes. Proanthocyanidins, a category of phenolic compounds, have exhibited chemoprotective attributes in countering oxygen-free radicals and oxidative stress (Herrera-Sotero et al. 2017). The most pronounced inhibitory effect was observed in colon ($CaCO_2$) and prostate (PC3) cancer cells, with viabilities falling below 30% when treated with an acidified extract derived from hybrid maize. Notably, the acidified extract from hybrid maize only reduced cell viability in HT29 cells to approximately 52%. Other flavonoids, such as isorhamnetin diglycosides, demonstrated comparable IC50 values against HT29 cells (Urias-Lugo et al. 2015b).

3.6.11 Health Benefits

The health benefits of maize are shown in Table 3.2.

3.6.12 Digestive Health and Prevention of Colon Cancer

Resistant starch (RS) is a kind of carbohydrate that evades breakdown in the small intestine and passes intact through to the colon. It functions as a prebiotic, encouraging the development of healthy gut flora. It also offers several health advantages, including aiding laxation and fermentation, improving mineral absorption, and alleviating diarrhea symptoms. The resistant starch in corn and goods produced from corn is a useful source (Xie et al. 2019). Moreover, the fermentation of RSs in the

Table 3.2 Pharmacological benefits of maize

Pharmacological properties	Action of mechanism	References
Digestive health and prevention of colon cancer	Resistant starch (RS) resists being broken down in the small intestine and instead travels to the colon, promoting the growth of beneficial intestinal flora, i.e., prebiotics which inhibit the growth of cancer cells, thus preventing cancer	Ai and Jane (2016), Siyuan et al. (2018), Xie et al. (2019)
Anti-inflammatory	Anthocyanins effectively inhibit pro-inflammatory cytokine signaling cascades and thus alleviate inflammation	Wu et al. (2021)
Antioxidant	Anthocyanins, present in colored kernels, have potential as functional ingredients due to their antioxidant properties. Proanthocyanidins and phenolic compounds counter oxygen-free radicals and reduce oxidative stress	Urias-Peraldí et al. (2013), Herrera-Sotero et al. (2017)
Antidiabetic effects	It increases the pancreatic cell's sensitivity to insulin, which will balance the body's blood sugar levels	Brobbey et al. (2017), Siyuan et al. (2018)
Prevention of CVDs	Anthocyanins are essential for scavenging free radicals, which may help avoid or alleviate chronic degenerative diseases like atherosclerosis	Saini et al. (2020), Siyuan et al. (2018)
Prevention of chronic kidney diseases	By decreasing the production of pro-inflammatory cytokines and limiting the infiltration of inflammatory cells into the renal tissue. Conditions such as glomerulonephritis and interstitial nephritis where inflammation plays a significant role	Parihar et al. (2022), Wu et al. (2021)

colon results in the formation of useful substances such as butyrate, which has been demonstrated to have anti-inflammatory characteristics and may help in reducing the risk of colon cancer (Ai and Jane 2016). The prebiotic impact of RSs also encourages the development of advantageous gut flora like Bifidobacterium and Lactobacillus, which can enhance immune system performance and gut health in general as shown in Fig. 3.2. In summary, resistant starch, which is abundant in maize and products produced from corn, offers several health advantages, including boosting laxation and fermentation, increasing mineral absorption, acting as prebiotics, and alleviating diarrheal symptoms. Corn products can be a simple and delectable method to increase dietary fiber intake and enhance overall gut health (Siyuan et al. 2018).

3.6.13 Antidiabetic Effects

High blood sugar levels are the major feature characteristic of the chronic disease known as diabetes. Maize can reduce blood sugar levels and enhance insulin sensitivity of the pancreatic cells helping to balance the blood sugar levels in the body, i.e., antidiabetic benefits. It is found that the presence of several substances, such as tannins, phenols, alkaloids, glycosides, flavonoids, and saponins is recognized to

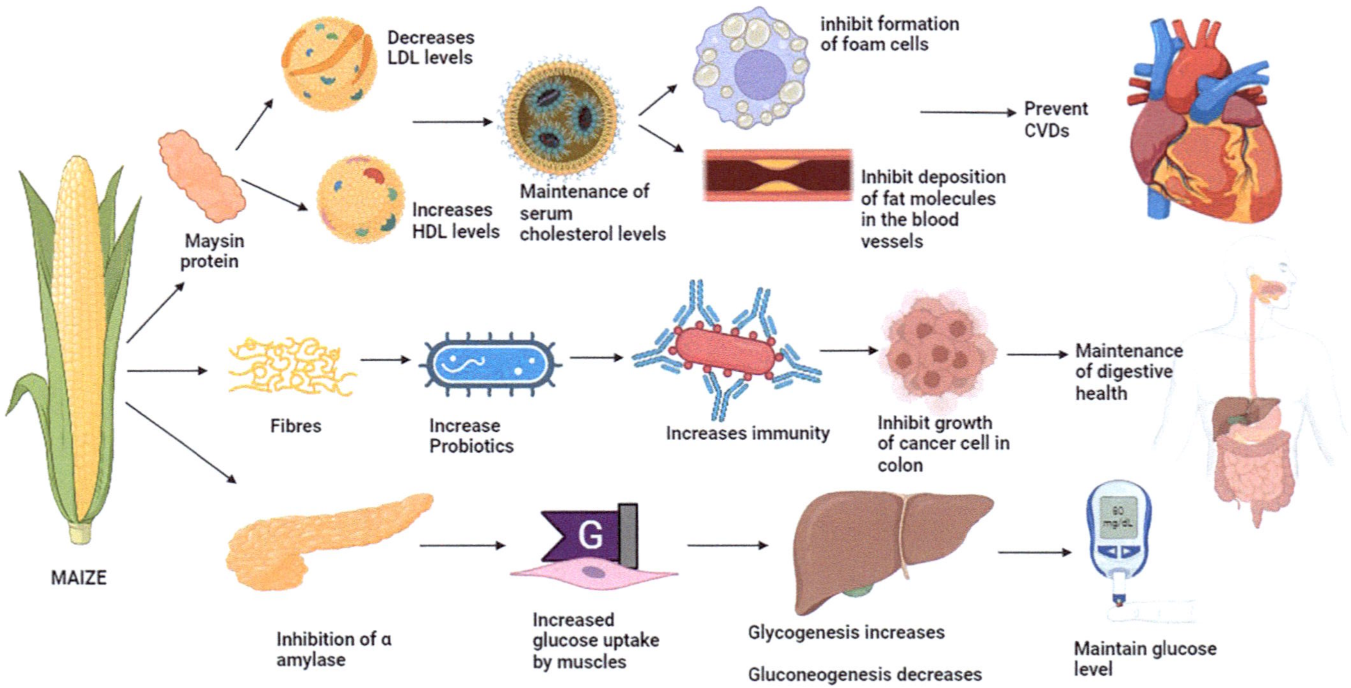

Fig. 3.2 Shows the action of the mechanism for the prevention of cardiovascular disease (CVDs), maintenance of digestive health, and blood glucose levels

have a variety of biological potential due to which it is frequently used for their therapeutic capabilities including alkaloids, glycosides, flavonoids, and saponins. Additionally, tannins and phenols are also found to have astringent and antioxidant properties present in the methanolic extract, making it highly potential and effective for the treatment of diabetes as shown in Fig. 3.2 (Brobbey et al. 2017).

3.7 Prevention of Chronic Kidney Diseases

Chronic, low-grade inflammation is an essential characteristic of chronic kidney disease (CKD), which has an important role in the pathophysiology of inflammation. Due to the presence of flavonoids, tannins, terpenoids, and alkaloids in its secondary metabolites, corn silk has been reported to have anti-inflammatory properties (Parihar et al. 2022). Maize extracts may help people with renal disorders like glomerulonephritis and interstitial nephritis where inflammation plays a prominent role by reducing the creation of pro-inflammatory cytokines and limiting the infiltration of inflammatory cells into the renal tissue (Wu et al. 2021).

3.7.1 Antioxidants

Antioxidants protect the body's cells from oxidative stress and damage caused by free radicals. Numerous disorders have oxidative stress, oxidative stress mediators, and inflammation as part of their pathophysiology, all of which play a role in the emergence of numerous disease consequences. Alpha terpineol, citronellol, and eugenol are a few of the main compounds linked to the antioxidant activity of maize. Antioxidant enzyme levels, such as glutathione peroxidase and superoxide dismutase, are efficiently increased. The amount of total phenol and flavonoids, such as fisetin, quercetin, naringin, and myricetin, also affects an object's ability to fight free radicals. According to research, maize extracts can neutralize free radicals and act as antioxidants, protecting tissue from oxidative damage (Murdia et al. 2016). Several antioxidants, including carotenoids (including lutein and zeaxanthin), phenolic compounds, and vitamin E, are found in maize (corn). Numerous health advantages of these antioxidants have been demonstrated, including a lower risk of chronic diseases like cancer, cardiovascular disease, and age-related macular degeneration as shown in Fig. 3.2 (D'Amato et al. 2019).

3.7.2 Purple Corn

Purple corn (*Zea mays L.*) belongs to the family Poaceae originated from Peru and has a purple cob, pericarp, and corn seeds. Anthocyanins, potent antioxidants that give purple maize its unique purple color, are abundant in purple maize especially in the aleurone layer and pericarp (Cristianini et al. 2020). Anthocyanins, which are present in purple maize, have been found to have anti-inflammatory, antioxidant, and cardiovascular benefits. Another substance present in purple maize called cyanidin 3-glucoside (C3G) has been demonstrated to enhance insulin sensitivity and glucose metabolism, making it advantageous for people with diabetes or metabolic syndrome (Kim et al. 2023). Purple maize also contains C3G and resistant starch, two substances that have been associated with several health advantages. A form of carbohydrate known as resistant starch ferments in the large intestine while defying digestion in the small intestine. Short-chain fatty acids are created during this fermentation process, and they can encourage the development of good bacteria in the stomach and enhance digestive health. Additionally connected to enhanced insulin sensitivity and a lower risk of colon cancer is resistant starch (Lao et al. 2017). The traditional Peruvian drink chicha is a well-liked way to savor purple maize. In Peru, chicha is frequently offered at celebratory events and special occasions. Chicha is a beverage made from purple corn that has been simmered with pineapple or other fruits, spices, and other ingredients before being fermented with yeast or bacteria (Vargas-Yana et al. 2020). Therefore, it has great potential in the food industry and pharmaceutical industries due to the presence of various bioactive components.

3.7.3 Sweet Corn

These are rich in fiber, vitamins, minerals, and antioxidants and also a good source of vitamins A, and B3 that help in metabolism, the neurological and digestive system maintenance. Moreover, it consists of vitamin C, protein, fiber, minerals, and folic acid and is high in carbohydrates and sugars. It consists of thiamine that acts as a neuroprotectant, whereas niacin present in it reduces the risk of cardiovascular diseases and prevents the deposition of cholesterol. Folic acid and pantothenic acid present are responsible for the prevention of anemia promoting healthy cell division and are necessary for protein, carbohydrate, and lipid metabolism in the body.

The dietary fiber present in it improves the functioning of the digestive system, and it may lower the incidence of colon cancer, type 2 diabetes, heart disease, and stroke. Additionally, fiber promotes satiety, making you feel fuller and preventing overeating. The carotenoids lutein and zeaxanthin, which are antioxidants that can help prevent free radical damage, are found in yellow sweet maize. Sweet corn also contains a lot of vitamin C, a powerful antioxidant that shields your cells from damage by reducing oxidative stress, thus protecting against cancer and heart disease (Swapna et al. 2020).

3.7.4 Value Addition

Maize is a crucial food crop for both humans and livestock due to its high energy content, vitamins, and relatively low protein content. The majority of maize production is used for local consumption with limited value-added processing. Value-added maize products encompass hot and cold pap, "tuwo," "massa," "couscous," "gwate," "nakia," "dambu," "dakuwa," popcorn, cooked and roasted maize, akamu, and boiled corn (Ater et al. 2018). Maize grain is employed to create value-added products like corn starch, dextrin, corn syrup, corn oil, sorbitol, and ethanol. These products find essential applications in manufacturing industrial items, including corn flakes, cake mixes, candies, paper adhesives, carbonated beverages, biodegradable plastics, antibiotics, and cosmetics (Shikha and Singh 2019).

Maize, an economical crop, holds promise as a primary option for impoverished communities. Scaling up the production of value-added maize food products on an industrial scale could enhance nutritional and economic security. The commercialization, promotion, and adoption of such products could lead to increased income for farmers, create job opportunities for women and youth, and improve dietary diversity for consumers (Murdia et al. 2016). The ingredients for milk-free soy-maize-sorghum-based ready-to-use therapeutic food consist of whole, non-defatted, non-dehulled soybean, maize, and sorghum grains/seeds. These ingredients undergo processing through extrusion cooking. Iron content was increased to counteract iron absorption inhibitors such as phytic acid and polyphenols. Iron fortification levels are based on WHO recommendations, with 100 mg of ferrous sulfate per 1000 Kcal, similar to fortification levels for therapeutic milk (Irena et al. 2015).

Corn bran, with its high dietary fiber content (up to 90%), was suitable for enhancing the fiber content of foods at low inclusion rates. When corn bran replaced flour at 18%, 20%, and 22% in white bread, the resulting product had reduced loaf volume compared to control bread and exhibited inferior microstructure and textural qualities. Defatted germ flour, when added to wheat flour, substantially boosts the protein content in the blends (Zhang et al. 2021). Ogi is a traditional fermented gruel or porridge produced from millet, sorghum, or maize. In countries such as Botswana and Southern Africa, Bogobe (ting) is crafted from fermented maize, sorghum, or pearl millet. In Benin, a similar product called gowe is prepared from fermented and malted sorghum, millet, and/or maize flour. Typically, it was consumed either as a porridge or a beverage after dilution with water and the addition of sugar, ice, and occasionally milk (Adebiyi et al. 2018).

Complementary foods (weaning food) were developed by blending fermented maize, beans, fishmeal, pawpaw, and sugar. These diets serve as effective alternatives to conventional weaning foods, enhancing the nutritional well-being of children and mitigating protein-energy malnutrition risk. They feature elevated levels of protein, carbohydrates, calories, and essential minerals, which are crucial for bone and teeth formation, muscle and nerve function, blood clotting, and hemoglobin synthesis in infants and young children (Tiencheu et al. 2016).

In South Africa, pregnant women were supplied with folic acid-fortified maize flour, which consisted of a daily portion of 30 g (dry weight) of maize meal porridge fortified with approximately 300 μg of folic acid per serving. The maize meal was fortified by adding 1 mg of crystalline folic acid per 30 kg of maize meal before the cooking process, resulting in a concentration of 33 parts per million (ppm) (Tablante et al. 2019). Gluten-free snacks, utilizing a blend of plantain-chickpea and maize, were created through extrusion and deep-frying techniques to produce snacks characterized by elevated dietary fiber content and a reduced predicted glycemic index. These snacks exhibited reduced fat content, increased dietary fiber levels, and a lower predicted glycemic index. They hold potential as an alternative for addressing issues related to excess weight and obesity in both the general population and the celiac community (Flores-Silva et al. 2015).

Blue maize flour is a valuable ingredient used in the formulation of gluten-free products because of its bioactive compounds and nutraceutical properties. In the formulation of composite gluten-free pasta, a mixture of white maize, blue maize, chickpea, and unripe plantain flour was employed. Notably, the inclusion of blue maize flour at a rate of 75% resulted in the highest retention of total phenolic content after both the extrusion and cooking processes (Camelo-Méndez et al. 2018). The current nutritional trend promotes the intake of cereal-based foods rich in resistant starch for improved health. In line with this trend, high-amylose maize starch flour was used in cookie formulation, resulting in increased dietary fiber and total starch content. Maize starch granules displayed various crystalline domains with varying resistance to digestive enzyme hydrolysis (Giuberti et al. 2016).

Various fermented maize-based products like ogi, mawe, banku, and kenkey in West Africa, togwa in East Africa, and mahewu in South Africa are well-documented. Non-alcoholic fermentation of maize using lactic acid bacteria (LAB) has been employed for centuries and remains crucial in the production of many contemporary fermented beverages. The consumption of these non-alcoholic beverages has the potential to mitigate adverse health and economic consequences associated with poor diets (Mashau et al. 2021). Corn syrup, derived from corn starch and primarily consisting of glucose, serves as a food syrup employed for texture modification, volume addition, sugar crystallization prevention, and flavor enhancement. Refined corn oil, valued for its high smoke point, is mainly utilized in cooking, particularly as a frying oil as shown in Fig. 3.3 (Saritha et al. 2020).

The formulation of ready-to-use therapeutic foods aimed to mitigate severe acute malnutrition and protein-energy deficits by utilizing maize grains, soybeans (pulses), and cooking banana fruits as ingredients. The optimization process concentrated on harnessing affordable plant protein sources like cereals, pulses, and soybeans for the effective feeding of children in need (Yazew 2022).

3.8 Applications

3.8.1 Biopolymer

Maize starch can be extracted that can be used in formation of the polymer which is easily biodegradable, non-toxic, and cost friendly consisting of linear amylose and branched amylopectin. This starch can be easily modified to make the bio-films and biocompatible, biodegradable polymers as shown in Fig. 3.3. These polymers exhibit a variety of characteristics properties, including increased mechanical strength, thermal stability, and chemical resistance. They may also provide benefits including lower costs, better processing, and the opportunity to customize features for certain applications (Hiremani et al. 2020).

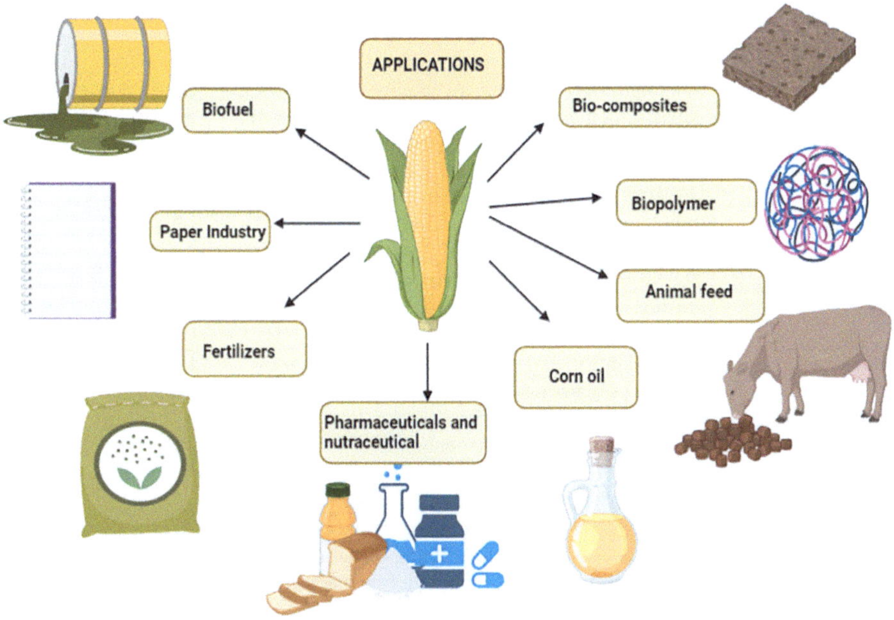

Fig. 3.3 Shows the application of maize

3.8.2 Paper Industry

In addition, paper, textiles, adhesives, and medications are all made with maize starch (Saeed et al. 2021). Bagasse is the residual crushed stalk from the cane sugar manufacturing process after the sugary juice has been drained away. The high pith ratio of the stalks, which is roughly 30% by weight of the stalk, is the main issue in bagasse pulping. Commonly, the pulp resembles hardwood pulps. Due to its high sucrose content, sugarcane is farmed in the majority of tropical and subtropical nations. The fibrous bagasse leftover after the cane has been processed through a number of crushes or diffusers to remove sugar juice is often burned in sugar mill boilers as shown in Fig. 3.3 (Varshney et al. 2019).

3.8.3 Fertilizers and Animal Feed

Nejayote is made from maize by a procedure called nixtamalization. It entails soaking maize kernels for a while in an alkaline solution, which is often created by scalding water with calcium hydroxide (lime). The hulls eventually loosen and separate from the kernels, which also causes the maize kernels to soften and swell. After draining the resulting mixture, the nejayote is the liquid that is still present. The solids of nejayote can be used as fertilizer or as animal feed, while the liquid can be used to cook with or as a foundation for soups and stews. The nutrient-rich

Table 3.3 Applications and utilization of maize

Product	Procedure	References
Fertilizers and animal feed	Soaking maize kernels in an alkaline solution for a while, which is commonly made by scalding water with lime or calcium hydroxide. Eventually, the hulls become less tight and separate from the kernels, causing the latter to swell and soften. The liquid that remains after draining the resultant mixture is called nejayote	Saeed et al. (2021), Acosta-Estrada et al. (2019)
Paper industry	The fibrous bagasse left behind after the cane is crushed and/or defused several times to make paper	Varshney et al. (2019), Saeed et al. (2021)
Biopolymer	It is possible to extract maize starch and use it to create a polymer made of branched and linear amylopectin that is inexpensive, non-toxic, and readily biodegradable. To create bio-films and biocompatible, biodegradable polymers, this starch is simply changed	Hiremani et al. (2020)
Nutraceuticals	Nejayote solids, when solid-state fermented with four distinct types of edible fungi, contain several beneficial components, including free phenolic compounds, antioxidant capacity, and soluble dietary fiber (SDF). The most effective strain for increasing SDF, antioxidant capacity, and phenolic component levels was found to be *P. ostreatus* Perla	Acosta-Estrada et al. (2019), Suri and Dutta (2022), Njideka et al. (2020), Saeed et al. (2021)
Pharmaceuticals	In the production of tablets and capsules, maize starch is utilized as a disintegrant, binder, and filler. Maize starch is an essential ingredient in the production of hard gelatin capsules. It makes it easier for a protective layer to form around therapeutic agents	Labib (2018), Builders and Arhewoh (2016)
Removal of toxicants	Maize products like activated carbon and biochar may effectively adsorb pollutants from water, while the husks and cobs of the grain have a great potential to adsorb colors, organic contaminants, and heavy metals	Sharma et al. (2019), de Santiago-Martín et al. (2020)

by-product of the nixtamalization method employed in traditional Mexican cooking is called nejayote as shown in Table 3.3. It is useful as a fertilizer and animal feed since it is nutrient-rich and contains calcium, potassium, and amino acids (Acosta-Estrada et al. 2019). Maize starch is a significant source of energy in animal feed and is frequently used in place of other grains like wheat and barley as shown in Fig. 3.3 (Saeed et al. 2021).

3.9 Nutraceuticals

3.9.1 Food Additives

However, a study discovered that nejayote solids consist of several bioactive components like free phenolic compounds, antioxidant capacity, and soluble dietary fiber (SDF) when solid-state fermented with four different kinds of edible fungi. *P. ostreatus* Perla was proved to be the most successful for boosting SDF, antioxidant capability, and phenolic component levels. These modifications significantly improve nejayote solids, indicating the possibility of using them as useful food additives (Acosta-Estrada et al. 2019).

3.9.2 Food Products

Maize starch is a versatile component that is employed in a range of industrial and food applications. It is frequently used in food products such as soups, sauces, and baked goods as a thickening, binder, and stabilizer. Maize is a valuable crop that has a wide range of applications and health advantages (Saeed et al. 2021). Oil is extracted from corn that is high in unsaturated fatty acids, i.e., PUFA and MUFA that are required for brain functioning and cognitive development as shown in Fig. 3.3 (Suri and Dutta 2022). Investigations have been carried out into the potential use of maize cob as a source of dietary fiber for high-fiber cookies (Njideka et al. 2020).

3.9.3 Pharmaceuticals

A common excipient in the pharmaceutical industry is maize starch. It is used in the formulation of tablets and capsules as a filler, disintegrant, and binder. A crucial component in the creation of firm gelatin capsules is maize starch. It facilitates the formation of a sturdy shell around medicinal substances. Drug delivery systems are developed using modified versions of maize starch. Among these can be formulations with controlled release, which improve the medication's absorption and effectiveness. Polylactic acid (PLA), one substance generated from maize, is becoming more and more popular in pharmaceutical packaging. A sustainable substitute, PLA is a biodegradable polymer made from maize starch as shown in Fig. 3.3 (Labib 2018).

Pharmaceutical companies use adhesives made from maize starch, especially when making transdermal patches and other sticky drug delivery devices. When making tablets, maize starch serves as a binder, strengthening the tablet's integrity and holding the active medicinal ingredients together. The pharmaceutical sector can employ maize starch as a lubricant. It facilitates the effective and seamless processing of pharmaceutical powders during the compression of tablets. Modified varieties of maize starch can be used to coat tablets, protecting the medication and regulating its release. To improve the overall composition and manufacturing process of oral dosage forms, maize starch can be utilized as a diluent. In conclusion, maize is a major ingredient in many medicinal products and helps with their formulation, production, and distribution. It is a useful ingredient in pharmaceutical applications because of its adaptability, biodegradability, and compatibility with medication formulations (Builders and Arhewoh 2016).

3.10 Removal of Toxicants

The potential of using maize to purify water has also been investigated. Researchers have discovered that products made from maize, such as activated carbon and biochar, can successfully adsorb pollutants from water. It has been demonstrated that the husks and cobs of maize have a significant potential for adsorbing colors, organic contaminants, and heavy metals (de Santiago-Martín et al. 2020). To increase their effectiveness, maize-based products have also been combined with other treatment techniques such as membrane filtration and electrochemical procedures. Maize has been investigated for its possible use in phytoremediation, which is the use of plants to remove toxins from soil or water, in addition to its application in water treatment. Heavy metals like lead and cadmium have been proven to be successfully removed from contaminated soil by maize (Sharma et al. 2019).

3.11 Conclusions

In summary, the scientific study of maize centers on the existence of phytochemicals and bioactive elements that support the crop's health advantages, rather than the grain's historical relevance or cultural influence. The identification of flavonoids, carotenoids, phenolic acids, and other bioactive substances among the phytochemicals found in maize highlights this crop's potential as a nutritional powerhouse. These substances give maize its vivid hues and tastes, but they also have several health advantages. Maize appears not just as a staple food but also as a source of useful substances that can improve human health, from antioxidant qualities that fight oxidative stress to possible anti-inflammatory benefits.

Furthermore, maize's uses go beyond its status as a staple grain. Numerous businesses, including pharmaceutical, nutraceutical, and even cosmetic ones, use its bioactive components. The potential of maize-derived chemicals as a useful resource for the creation of functional foods and health-promoting products is demonstrated by their diversity, which creates opportunities for study and innovation. Maize represents a promising future in the pursuit of holistic health and sustainable innovation, as its genetic makeup is unveiled and its bioactive components are fully utilized.

References

Acosta-Estrada BA, Villela-Castrejón J, Perez-Carrillo E, Gómez-Sánchez CE, Gutiérrez-Uribe JA (2019) Effects of solid-state fungi fermentation on phenolic content, antioxidant properties and fiber composition of lime cooked maize by-product (nejayote). J Cereal Sci 90:102837

Adebiyi JA, Obadina AO, Adebo OA, Kayitesi E (2018) Fermented and malted millet products in Africa: expedition from traditional/ethnic foods to industrial value-added products. Crit Rev Food Sci Nutr 58(3):463–474. https://doi.org/10.1080/10408398.2016.1188056

Ai Y, Jane JL (2016) Macronutrients in corn and human nutrition. Compr Rev Food Sci Food Saf 15(3):581–598

Appendini K, Quijada MG (2016) Consumption strategies in Mexican rural households: pursuing food security with quality. Agric Hum Values 33:439–454. https://doi.org/10.1007/s10460-015-9614-y

Assefa Y, Carter P, Hinds M, Bhalla G, Schon R, Jeschke M et al (2018) Analysis of long term study indicates both agronomic optimal plant density and increase maize yield per plant contributed to yield gain. Sci Rep 8(1):4937. https://doi.org/10.1038/s41598-018-23362-x

Ater PI, Aye GC, Daniel A (2018) Analysis of maize value addition among entrepreneurs in Taraba State, Nigeria. Int J Environ Agric Biotechnol 3(6):2011–2019. https://doi.org/10.22161/ijeab/3.6.8

Azam FMS, Lian T, Liang Q, Wang W, Zhang C, Jiang L (2022) Variation of vitamin B contents in maize inbred lines: potential genetic resources for biofortification. Front Nutr 9:1029119. https://doi.org/10.3389/fnut.2022.1029119

Bathla S, Jaidka M, Kaur R (2019) Nutritive value. In: Hossain A (ed) Maize-production and use. IntechOpen, London

Bedoya CA, Dreisigacker S, Hearne S, Franco J, Mir C, Prasanna BM et al (2017) Genetic diversity and population structure of native maize populations in Latin America and the Caribbean. PLoS One 12(4):e0173488. https://doi.org/10.1371/journal.pone.0173488

Beta T, Hwang T (2018) Influence of heat and moisture treatment on carotenoids, phenolic content, and antioxidant capacity of orange maize flour. Food Chem 246:58–64. https://doi.org/10.1016/j.foodchem.2017.10.150

Bojtor C, Mousavi SMN, Illés Á, Golzardi F, Széles A, Szabó A et al (2022) Nutrient composition analysis of maize hybrids affected by different nitrogen fertilisation systems. Plan Theory 11(12):1593. https://doi.org/10.3390/plants11121593

Bonavia D (2013) Maize: origin, domestication, and its role in the development of culture. Cambridge University Press, Cambridge

Brandenburg JT, Mary-Huard T, Rigaill G, Hearne SJ, Corti H, Joets J et al (2017) Independent introductions and admixtures have contributed to adaptation of European maize and its American counterparts. PLoS Genet 13(3):e1006666. https://doi.org/10.1371/journal.pgen.1006666

Brobbey AA, Somuah-Asante S, Asare-Nkansah S, Boateng FO, Ayensu I (2017) Preliminary phytochemical screening and scientific validation of the anti-diabetic effect of the dried husk of Zea mays L. (Corn, Poaceae). Int J Phytopharm 7(1):1–5

Builders PF, Arhewoh MI (2016) Pharmaceutical applications of native starch in conventional drug delivery. Starch-Stärke 68(9–10):864–873

Camelo-Méndez GA, Tovar J, Bello-Pérez LA (2018) Influence of blue maize flour on gluten-free pasta quality and antioxidant retention characteristics. J Food Sci Technol 55:2739–2748. https://doi.org/10.1007/s13197-018-3196-9

Capocchi A, Bottega S, Spanò C, Fontanini D (2017) Phytochemicals and antioxidant capacity in four Italian traditional maize (Zea mays L.) varieties. Int J Food Sci Nutr 68(5):515–524. https://doi.org/10.1080/09637486.2016.1261809

Chavarín-Martínez CD, Gutiérrez-Dorado R, Perales-Sánchez JXK, Cuevas-Rodríguez EO, Milán-Carrillo J, Reyes-Moreno C (2019) Germination in optimal conditions as effective strategy to improve nutritional and nutraceutical value of underutilized Mexican blue maize seeds. Plant Foods Hum Nutr 74:192–199. https://doi.org/10.1007/s11130-019-00717-x

Chukwudi UP, Kutu FR, Mavengahama S (2021) Influence of heat stress, variations in soil type, and soil amendment on the growth of three drought–tolerant maize varieties. Agronomy 11(8):1485. https://doi.org/10.3390/agronomy11081485

Cristianini M, Guillén Sánchez JS (2020) Extraction of bioactive compounds from purple corn using emerging technologies: a review. J Food Sci 85(4):862–869

Cuello JP, Hwang HY, Gutierrez J, Kim SY, Kim PJ (2015) Impact of plastic film mulching on increasing greenhouse gas emissions in temperate upland soil during maize cultivation. Appl Soil Ecol 91:48–57. https://doi.org/10.1016/j.apsoil.2015.02.007

D'Amato R, De Feudis M, Guiducci M, Businelli D (2019) Zea mays L. grain: increase in nutraceutical and antioxidant properties due to Se fortification in low and high water regimes. J Agric Food Chem 67(25):7050–7059. https://doi.org/10.1021/acs.jafc.9b02446

Dabija A, Ciocan ME, Chetrariu A, Codină GG (2021) Maize and sorghum as raw materials for brewing, a review. Appl Sci 11(7):3139. https://doi.org/10.3390/app11073139

Dall'Asta C, Giorni P, Cirlini M, Reverberi M, Gregori R, Ludovici M et al (2015) Maize lipids play a pivotal role in the fumonisin accumulation. World Mycotoxin J 8(1):87–97. https://doi.org/10.3920/WMJ2014.1754

Das AK, Singh V (2015) Antioxidative free and bound phenolic constituents in pericarp, germ and endosperm of Indian dent (Zea mays var. indentata) and flint (Zea mays var. indurata) maize. J Funct Foods 13:363–374. https://doi.org/10.1016/j.jff.2015.01.012

de Santiago-Martín A, Meffe R, Teijón G, Hernández VM, Lopez-Heras I, Alonso CA et al (2020) Pharmaceuticals and trace metals in the surface water used for crop irrigation: risk to health or natural attenuation? Sci Total Environ 705:135825

Deribe Y, Kassa E (2020) Value creation and sorghum-based products: what synergetic actions are needed? Cogent Food Agric 6(1):1722352. https://doi.org/10.1080/23311932.2020.1722352

Dong WL, Wang XB, Jun YANG (2015) Future perspective of China's feed demand and supply during its fast transition period of food consumption. J Integr Agric 14(6):1092–1100. https://doi.org/10.1016/S2095-3119(14)60992-8

Eleazu CO, Eleazu KF, Ukamaka G, Adeolu T, Ezeorah V, Ezeorah B et al (2020) Nutrient and antinutrient composition and heavy metal and phenolic profiles of maize (Zea mays) as affected by different processing techniques. ACS Food Sci Technol 1(1):113–123. https://doi.org/10.1021/acsfoodscitech.0c00045?ref=pdf

Emerson TE, Hedman KM, Simon ML, Fort MA, Witt KE (2020) Isotopic confirmation of the timing and intensity of maize consumption in greater Cahokia. Am Antiq 85(2):241–262. https://doi.org/10.1017/aaq.2020.7

Enyisi IS, Umoh VJ, Whong CMZ, Alabi O, Abdullahi IO (2014) Chemical and nutritional values of maize and maize products obtained from selected markets in Kaduna. J Pharm Allied Sci 11(2):2106–2113. https://doi.org/10.14303/ajfst.2014.029

Erenstein O, Jaleta M, Sonder K, Mottaleb K, Prasanna BM (2022) Global maize production, consumption and trade: trends and R&D implications. Food Sec 14(5):1295–1319. https://doi.org/10.1007/s12571-022-01288-7

FAO Stat (2021) FAO Stat. FAO, Rome. http://www.fao.org/faostat

Félix-Medina JV, Gutiérrez-Dorado R, López-Valenzuela JA, López-Ángulo G, Quintero-Soto MF, Perales-Sánchez JXK, Montes-Ávila J (2021) Nutritional, antioxidant and phytochemical characterization of healthy ready-to-eat expanded snack produced from maize/common bean mixture by extrusion. LWT 142:111053. https://doi.org/10.1016/j.lwt.2021.111053

Flores-Silva PC, Rodriguez-Ambriz SL, Bello-Pérez LA (2015) Gluten-free snacks using plantain–chickpea and maize blend: chemical composition, starch digestibility, and predicted glycemic index. J Food Sci 80(5):C961–C966. https://doi.org/10.1111/1750-3841.12865

Gałązka A, Gawryjołek K, Grządziel J, Frąc M, Księżak J (2017) Microbial community diversity and the interaction of soil under maize growth in different cultivation techniques. Plant Soil Environ 63(6):264–270. https://doi.org/10.17221/171/2017-PSE

Giuberti G, Gallo A, Fortunati P, Rossi F (2016) Influence of high-amylose maize starch addition on in vitro starch digestibility and sensory characteristics of cookies. Starch-Stärke 68(5–6):469–475. https://doi.org/10.1002/star.201500228

Goredema-Matongera N, Ndhlela T, Magorokosho C, Kamutando CN, van Biljon A, Labuschagne M (2021) Multinutrient biofortification of maize (Zea mays L.) in Africa: current status, opportunities and limitations. Nutrients 13(3):1039. https://doi.org/10.3390/nu13031039

Grote U, Fasse A, Nguyen TT, Erenstein O (2021) Food security and the dynamics of wheat and maize value chains in Africa and Asia. Front Sustain Food Syst 4:617009. https://doi.org/10.3389/fsufs.2020.617009

Gwirtz JA, Garcia-Casal MN (2014) Processing maize flour and corn meal food products. Ann N Y Acad Sci 1312(1):66–75. https://doi.org/10.1111/nyas.12299

Hammad HM, Abbas F, Saeed S, Fahad S, Cerdà A, Farhad W et al (2018) Offsetting land degradation through nitrogen and water management during maize cultivation under arid conditions. Land Degrad Dev 29(5):1366–1375

Hançerlioğulları BZ, Toprak U, Yılmaz R (2024) Analyses of metabolites in microwave-treated maize flours. Food Bioprocess Technol 17:686–696. https://doi.org/10.1007/s11947-023-03164-4

Herrera-Sotero MY, Cruz-Hernández CD, Trujillo-Carretero C, Rodríguez-Dorantes M, García-Galindo HS, Chávez-Servia JL et al (2017) Antioxidant and antiproliferative activity of blue corn and tortilla from native maize. Chem Cent J 11(1):1–8. https://doi.org/10.1186/s13065-017-0341-x

Hiremani VD, Sataraddi S, Bayannavar PK, Gasti T, Masti SP, Kamble RR, Chougale RB (2020) Mechanical, optical and antioxidant properties of 7-hydroxy-4-methyl coumarin doped polyvinyl alcohol/oxidized maize starch blend films. SN Appl Sci 2:1–18

Hossain F, Jaiswal SK, Muthusamy V, Zunjare RU, Mishra SJ, Chand G et al (2023) Enhancement of nutritional quality in maize kernel through marker-assisted breeding for vte4, crtRB1, and opaque2 genes. J Appl Genet 64:431–443. https://doi.org/10.1007/s13353-023-00768-6

Huang Y, Huang W, Meng Z, Braz GT, Li Y, Wang K et al (2021) Megabase-scale presence-absence variation with Tripsacum origin was under selection during maize domestication and adaptation. Genome Biol 22:1–28. https://doi.org/10.1186/s13059-021-02448-2

Hussain M, Saeed F, Niaz B, Afzaal M, Ikram A, Hussain S et al (2021) Biochemical and nutritional profile of maize bran-enriched flour in relation to its end-use quality. Food Sci Nutr 9(6):3336–3345. https://doi.org/10.1002/fsn3.2323

Hwang T, Ndolo VU, Katundu M, Nyirenda B, Bezner-Kerr R, Arntfield S, Beta T (2016) Provitamin A potential of landrace orange maize variety (Zea mays L.) grown in different geographical locations of Central Malawi. Food Chem 196:1315–1324. https://doi.org/10.1016/j.foodchem.2015.10.067

Irena AH, Bahwere P, Owino VO, Diop EI, Bachmann MO, Mbwili-Muleya C et al (2015) Comparison of the effectiveness of a milk-free soy-maize-sorghum-based ready-to-use therapeutic food to standard ready-to-use therapeutic food with 25% milk in nutrition management of severely acutely malnourished Z ambian children: an equivalence non-blinded cluster randomised controlled trial. Matern Child Nutr 11:105–119. https://doi.org/10.1111/mcn.12054

Jat HS, Singh G, Singh R, Choudhary M, Jat ML, Gathala MK, Sharma DK (2015) Management influence on maize–wheat system performance, water productivity and soil biology. Soil Use Manag 31(4):534–543. https://doi.org/10.1111/sum.12208

Jeon J, Lee SH, Choi YM, Choung MG (2021) Thiamine, riboflavin, and niacin content in Korean minor cereals. J Korean Soc Food Sci Nutr 50(12):1299–1307. https://doi.org/10.3746/jkfn.2021.50.12.1299

Jun XUE, XIE RZ, ZHANG WF, WANG KR, Peng HOU, Bo MING et al (2017) Research progress on reduced lodging of high-yield and-density maize. J Integr Agric 16(12):2717–2725. https://doi.org/10.1016/S2095-3119(17)61785-4

Kang'Ethe EK, Gatwiri M, Sirma AJ, Ouko EO, Mburugu-Musoti CK, Kitala PM et al (2017) Exposure of Kenyan population to aflatoxins in foods with special reference to Nandi and Makueni counties. Food Qual Saf 1(2):131–137. https://doi.org/10.1093/fqs/fyx011

Kaplan M, Kale H, Karaman K, Ünlükara A (2017) Influence of different irrigation and nitrogen levels on crude oil and fatty acid composition of maize (Zea mays L.). Grasas y Aceites 68(3):e207. https://doi.org/10.3989/gya.0222171

Kataria R (2014) Proximate nutritional evaluation of maize and rice-gluten free cereal. IOSR J Nurs Health Sci 3(2):1–6
Kim HY, Lee KY, Kim M, Hong M, Deepa P, Kim S (2023) A review of the biological properties of purple corn (Zea mays L.). Sci Pharm 91(1):6
Krämer-Schmid M, Lund P, Weisbjerg MR (2016) Importance of NDF digestibility of whole crop maize silage for dry matter intake and milk production in dairy cows. Anim Feed Sci Technol 219:68–76. https://doi.org/10.1016/j.anifeedsci.2016.06.007
Krimmer M, Farber C, Kurouski D (2019) Rapid and noninvasive typing and assessment of nutrient content of maize kernels using a handheld raman spectrometer. ACS Omega 4(15):16330–16335. https://doi.org/10.1021/acsomega.9b01661
Kumar A, Jat SL, Kumar R, Yadav OP (2013) Maize production systems for improving resource-use efficiency and livelihood security. Directorate of Maize Research, New Delhi
Labib G (2018) Overview on zein protein: a promising pharmaceutical excipient in drug delivery systems and tissue engineering. Expert Opin Drug Deliv 15(1):65–75
Lao F, Sigurdson GT, Giusti MM (2017) Health benefits of purple corn (Zea mays L.) phenolic compounds. Compr Rev Food Sci Food Saf 16(2):234–246. https://doi.org/10.1111/1541-4337.12249
Li Y, Bingxin Y, Yiming Y, Xiantao H, Quanwei L, Zhijie L et al (2016) Global overview of research progress and development of precision maize planters. Int J Agric Biol Eng 9(1):9–26
Li C, Wang C, Wen X, Qin X, Liu Y, Han J et al (2017) Ridge–furrow with plastic film mulching practice improves maize productivity and resource use efficiency under the wheat–maize double–cropping system in dry semi–humid areas. Field Crop Res 203:201–211. https://doi.org/10.1016/j.fcr.2016.12.029
Lopez-Martinez LX, Garcia HS (2015) Processing of corn (maize) and compositional features. In: Processing and impact on active components in food. Academic Press, London, pp 329–336. https://doi.org/10.1016/B978-0-12-404699-3.00039-1
Mageto EK, Makumbi D, Njoroge K, Nyankanga R (2017) Genetic analysis of early-maturing maize (Zea Mays L.) inbred lines under stress and nonstress conditions. J Crop Improv 31(4):560–588. https://doi.org/10.1080/15427528.2017.1315625
Mankambou JG, Fagbohoun JB, Ya KC, Blei SK, Kouame LP (2021) Assessment of minerals, vitamins and functional properties of flours from germinated yellow maize (Zea mays L.) seeds from Daloa (Côte d'Ivoire). Int J Food Sci Nutr Eng 11(2):35–42. https://doi.org/10.5923/j.food.20211102.01
Mashau ME, Maliwichi LL, Jideani AIO (2021) Non-alcoholic fermentation of maize (Zea mays) in Sub-Saharan Africa. Fermentation 7(3):158. https://doi.org/10.3390/fermentation7030158
Meena KK, Taneja NK, Jain D, Ojha A, Saravanan C, Mudgil D (2022) Bioactive components and health benefits of maize-based fermented foods: a review. Biointerface Res Appl Chem 13(4):338. https://doi.org/10.33263/BRIAC134.338
Mukri G, Patil MS, Motagi BN, Bhat JS, Singh C, Jeevan Kumar SP et al (2022) Genetic variability, combining ability and molecular diversity-based parental line selection for heterosis breeding in field corn (Zea mays L.). Mol Biol Rep 49(6):4517–4524. https://doi.org/10.1007/s11033-022-07295-3
Murdia LK, Wadhwani R, Wadhawan N, Bajpai P, Shekhawat S (2016) Maize utilization in India: an overview. Am J Food Nutr 4(6):169–176. https://doi.org/10.12691/ajfn-4-6-5
Nazari M, Riebeling S, Banfield CC, Akale A, Crosta M, Mason-Jones K et al (2020) Mucilage polysaccharide composition and exudation in maize from contrasting climatic regions. Front Plant Sci 11:587610. https://doi.org/10.3389/fpls.2020.587610
Ndukwe OK, Edeoga HO, Omosun G (2015) Varietal differences in some nutritional composition of ten maize (Zea mays L.) varieties grown in Nigeria. Int J Acad Res Reflect 3(5):1–11
Njideka BE, Chijioke MB, Chioma ON, Nnennaya AN, Iheduzaju PIA (2020) Maize cob as dietary fiber source for high-fiber biscuit. GSC Biol Pharm Sci 12(1):138–144
Nuss ET, Tanumihardjo SA (2010) Maize: a paramount staple crop in the context of global nutrition. Compr Rev Food Sci Food Saf 9(4):417–436. https://doi.org/10.1111/j.1541-4337.2010.00117.x

Oas SE, Adams KR (2022) The nutritional content of five Southwestern US indigenous maize (Zea Mays L.) landraces of varying endosperm type. Am Antiq 87(2):284–302. https://doi.org/10.1017/aaq.2021.131

Okonkwo C, Agharandu U (2017) Proximate and vitamin composition of selected cereals commonly used for weaning babies' food preparation in South-Eastern Nigeria. J Biol Agric Healthc 7:22

Oladapo AS, Adepeju AB, Akinyele AA, Adepeju DM (2017) The proximate, functional and anti-nutritional properties of three selected varieties of maize (yellow, white and pop corn) flour. Int J Sci Eng Sci 1(2):23–26

Oladeji BS, Irinkoyenikan OA, Gbadamosi OS, Ibironke SI, Akanbi CT, Taiwo KA (2016) Comparative analysis of physico-chemical properties and amino acids profile of three tropical maize hybrid cultivars in Nigeria. Nutr Food Sci 46(5):695–705. https://doi.org/10.1108/NFS-10-2015-0120

Oluba OM, Oredokun-Lache AB (2018) Nutritional composition and glycemic index analyses of vitamin A-biofortified maize in healthy subjects. Food Sci Nutr 6(8):2285–2292. https://doi.org/10.1002/fsn3.801

Palacios-Rojas N, McCulley L, Kaeppler M, Titcomb TJ, Gunaratna NS, Lopez-Ridaura S, Tanumihardjo SA (2020) Mining maize diversity and improving its nutritional aspects within agro-food systems. Compr Rev Food Sci Food Saf 19(4):1809–1834. https://doi.org/10.1111/1541-4337.12552

Parihar PS, Jindal P, Nalwaya N (2022) Corn silk-a natural therapy for curing diseases and its benefits. Int Res J Eng Technol 9:1361–1367

Pingali P (2015) Agricultural policy and nutrition outcomes–getting beyond the preoccupation with staple grains. Food Sec 7:583–591. https://doi.org/10.1007/s12571-015-0461-x

Mt. Pleasant J (2016) Food yields and nutrient analyses of the three sisters: a Haudenosaunee cropping system. Ethnobiol Lett 7(1):87–98

Pradhan A, Chan C, Roul PK, Halbrendt J, Sipes B (2018) Potential of conservation agriculture (CA) for climate change adaptation and food security under rainfed uplands of India: a transdisciplinary approach. Agric Syst 163:27–35. https://doi.org/10.1016/j.agsy.2017.01.002

Prasanthi PS, Naveena N, Vishnuvardhana Rao M, Bhaskarachary K (2017) Compositional variability of nutrients and phytochemicals in corn after processing. J Food Sci Technol 54:1080–1090. https://doi.org/10.1007/s13197-017-2547-2

Ramirez-Cabral NY, Kumar L, Shabani F (2017) Global alterations in areas of suitability for maize production from climate change and using a mechanistic species distribution model (CLIMEX). Sci Rep 7(1):5910. https://doi.org/10.1038/s41598-017-05804-0

Ramos-Madrigal J, Smith BD, Moreno-Mayar JV, Gopalakrishnan S, Ross-Ibarra J, Gilbert MTP, Wales N (2016) Genome sequence of a 5,310-year-old maize cob provides insights into the early stages of maize domestication. Curr Biol 26(23):3195–3201. https://doi.org/10.1016/j.cub.2016.09.036

Ranum P, Peña-Rosas JP, Garcia-Casal MN (2014) Global maize production, utilization, and consumption. Ann N Y Acad Sci 1312(1):105–112. https://doi.org/10.1111/nyas.12396

Rao CS, Lal R, Prasad JV, Gopinath KA, Singh R, Jakkula VS et al (2015) Potential and challenges of rainfed farming in India. Adv Agron 133:113–181. https://doi.org/10.1016/bs.agron.2015.05.004

Revilla P, Alves ML, Andelković V, Balconi C, Dinis I, Mendes-Moreira P et al (2022) Traditional foods from maize (Zea mays L.) in Europe. Front Nutr 8:683399. https://doi.org/10.3389/fnut.2021.683399

Rodriguez-Salinas PA, Zavala-Garcia F, Urias-Orona V, Muy-Rangel D, Heredia JB, Nino-Medina G (2020) Chromatic, nutritional and nutraceutical properties of pigmented native maize (Zea mays L.) genotypes from the northeast of Mexico. Arab J Sci Eng 45:95–112. https://doi.org/10.1007/s13369-019-04086-0

Ronquest-Ross LC, Vink N, Sigge GO (2015) Food consumption changes in South Africa since 1994. S Afr J Sci 111(9–10):1–12

Rouf Shah T, Prasad K, Kumar P (2016) Maize—a potential source of human nutrition and health: a review. Cogent Food Agric 2(1):1166995. https://doi.org/10.1080/23311932.2016.1166995

Rutkowska B, Szulc W, Sosulski T, Skowrońska M, Szczepaniak J (2018) Impact of reduced tillage on CO_2 emission from soil under maize cultivation. Soil Tillage Res 180:21–28. https://doi.org/10.1016/j.still.2018.02.012

Saeed MS, Saeed A (2020) Health benefits of maize crop—an overview. Curr Res Agric Farming 1:5–8. https://doi.org/10.18782/2582-7146.114

Saeed F, Hussain M, Arshad MS, Afzaal M, Munir H, Imran M et al (2021) Functional and nutraceutical properties of maize bran cell wall non-starch polysaccharides. Int J Food Prop 24(1):233–248

Sah RP, Chakraborty M, Prasad K et al (2020) Impact of water deficit stress in maize: phenology and yield components. Sci Rep 10:2944. https://doi.org/10.1038/s41598-020-59689-7

Saini RK, Rengasamy KR, Ko EY, Kim JT, Keum YS (2020) Korean maize hybrids present significant diversity in fatty acid composition: an investigation to identify PUFA-rich hybrids for a healthy diet. Front Nutr 7:578761. https://doi.org/10.3389/fnut.2020.578761

Sanjeev P, Chaudhary DP, Sreevastava P, Saha S, Rajenderan A, Sekhar JC, Chikkappa GK (2014) Comparison of fatty acid profile of specialty maize to normal maize. J Am Oil Chem Soc 91(6):1001–1005. https://doi.org/10.1007/s11746-014-2429-y

Saritha A, Ramanjaneyulu AV, Sainath N, Umarani E (2020) Nutritional importance and value addition in maize. Biot Res Today 2(9):974–977

Serna-Saldivar SO, Carrillo EP (2019) Food uses of whole corn and dry-milled fractions. In: Corn. AACC International Press, Oxford, pp 435–467. https://doi.org/10.1016/B978-0-12-811971-6.00016-4

Sharma A, Tomer A, Singh J, Chhikara BS (2019) Biosorption of metal toxicants and other water pollutants by corn (maize) plant: a comprehensive review. J Integr Sci Technol 7(2):19–28

Shikha K, Singh S (2019) Maize utilization and value addition. Agric Food e-newslett 32:63

Singh VK, Dwivedi BS, Singh SK, Majumdar K, Jat ML, Mishra RP, Rani M (2016) Soil physical properties, yield trends and economics after five years of conservation agriculture based rice-maize system in North-Western India. Soil Tillage Res 155:133–148. https://doi.org/10.1016/j.still.2015.08.001

Singh N, Singh S, Shevkani K (2019) Maize: composition, bioactive constituents, and unleavened bread. In: Flour and breads and their fortification in health and disease prevention. Academic Press, London, pp 111–121. https://doi.org/10.1016/B978-0-12-814639-2.00009-5

Siyuan S, Tong L, Liu R (2018) Corn phytochemicals and their health benefits. Food Sci Human Wellness 7(3):185–195. https://doi.org/10.1016/j.fshw.2018.09.003

Stitzer MC, Ross-Ibarra J (2018) Maize domestication and gene interaction. New Phytol 220(2):395–408. https://doi.org/10.1111/nph.15350

Sun X, Ma L, Lux PE, Wang X, Stuetz W, Frank J, Liang J (2022) The distribution of phosphorus, carotenoids and tocochromanols in grains of four Chinese maize (Zea mays L.) varieties. Food Chem 367:130725. https://doi.org/10.1016/j.foodchem.2021.130725

Suri S, Dutta A (2022) Bioactive compounds, antioxidant properties, and health benefits of whole maize and its components. In: Purewal SS, Kaur P, Bangar SP, Sandhu KS, Singh SK, Kaur M (eds) Maize: nutritional composition, processing, and industrial uses, vol 211. CRC Press, London

Suri DJ, Tanumihardjo SA (2016) Effects of different processing methods on the micronutrient and phytochemical contents of maize: from A to Z. Compr Rev Food Sci Food Saf 15(5):912–926. https://doi.org/10.1111/1541-4337.12216

Suriano S, Balconi C, Valoti P, Redaelli R (2021) Comparison of total polyphenols, profile anthocyanins, color analysis, carotenoids and tocols in pigmented maize. LWT 144:111257. https://doi.org/10.1016/j.lwt.2021.111257

Swapna G, Jadesha G, Mahadevu P (2020) Sweet corn—a future healthy human nutrition food. Int J Curr Microbiol App Sci 9(7):3859–3865. https://doi.org/10.20546/ijcmas.2020.907.452

Tablante EC, Pachón H, Guetterman HM, Finkelstein JL (2019) Fortification of wheat and maize flour with folic acid for population health outcomes. Cochrane Database Syst Rev 7:CD012150. https://doi.org/10.1002/14651858.CD012150.pub2

Tanumihardjo SA, McCulley L, Roh R, Lopez-Ridaura S, Palacios-Rojas N, Gunaratna NS (2020) Maize agro-food systems to ensure food and nutrition security in reference to the sustainable development goals. Glob Food Sec 25:100327. https://doi.org/10.1016/j.gfs.2019.100327

Tien Lea D, Duc Chua H, Quynh Lea N (2016) Improving nutritional quality of plant proteins through genetic engineering. Curr Genom 17(3):220–229. https://doi.org/10.2174/1389202917666160202215934

Tiencheu B, Achidi AU, Fossi BT, Tenyang N, Flore E, Ngongang T, Womeni HM (2016) Formulation and nutritional evaluation of instant weaning foods processed from maize (Zea mays), pawpaw (Carica papaya), red beans (Phaseolus vulgaris) and mackerel fish meal (Scomber scombrus). Am J Food Sci Technol 4(5):149–159. https://doi.org/10.12691/ajfst-4-5-5

Trinidad-Calderon PA, Acosta-Cruz E, Rivero-Masante MN, Diaz-Gomez JL, Garcia-Lara S, Lopez-Castillo LM (2021) Maize bioactive peptides: from structure to human health. J Cereal Sci 100:103232. https://doi.org/10.1016/j.jcs.2021.103232

Urias-Lugo DA, Heredia JB, Serna-Saldivar SO, Muy-Rangel MD, Valdez-Torres JB (2015a) Total phenolics, total anthocyanins and antioxidant capacity of native and elite blue maize hybrids (Zea mays L.). CyTA-J Food 13(3):336–339. https://doi.org/10.1080/19476337.2014.980324

Urias-Lugo DA, Heredia JB, Muy-Rangel MD, Valdez-Torres JB, Serna-Saldívar SO, Gutierrez-Uribe JA (2015b) Anthocyanins and phenolic acids of hybrid and native blue maize (Zea mays L.) extracts and their antiproliferative activity in mammary (MCF7), liver (HepG2), colon (Caco2 and HT29) and prostate (PC3) cancer cells. Plant Foods Hum Nutr 70:193–199. https://doi.org/10.1007/s11130-015-0479-4

Urias-Peraldí M, Gutiérrez-Uribe JA, Preciado-Ortiz RE, Cruz-Morales AS, Serna-Saldívar SO, García-Lara S (2013) Nutraceutical profiles of improved blue maize (Zea mays) hybrids for subtropical regions. Field Crop Res 141:69–76. https://doi.org/10.1016/j.fcr.2012.11.008

Vargas-Yana D, Aguilar-Morón B, Pezo-Torres N, Shetty K, Ranilla LG (2020) Ancestral Peruvian ethnic fermented beverage "Chicha" based on purple corn (Zea mays L.): unraveling the health-relevant functional benefits. J Ethnic Foods 7:1–12

Varshney D, Mandade P, Shastri Y (2019) Multi-objective optimization of sugarcane bagasse utilization in an Indian sugar mill. Sustain Product Consumpt 18:96–114

Vaswani S, Kumar R, Kumar V, Roy D, Kumar M (2016) Nutritional and mineral composition of different varieties of normal and high quality protein maize fodder at post-cob stage. Int J Sci Environ Technol 5(5):2719–2727

Wang H, Studer AJ, Zhao Q, Meeley R, Doebley JF (2015) Evidence that the origin of naked kernels during maize domestication was caused by a single amino acid substitution in tga1. Genetics 200(3):965–974. https://doi.org/10.1534/genetics.115.175752

Wang L, Beissinger TM, Lorant A, Ross-Ibarra C, Ross-Ibarra J, Hufford MB (2017) The interplay of demography and selection during maize domestication and expansion. Genome Biol 18(1):1–13. https://doi.org/10.1186/s13059-017-1346-4

Worku IH, Dereje M, Minten B, Hirvonen K (2017) Diet transformation in Africa: the case of Ethiopia. Agric Econ 48(S1):73–86. https://doi.org/10.1111/agec.12387

Wu B, Chang H, Marini R, Chopra S, Reddivari L (2021) Characterization of maize near-isogenic lines with enhanced flavonoid expression to be used as tools in diet-health complexity. Front Plant Sci 11:619598

Xie Z, Wang S, Wang Z, Fu X, Huang Q, Yuan Y et al (2019) In vitro fecal fermentation of propionylated high-amylose maize starch and its impact on gut microbiota. Carbohydr Polym 223:115069

Yadav V, Supriya P (2014) Value addition in maize. In: Chaudhary D, Kumar S, Langyan S (eds) Maize: nutrition dynamics and novel uses. Springer, New Delhi. https://doi.org/10.1007/978-81-322-1623-0_12

Yazew T (2022) Therapeutic food development from maize grains, pulses, and cooking banana fruits for the prevention of severe acute malnutrition. ScientificWorldJournal 2022:3547266. https://doi.org/10.1155/2022/3547266

Yu JK, Moon YS (2021) Corn starch: quality and quantity improvement for industrial uses. Plants 11(1):92. https://doi.org/10.3390/plants11010092

Zhang Z, Zheng X, Yang J, Messing J, Wu Y (2016) Maize endosperm-specific transcription factors O_2 and PBF network the regulation of protein and starch synthesis. Proc Natl Acad Sci 113(39):10842–10847. https://doi.org/10.1073/pnas.1613721113

Zhang Z, Liang Z, Yin L, Li QX, Wu Z (2018) Distribution of four bioactive flavonoids in maize tissues of five varieties and correlation with expression of the biosynthetic genes. J Agric Food Chem 66(40):10431–10437. https://doi.org/10.1021/acs.jafc.8b03865

Zhang S, Ji J, Zhang S, Xiao W, Guan C, Wang G, Wang Y (2020) Changes in the phenolic compound content and antioxidant activity in developmental maize kernels and expression profiles of phenolic biosynthesis-related genes. J Cereal Sci 96:103113. https://doi.org/10.1016/j.jcs.2020.103113

Zhang R, Ma S, Li L, Zhang M, Tian S, Wang D et al (2021) Comprehensive utilization of corn starch processing by-products: a review. Grain Oil Sci Technol 4(3):89–107. https://doi.org/10.1016/j.gaost.2021.08.003

Zhu F (2018) Anthocyanins in cereals: composition and health effects. Food Res Int 109:232–249. https://doi.org/10.1016/j.foodres.2018.04.015

Oats

4

Vaishnavi and Mamta

Abstract

This chapter provides a comprehensive overview of oats (*Avena sativa*) from various dimensions, encompassing their origin and historical significance, production and cultivation practices, global consumption patterns, and chemical composition. It delves into the presence of antinutrients and toxic compounds in oats and explores the bioactive and functional components that contribute to their nutritional value. Additionally, the health implications associated with regular oat consumption, including their role in heart health and digestive well-being, are discussed. The chapter offers insights into the processing of oats, transforming them into diverse products, and explores value addition opportunities in the food industries. Safety considerations related to gluten contamination are examined, highlighting the rise of gluten-free oat varieties and production facilities. Looking ahead, the chapter addresses future challenges and opportunities in oat production, as well as the potential for innovative oat-based products and ingredients. Oats emerge as a versatile, nutritious, and adaptable grain with a promising role in global agriculture and the evolving food landscape.

Keywords

Oats · Nutrients · Functional · β-glucan · Health

Vaishnavi (✉)
Department of Food Science, Nutrition and Technology, CSK Himachal Pradesh Krishi Vishvavidalaya, Palampur, Himachal Pradesh, India

Mamta
School of Bioengineering and Food Technology, Shoolini University, Solan, Himachal Pradesh, India

J. Singh et al. (eds.), *Cereals and Nutraceuticals*,
https://doi.org/10.1007/978-981-97-2542-7_4

4.1 Origin and Background

Oats, scientifically referred to as *Avena sativa*, is a member of the grass family Poaceae. It is an annual cereal plant cultivated for its edible seeds, which are commonly referred to as oat grains or oats. *Avena sativa*, commonly known as cultivated oats, is a distinctive plant with a typical height range of 60–120 cm (2–4 ft). Its botanical features include long, narrow leaves that grow alternately along the stem, displaying parallel venation. The stem is characterized by its hollow, rounded, and erect structure, featuring nodes where leaves and branches emerge (Alemayehu et al. 2023). Oats produce terminal panicles, branching clusters of flowers known as inflorescence. These panicles house numerous small spikelets containing the flowers, which are typically small and bear two specialized bracts called glumes. Oat flowers, which are wind-pollinated, exhibit a diverse range of colors, including green, yellow, or reddish hues. After successful pollination, the flower's ovaries develop into elongated seeds, commonly referred to as oat grains. These seeds are encased in a fibrous outer husk known as the hull. Oats are highly valued for their nutritious grains and are cultivated worldwide for a variety of culinary and agricultural purposes (Butt et al. 2008).

The history and origin of oats (*Avena sativa*) can be traced back thousands of years. Oats are believed to have originated in the Fertile Crescent of the Near East, an area that includes parts of modern-day Iraq, Syria, and Iran. They were among the earliest crops cultivated by ancient civilizations in this region. Oats were cultivated by ancient Egyptians, Greeks, and Romans. They were primarily grown as a source of food for both humans and livestock. The use of oats as a food source can be traced back to around 2000 BCE. Oats became especially popular in northern Europe, including regions such as Scotland and Scandinavia. These areas had climates conducive to oat cultivation, and oats became a staple food for many people in these regions (Chandel 2013). Oats could withstand colder and wetter conditions better than some other cereal grains. Oats continued to be a significant food source during the Middle Ages. They were often used to make porridge and gruel, which were staple foods for peasants and commoners. Oats were brought to North America by European colonists. They were grown in various regions of the continent, but oats did not become as prominent as crops like wheat and corn in the United States. Oats gained popularity as a food source for horses in the nineteenth century, particularly in North America (Gazal et al. 2014). They were used extensively as horse feed due to their nutritional value and suitability for equine diets. In twentieth century, oats remained an important crop for animal feed, and they also gained popularity as a human food source for breakfast cereals and other products. This surge was largely driven by advancements in processing techniques, leading to the widespread availability of oatmeal and oat flour. Rolled oats and oatmeal became common breakfast foods. Today, oats are cultivated worldwide and are known for their nutritional benefits, particularly their high-fiber content and various vitamins and minerals. They are used in a wide range of culinary applications, from breakfast cereals and granola bars to oat flour for baking. Oats also have a reputation for being a heart-healthy food due to their cholesterol-lowering properties, which have contributed to their continued popularity in modern diets (Sangwan et al. 2014).

4.2 Production and Cultivation Practices

Oats are a crucial component of the global cereal crop industry, valued for their nutritious grains, which serve various purposes, including human consumption, livestock feed, and industrial applications. Despite their significance, oats often remain in the shadow of more prominent grains such as wheat, rice, and maize. In 2020, the Food and Agriculture Organization (FAO) reported a global oat production of approximately 23 million metric tons.

This production is distributed across numerous countries, with Russia, Canada, and the United States ranking among the world's leading oat producers. Oats thrive in regions characterized by cool, temperate climates, making them particularly well-suited for cultivation in countries like Canada, where these conditions are optimal. These regions provide the ideal combination of temperature, moisture, and soil quality necessary for robust oat crops (Popa et al. 2012).

The successful cultivation of oats involves several critical stages, each contributing to the overall health and yield of the crop. These stages include land preparation, planting, fertilization, and pest management. Land preparation is a pivotal initial step in creating a suitable environment for oat growth. It encompasses soil testing and nutrient adjustment to ensure optimal oat development. Adequate soil drainage and maintenance of appropriate pH levels are also vital considerations during this phase (Mostafalou and Abdollahi 2014).

Oat planting typically occurs in early spring or late summer, depending on regional climate conditions. The choice of planting method, whether broadcast seeding or drilling, depends on local practices and the availability of suitable equipment. Fertilization plays a critical role in providing oats with the necessary nutrients for healthy growth. Oats require fewer nutrients (N—Nitrogen, P—Phosphorus, and K—Potassium) compared to crops like wheat or maize. Nitrogen, phosphorus, and potassium are key elements in oat nutrition, with their application adjusted based on soil test results and local recommendations (Schwarz et al. 2014).

Effective pest management strategies are essential to protect oat crops from various pathogens and pests. Integrated pest management (IPM) approaches, including crop rotation, the utilization of disease-resistant oat varieties, and timely pesticide application, are employed to minimize yield losses. Oats are typically harvested when they reach maturity, a stage achieved after approximately 90–120 days of growth, depending on environmental conditions and the oat variety. Combine harvesters are commonly used for efficient and timely oat harvesting. Given that oats require more moisture to produce a given unit of dry matter than all other cereals except rice, they thrive particularly well in cool and moist climates (Ti et al. 2016).

4.3 Consumption Pattern

The consumption pattern of oats has undergone a remarkable transformation, transitioning from a modest staple to a globally acclaimed superfood. Oat (*Avena sativa L.*) consumption in human diets has witnessed significant growth, driven by the well-documented health benefits associated with its well-balanced nutritional composition. Oats are cultivated for both cattle feed (70%) and human consumption (30%). In recent years, whole grain consumption has surged in popularity due to the health advantages they offer, presenting diverse opportunities for incorporating flavorful cereal grains into diets.

The American Association of Cereal Chemists (AACC) International has defined whole grains as "the intact ground, cracked, or flaked caryopsis, whose principal anatomical components, the starchy endosperm, germ, and bran, are present in substantially the same relative proportion as they exist in the intact caryopsis." Whole grain oats contain a wealth of valuable nutrients, including proteins, starch, unsaturated fatty acids, and dietary fiber in both soluble and insoluble fractions. Additionally, oats boast micronutrients such as vitamin E, folates, zinc, iron, selenium, copper, manganese, carotenoids, betaine, choline, sulfur-containing amino acids, phytic acid, lignins, lignane, and alkyl resorcinols. While wheat and rice are consumed in significantly larger quantities worldwide compared to oats, oats hold the advantage of being consumed as a whole grain cereal more frequently than their processed products. The recognition of the health benefits associated with the consumption of whole grain cereals is on the rise. Oats are utilized in various forms for food production, including whole grains, rolled oats, crushed oatmeal, and oat flour. Oats are best known as a breakfast cereal, whether consumed whole or in the form of rolled oats. Oatmeal is predominantly used for porridge and in the preparation of various baked goods such as oatcakes, oatmeal cookies, and oat bread. In recent times, several innovative oat-based products have been developed. The increasing popularity of oat-based foods can be attributed to their numerous health benefits. Consumption of oat products has been linked to reduced serum cholesterol levels, a lower risk of cardiovascular disease (CVD), decreased risk of obesity, hypertension, cancer, diabetes, and gastrointestinal disorders. The evolving consumption patterns of oats reflect a convergence of nutritional awareness, culinary innovation, and global food trends, positioning oats as a versatile and celebrated superfood on the world stage.

4.4 Chemical Composition

Nutrient composition of oats encompasses macronutrients, micronutrients, dietary fiber, and bioactive compounds.

4.4.1 Macronutrient Composition

Carbohydrates Oats are predominantly composed of carbohydrates, making up approximately 65–70% of their dry weight. The primary carbohydrate component in oats is starch, serving as a substantial source of energy. What sets oat starch apart are its distinctive characteristics, including short amylose, relatively high crystallinity, and small granule surface, making it unique among cereal starches (Zhu 2017).

Proteins and Amino Acids Oats contain approximately 11–17% protein by dry weight. Unlike many other cereals that rely heavily on prolamins as their main storage proteins, oats have a distinctive protein composition. Globulins, which are salt-water-soluble and account for roughly 55% of the total Osborn protein solubility classification, serve as the main storage proteins in oats, with prolamins constituting a minor portion (Chen et al. 2021). Avenins also play a role, accounting for 10–13% of the total protein content. Notably, oat protein has a profile rich in certain essential amino acids such as glutamine, lysine, and threonine and is lower in proline compared to other cereal grains. Amino acids are particularly concentrated in the embryonic axis and the scutellum within the oat kernel (Kumar et al. 2021).

Fats Oats boast the highest fat content among cereals. Their lipid profile is characterized by a high content of linoleic acid and low levels of saturated fat, contributing to the reduction of the risk of heart and vascular diseases. Monounsaturated fatty acids (MUFA, C 18:1) and polyunsaturated fatty acids (PUFA, C18:2) are the predominant fatty acids in oats, followed by saturated fatty acids (C16:0) (Van Den Broeck et al. 2015). Oats also contain triglycerides as the primary component of lipids, along with significant amounts of phospholipids, glycolipids, and sterols. This rich lipid content makes oats a valuable functional food ingredient across various industries (Singh et al. 2013).

4.4.2 Micronutrient Composition

Vitamins and Minerals Micronutrients, comprising vitamins and minerals, are vital for the body's proper functioning and tissue maintenance. These nutrients, although required in small amounts, are non-negotiable, and deficiencies can have severe consequences. Minerals are categorized into two groups: major and minor. Major minerals, including sodium, potassium, magnesium, calcium, phosphorus, chlorine, and sulfur, are needed in daily quantities exceeding 100 mg for adults. In contrast, minor or trace elements, such as iron, zinc, copper, chromium, cobalt, molybdenum, selenium, nickel, manganese, fluorine, iodine, silicon, tin, and vanadium, are required in daily amounts less than 100 mg. Oats, like other cereal grains, contain minerals in the range of 2–3%. (Kaur et al. 2019) For instance, the iron and zinc content in oats can vary from 1.8 to 6.8 mg/100 g and 6.5 to 10.2 mg/100g, respectively. In a 100 g sample of oat flour, you may find 60 mg of calcium, 372 mg of phosphorus, 3.8 mg of iron, and 3.9 mg of zinc (De Oliveira Maximino et al. 2021). Oats are also rich in vitamins, including thiamine (0.76), riboflavin (0.14), niacin (0.96), pantothenic acid (1.35), vitamin B-6 (0.12), and total folate (56) in milligrams per 100 g. Additionally, vitamin C (0.1), thiamine (0.44–0.53), riboflavin (0.40–0.60), and vitamin E (0.13–0.87) have been reported in milligrams per 100 g of oats. Furthermore, studies have shown vitamin C (0.1), niacin (0.68), vitamin B-6 (0.18), and vitamin E (1.32) content in oats (mg/100 g). Oats are a valuable source of essential vitamins and minerals, contributing to a balanced and nutritious diet (Gabrovska et al. 2004).

4.4.3 Enzymes

Enzymes play a vital role in oats' metabolic activities, and they are abundant in this cereal grain. Various enzymes have been identified in oats, including maltase, proteases, phenoxyacetylase, hydroxylase, α-amylase, lichenase, tyrosinase, phosphatase, and lipase. These enzymes contribute to the versatility and utility of oats in food processing and nutrition.

Oats, with their unique macronutrient composition, stand out as a nutritional powerhouse, offering a rich source of energy, proteins, and healthy fats, making them a versatile and health-conscious choice for various dietary applications (Sangwan et al. 2014).

4.4.4 Dietary Fiber

Oats are renowned for their remarkable dietary fiber content, with approximately 60% of this fiber being insoluble and the remaining 40% soluble. This unique composition contributes significantly to the health benefits associated with oat consumption (Rezende et al. 2021). Two primary sources of dietary fiber in oats are mixed-linkage (1–3), (1–4)-β-D-glucans, commonly referred to as β-glucans, and arabinoxylans. Soluble β-glucans, found within the subaleurone cell walls of oats, have garnered extensive research attention for their potential health advantages. β-glucan is a polysaccharide composed of glucose units linked together. What sets oat β-glucan apart is its distinctive composition. It consists of linear polymers of glucose molecules linked by approximately 30% (1–3) linkages and 70% (1–4) linkages. These linkages, rather than being randomly arranged, exhibit specific patterns: (1–4) links tend to appear in groups of two to four, while (1–3) links appear singly. The presence of (1–3) links introduces flexibility into the molecular chains, facilitating the penetration of water and the solubilization of the fiber. However, it is important to note that adjacent (1–4) links can engage in interchain aggregation through strong hydrogen bonds, which may reduce the solubility of β-glucans (Menon et al. 2016).

In essence, the intricate composition of oat fiber, specifically the unique interplay between (1–3) and (1–4) linkages in β-glucans, contributes to its exceptional solubility and potential health-promoting properties. This complexity underscores the significance of oats as a valuable dietary component for supporting various aspects of human health (Yoo et al. 2020).

4.5 Antinutrients and Toxic Compounds

4.5.1 Antinutrients

While oats (*Avena sativa*) are known for their nutritional advantages, it is essential to recognize that they also contain specific antinutrients and potentially toxic compounds that can impact human health negatively. These substances, present in various foods, can have adverse effects on the body (Ratnasari et al. 2017).

Antinutrients are components of food that, when consumed, exert adverse effects on the body's nutritional status. They often reduce the bioavailability of essential nutrients, potentially leading to deficiencies that can negatively impact health. Plants produce antinutrients as part of their defense mechanisms, protecting themselves from herbivores, insects, pathogens, and adverse environmental conditions. The biological effects of antinutrients can vary widely, depending on the chemical structure of each compound. Oats, like many other plant-based foods, contain antinutrients and potentially toxic compounds that warrant attention. Understanding these substances is crucial for making informed dietary choices and ensuring the optimal utilization of oat-derived nutrients. In the following sections, we will delve into specific antinutritional and toxic compounds found in oats and their potential implications for human health (Graf et al. 2018).

4.5.1.1 Phytates (Phytic Acid)

- **Occurrence:** Abundant in oat grains, with a predilection for the bran layer.
- **Effect:** Adept at binding essential minerals—iron, zinc, calcium, and magnesium—thus diminishing their absorptive capacity within the digestive tract.
- **Management:** Employing techniques such as soaking, sprouting, or fermentation can effectively curtail phytate content, augmenting mineral bioavailability for absorption (Graf et al. 2018).

4.5.1.2 Tannins

- **Occurrence:** Oats harbor tannins, compounds contributing to astringency and bitterness.
- **Effect:** Proficient at hindering digestive enzymes, leading to potential disruptions in protein and carbohydrate digestion.
- **Management:** The application of cooking and processing methodologies stands as a dependable means of reducing tannin levels and attenuating their adverse influence (Popa et al. 2012).

4.5.1.3 Oxalates

- **Occurrence:** Though not as pervasive as in some other dietary sources, oats contain oxalates.
- **Effect:** Elevated oxalate intake may contribute to the formation of kidney stones, particularly in individuals predisposed to such conditions.
- **Management:** Preventing oxalate-related risks involves maintaining adequate hydration and diversifying one's diet (Chen et al. 2021).

4.5.1.4 Protease Inhibitors

- **Occurrence:** Oats house protease inhibitors, compounds known to impede enzymes responsible for protein digestion.
- **Effect:** The presence of these inhibitors may curtail protein digestibility.
- **Management:** Adequate cooking and processing effectively neutralize protease inhibitors, preserving protein bioavailability (Kumar et al. 2021).

4.5.1.5 Lectins

- **Occurrence:** While lectin concentrations in oats are relatively modest, they are present.
- **Effect**: Elevated lectin consumption can potentially disrupt nutrient absorption and digestive processes.
- **Management:** Diligent cooking and thorough processing techniques serve to diminish lectin content, reducing their impact (Butt et al. 2008).

4.5.1.6 Avenin (Gluten-Like Protein)

- **Occurrence:** Oats contain avenin, a protein akin to gluten.
- **Effect:** Individuals with gluten sensitivity or celiac disease may experience digestive discomfort upon exposure to avenin.
- **Management:** For individuals with gluten sensitivity, certified gluten-free oats serve as a suitable and safe alternative (Van Den Broeck et al. 2015).

4.5.1.7 Alkaloids

- **Occurrence:** Oats may contain trace amounts of alkaloids.
- **Effect:** Excessive consumption of specific alkaloids could potentially exert toxic effects on the nervous system.
- **Management:** In oat production, rigorous quality control measures are essential to minimize alkaloid levels and ensure product safety (Menon et al. 2016).

4.5.1.8 Here Are Some Examples of How Processing Affects Antinutrients

- Germination: Germination is a traditional processing method that plays a significant role in reducing antinutrients, especially phytic acid levels in pulses and cereals.
- Soaking: Soaking is another effective traditional method that can help to reduce antinutrients. For instance, soaking lentils was found to reduce phytic acid con-

tent and increase catechins and tannins, while not affecting trypsin inhibitor activity.

- Cooking: Cooking, especially after soaking, can further reduce antinutrient levels. Cooking presoaked seeds, for example, led to a reduction in phytic acid content and complete removal of trypsin inhibitor.
- Heat Treatment: Heat treatment denatures the structure of protein inhibitor enzymes, effectively inactivating them. This method has been found to be useful in beans.
- Dehulling: Dehulling, the removal of the outer shell or husk, can also reduce antinutrient levels.
- Pressure cooking: Pressure cooking is one of the more effective methods for reducing antinutrient content in various foods.
- Sprouting: Sprouting is another process that can significantly decrease antinutrient levels.
- Ordinary cooking: Regular cooking methods can also contribute to antinutrient reduction.

Comprehensive knowledge regarding the presence and implications of these antinutrients empowers individuals to make informed dietary choices and employ appropriate food preparation techniques. Different processing methods can have varying effects on antinutrient reduction, and the choice of method may depend on the specific food and desired nutritional outcomes. These processing techniques are important for improving the nutritional quality of foods and making them more digestible and nutritious (Bora 2014).

4.5.2 Toxic Compounds

1. **Avenin:** Avenin is a protein found in oats and structurally resembles gluten, the protein responsible for gluten intolerance and celiac disease. While oats are generally considered safe for most individuals with celiac disease, a small subset may still react to avenin, necessitating caution (Comino et al. 2015).
2. **Mycotoxins:** Oats, like other grains, can be vulnerable to mycotoxin contamination, particularly by molds such as Fusarium and Penicillium. Mycotoxins produced by these molds, such as deoxynivalenol, can have adverse health effects if ingested in significant quantities (Yu and Pedroso 2023).
3. **Pesticide residues:** Oats are susceptible to pesticide residues, depending on agricultural practices. The presence of pesticide residues in oats can pose health risks if not properly controlled through cultivation and processing (Lee et al. 2009).

4.5.2.1 Mitigation and Safe Consumption

Many antinutrients and contaminants can be reduced through processing methods such as soaking, sprouting, and heat treatment. These methods can help improve the nutritional quality of oats by reducing the levels of phytic acid, tannins, mycotoxins, and pesticide residues. For individuals with celiac disease or gluten sensitivity, it is essential to choose oats labelled as "certified gluten-free" to minimize the risk of cross-contamination with gluten-containing grains. Understanding and addressing these potentially toxic compounds in oats are essential for ensuring their safe consumption and maximizing their nutritional benefits (Alemayehu et al. 2023).

4.6 Bioactive and Functional Components

Oats (*Avena sativa*) are not only a rich source of macronutrients but also contain bioactive and functional components that contribute to their reputation as a health-promoting food.

4.6.1 Bioactive Components

Oats (*Avena sativa*) are well-known for their rich content of bioactive compounds, which contribute to their reputation as a healthy and nutritious food. These bioactive compounds offer a range of health benefits and have garnered significant scientific interest. Following are various bioactive compounds found in oats, their roles in promoting health, and the scientific evidence supporting their benefits.

Beta-Glucans Beta-glucans are a group of polysaccharides primarily found in the endosperm cell walls of oats. They are composed of glucose molecules linked by both (1 → 3) and (1 → 4) glycosidic bonds, with a higher proportion of (1 → 4) linkages.

Beta-glucans are renowned for their cholesterol-lowering effects. These soluble fibers form a gel-like matrix in the digestive tract, which binds to dietary cholesterol and reduces its absorption. Multiple studies have shown that regular consumption of beta-glucan-rich oats can significantly reduce LDL cholesterol levels, which is a key risk factor for cardiovascular diseases. A meta-analysis of randomized controlled trials demonstrated the efficacy of oat beta-glucan in lowering cholesterol. This property makes oats a valuable dietary choice for heart health (Ratnasari et al. 2017).

Avenanthramides Avenanthramides are polyphenolic compounds unique to oats. They are primarily found in the oat bran and are responsible for the grain's antioxidant properties. Avenanthramides exhibit potent antioxidant and anti-inflammatory activities. They have the ability to scavenge free radicals and reduce oxidative stress in the body. Additionally, avenanthramides have been associated with improved cardiovascular health. They may help protect against atherosclerosis, reduce blood pressure, and enhance overall vascular function. These properties make oats a potential dietary strategy for preventing cardiovascular diseases (Tripathi et al. 2018).

Phenolic Acids Oats contain various phenolic acids, with ferulic acid and caffeic acid being notable examples. These phenolic compounds are distributed throughout the grain. Phenolic acids contribute significantly to oats' antioxidant capacity. They play a crucial role in reducing the risk of chronic diseases, including cardiovascular diseases and certain cancers. For instance, ferulic acid, a common phenolic acid in oats, has been associated with anti-inflammatory and anticarcinogenic properties. Long-term oat bran consumption has shown potential in reducing cholesterol levels in hypercholesterolemic individuals, demonstrating the cholesterol-lowering effect of phenolic compounds in oats (Anderson et al. 2018).

These bioactive compounds in oats are not only responsible for the grain's nutritional value but also offer a wide range of health benefits. Understanding the structural characteristics and health-promoting properties of beta-glucans, avenanthramides, and phenolic acids is essential for harnessing the full potential of oats as a functional food. Incorporating oats into one's diet can contribute to improved cardiovascular health, antioxidant effects, and protection against chronic diseases, making them a valuable dietary choice for overall well-being.

4.6.2 Functional Components

The functional components of oats refer to the specific compounds and nutrients found in oats that contribute to their health benefits and functional properties. These components make oats a valuable addition to a healthy diet and have various applications in nutrition and food products. Here are some key functional components of oats:

Dietary Fiber Oats are rich in dietary fiber, with a unique composition of both soluble and insoluble fibers. Soluble fiber, particularly beta-glucans, is found in the bran and endosperm layers, while insoluble fiber is primarily located in the outer bran layers. Dietary fiber plays a pivotal role in promoting digestive health. Insoluble fiber adds bulk to stool, preventing constipation and promoting regular bowel movements. Soluble fiber, mainly beta-glucans, offers several health benefits. It is well-known for its cholesterol-lowering effects by binding to dietary cholesterol in the digestive tract. Additionally, beta-glucans contribute to improved glycemic control by slowing down the absorption of carbohydrates. The feeling of fullness induced by soluble fiber also aids in weight management (Punia et al. 2020).

Proteins Oat proteins are of high quality and contain essential amino acids, making them a valuable component of a balanced diet. Oat proteins are essential for muscle maintenance and repair. They have gained popularity in various dietary contexts, including plant-based diets, sports nutrition, and the development of gluten-free products. The well-balanced amino acid profile of oat proteins makes them suitable for meeting diverse dietary needs (Butt et al. 2008).

Minerals Oats provide essential minerals such as manganese, phosphorus, magnesium, and iron. These minerals play crucial roles in various physiological processes. Manganese contributes to bone health and wound healing. Phosphorus is essential for energy metabolism, including the conversion of food into energy. Magnesium is crucial for muscle and nerve function, maintaining a steady heartbeat and supporting the immune system. Iron is necessary for oxygen transport in the blood, preventing anemia and fatigue (Godswill et al. 2020).

Vitamins Oats contain several vitamins, including thiamine (vitamin B1), riboflavin (vitamin B2), niacin (vitamin B3), pantothenic acid (vitamin B5), vitamin B6, and folate (vitamin B9). These B vitamins are essential for energy production, nerve function, and DNA synthesis. They play vital roles in maintaining proper bodily functions, including metabolism, cognition, and overall well-being (Tripathi et al. 2018).

Phenolic Compounds Oats contain phenolic compounds, including avenanthramides and phenolic acids such as ferulic acid and caffeic acid. Phenolic compounds are known for their antioxidant properties. They help protect cells from oxidative stress and inflammation. Consuming oats with phenolic compounds has been associated with reducing the risk of chronic diseases like cardiovascular disease and certain types of cancer (Lee et al. 2009).

Lignans Oats are a source of lignans, a class of phytochemicals. Lignans also possess antioxidant properties and may have hormone-balancing effects. Research is ongoing to understand their potential role in reducing the risk of hormone-related cancers and other health conditions.

These functional components collectively make oats a versatile and nutritious food choice. They offer a wide range of health benefits, from supporting heart health and digestion to providing essential nutrients for overall well-being. Incorporating oats into your diet can be a valuable step toward a healthier lifestyle (Peterson et al. 2010).

4.7 Health Implications

Due to the several significant health advantages of including oats in our daily diet, oats are becoming more and more well-liked as a component of a balanced diet. Oat grains are rich in nutrients and offer significant health advantages. Oat bran is an excellent source of β-glucan and has been shown to be useful in decreasing plasma cholesterol, and postprandial blood sugar levels also aid in lowering coronary heart disease risk. The FDA's suggested daily allowance of oat β-glucan for reaching

0.75% of each serving, or 3 g/day, should be ingested for health effects. Numerous positive effects of oat consumption in diabetes mellitus patients have been documented, including cardiovascular disorders, hypertension, and infections. Diabetes patients should pay special attention to the low glycemic index of oat products, and it has been observed that consuming viscous foods high in β-glucan affects the amount of fat absorbed in the gastrointestinal tract and lowers lipase activity. Low molecular fatty acids are released during β-glucan fermentation in the colon, which preconditions its potential anticarcinogenic activity (Alemayehu et al. 2023). The following is a list of some of oat's health benefits:

4.7.1 Cholesterol and Oats

Oats are a key source of dietary fiber (both soluble and insoluble) and other vital elements that are required by the human body. The soluble fiber present in oats includes β-glucan, which has been shown to reduce blood cholesterol. Oat β-glucan consumption may lower LDL cholesterol and total blood cholesterol levels. As it moves through the digestive system, oat-soluble fiber-glucan breaks down, generating the gel that retains some elements connected to cholesterol-rich bile acids (Ryan et al. 2007). The absorption of cholesterol into the bloodstream is decreased by this trapping. Dietary oat β-glucan forming a viscous coating in the small intestine is one of the main proposed mechanisms. The viscous layer reduces both the reabsorption of bile acids (which the body produces from cholesterol) and intestinal absorption of ingested cholesterol. In reaction, the body creates fresh bile acids by drawing on the pool of circulating cholesterol. Less dietary cholesterol is absorbed from the gut, and more body cholesterol is produced to make bile, which lowers the amount of circulating cholesterol in the blood. A daily intake of at least 3 g of β-glucan may lower total plasma and LDL cholesterol levels by 5–10%, respectively (Othman et al. 2011).

One of the richest sources of the substances known as tocotrienols is the grain of oats. These are the antioxidants that makeup vitamin E, together with tocopherols. These tocotrienols have been discovered to reduce blood cholesterol levels by inhibiting the manufacture of cholesterol. Numerous forms of cardiovascular disease are linked to cholesterol buildup. When oats are included in a low-fat, high-fiber diet, they are more effective at lowering cholesterol. Additionally, according to recent studies, oat antioxidants lower cholesterol by lowering blood cells' tendency to adhere to arterial walls (Maki et al. 2007).

4.7.2 Diabetes and Oats

The oat β-glucan has positive effects on blood sugar or diabetes. When offered to diabetics, oat fiber-rich diets exhibit much lower blood sugar spikes. Therefore, consuming oats can lower blood sugar levels for a longer length of time. Controlling blood sugar and insulin levels is crucial for avoiding many of the complications linked to diabetes (Jenkins et al. 2002). The rise in blood sugar levels after meals and the delay in their return to pre-meal levels are both slowed by oat-glucan. The viscosity of the contents of the stomach and small intestine increases as the soluble oat fiber (β-glucan) is digested and forms a gel. As a result, digestion is slowed, and the body takes longer to absorb carbohydrates into the bloodstream. This means that dramatic changes in blood sugar levels are avoided (Tosh et al. 2013).

4.7.3 Cancer and Oats

Oats contain specific phytochemicals (plant compounds) in their hundreds, which, when taken, can lower the risk of cancer. Muesli is recommended by many cancer-related organizations as an advantageous component of a healthy diet for reducing the risk of cancer. The phytoestrogen compounds (known as lignans) present in oats have been linked to a decreased risk of diabetes, hypertension, obesity, hormone-related diseases like breast cancer, and many other compounds are present in oats which prevent different types of cancers and act as anticarcinogenic agents. Fiber in oats lowers the risk of colon cancer (Hsueh et al. 2011). Oats include vitamins and minerals that help the body fight against cancer cells. Oat eating will lower the number of polyps in colon cancer patients, according to a study conducted by Harvard. According to international studies, women who consume more dietary fiber have lower levels of circulating estrogen, which is a risk factor for breast cancer. The insoluble fibers in oats may also lessen carcinogens in the digestive system (Ratnasari et al. 2017).

4.7.4 Blood Pressure and Oats

A systolic blood pressure of more than 140 mmHg or a diastolic blood pressure of more than 90 mmHg is considered high blood pressure. Additionally, oats support the host's blood pressure regulation. Oat-soluble fiber consumption can lower hypertension or high blood pressure and hence lessen the requirement for hypertensive medication. By enhancing the glycemic and insulinemic profiles, oats may aid in the maintenance of healthy blood pressure. Oat cereals dramatically lower both systolic and diastolic blood pressure in hypertensive individuals when added to their regular diet. Whole oats that are high in soluble fiber may be a useful dietary therapy in the management and prevention of hypertension. Dietary consumption of oats is associated with the DASH (Dietary Approaches to Stop Hypertension) recommendations and may confer a health benefit due to its fiber content (Maki et al. 2007).

4.7.5 Oat Improves Bowel Function

Dietary fiber, which is abundant in oats, is important for maintaining regular bowel movements. Both soluble and insoluble fibers are abundant in oats. Water does not cause the insoluble fibers to dissolve, but they are spongy and can absorb many times their own weight in liquid. Constipation is relieved by the insoluble fiber, which makes stools bulkier and hastens their transit through the digestive tract (Khalid et al. 2022).

4.7.6 Oat Helps in Weight Control

The feeling of satiety alerts the stomach that it has had enough food and that it is time to quit eating. Oat-soluble fibers, β-glucans, are broken down in the stomach and produce a gel that makes the stomach and small intestine's contents more viscous. This enhances fullness and helps curb hunger. The gel postpones stomach emptying, prolonging your feeling of fullness and assisting in weight loss. When combined with a good, balanced diet and enough exercise, the urge for food is lessened, which aids with weight control. According to reports, youngsters who ate muesli had a 50% lower risk of becoming overweight than those who didn't eat it. There is some evidence that children between the ages of 2 and 18 who regularly consume muesli reduce their chance of becoming obese. Oats are primarily valued for their ability to supply calories for energy needs, much as other cereal grains. Regular consumption of oats has beneficial impacts on athletes' performance. Additionally, it has been discovered that oat protein is more satiating than isoenergetic doses of fat or carbohydrate. When consumed 45 to 1 h prior to moderately intense exercise, oats have been demonstrated in trials to favorably modify metabolism and improve performance (Karabudak, et al. 2019).

4.7.7 General Health and Longevity

In comparison to other cereals, oats have a larger percentage of balanced protein. They also include a good balance of vital fatty acids, which are associated with long life and general health. Oats include phytochemicals that have been linked to defense against chronic illnesses including cancer.

The best amino acid profiles of any grain are found in oats; these vital proteins aid in aiding the body's optimal operation. Oat β-glucan also seems to speed up the immune system's reaction to infection, which could lead to quicker healing (Chen et al. 2021).

4.8 Processing

Agricultural commodities must be processed into edible, tasty, and appealing foods before they can be distributed to customers. Food processing can extend shelf life, make nutrients more bioavailable, stable color and flavor, boost economic value, and make preparing raw food ingredients easier through a variety of physical and chemical processes (Decker et al. 2014) (Fig. 4.1).

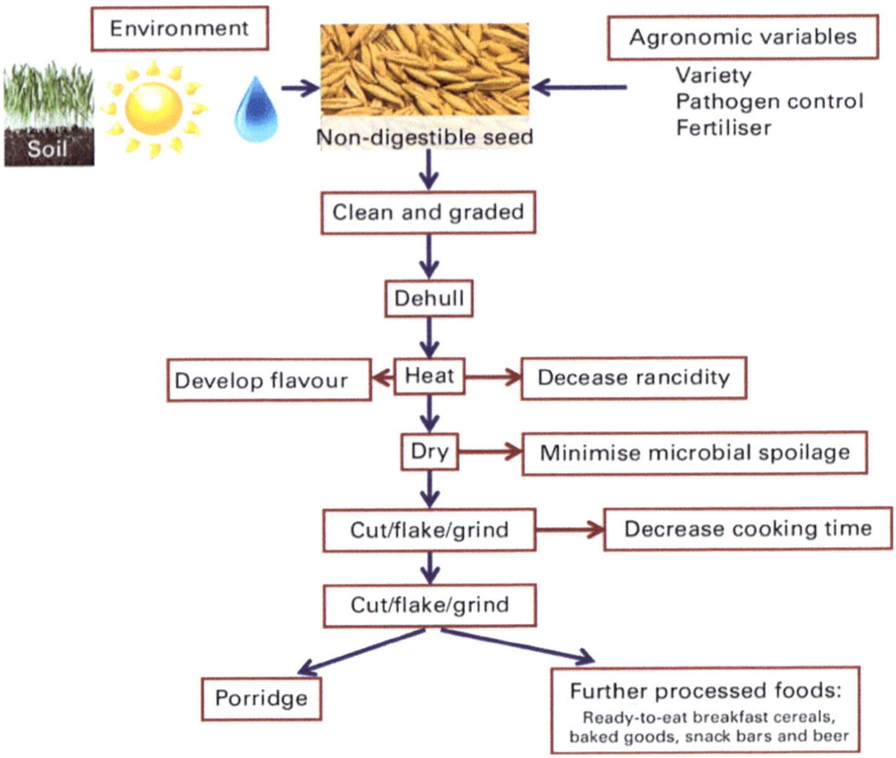

Fig. 4.1 A flow diagram showing the production of oats on the farm through typical oat-processing operations and production of human foods. The impact of processing operations on the farm variables is included in other sections in this special addition. How the quality of the oats is affected by some processing steps are included in the side boxes. Available at: http://www.journals.cambridge.org/bjn

4.8.1 Cleaning

Oats will be mixed with other ingredients in the field and during transit after being harvested, as is the case with many agricultural products. These foreign substances must be eliminated in order to make oats fit for human consumption. The main method for achieving this is screening. As soon as the oats reach the mill, they pass through a magnetic separator to get rid of any unwanted metal items, which is something that many food-processing facilities do. The oats are subsequently exposed to a series of revolving or oscillating screens that have the ability to both retain large materials (such as straw, twigs, and stone) and allow small objects to pass through, like immature oats, dirt, weed seeds, and dust. The remaining light components are then aspirated out of the retained oat stream. The next step is a dry stoner, which eliminates high-density but comparable-sized particles like pebbles and other grains like maize. Oats may occasionally be clipped before cleaning, in which the tip of the oat is removed to improve the efficiency of the subsequent dehulling. In order to remove the clipped-off piece before continuing to mill, clipping is done before cleaning. A mesh screen is used for clipping, and the oat's narrow end can fit through it. The oat is then displaced from the mesh by a spinning bar, breaking off the tip, and the cut-off tips are aspirated away (Girardet and Webster 2011).

Additionally, the oats can be divided into various size classes using a rotating separator, which can improve milling effectiveness by removing small kernels that contain a small proportion of growth. For this procedure, a rotational drum with indentations is used. As the name suggests, an indent separator contains indents on the inner face of a revolving cylinder that are the size of the seeds that are to be removed; for example, an indent separator designed to remove weed seeds has indents that are smaller than oat seeds. The weed seeds are carried high up the side of the rotating cylinder by fitting securely into the indents. The tight fit of the seed in the indentation allows it to be carried far up the side of the cylinder until the pull of gravity drops the weed seed into a trough that is in the center of the cylinder. The larger oats, which cannot be hoisted high enough up the side walls of the cylinder, continue to travel out of the bottom of the cylinder while the seeds in the trough are subsequently removed by a screw conveyor. By altering the height of the catch trough, the size of the particle removed by the screw conveyer can be changed. Because even particles that do not precisely fit in the indents will ascend up the sides of the cylinder due to friction, the lower the trough, the larger the particles that are separated. This method can be used to remove some small oats, including light, double, and pin oats, which are then fed to animals (Fellows 2009).

4.8.2 Grading

The medium and small oats then flow into a second perforated cylinder, which has smaller pores than the first. The isolated oats must be separated into various sizes in order to run the oat mill as efficiently as possible, which is accomplished using the variation in the weight and density of the fraction. It is more accurate to separate oats with comparable weights based on width, this method of separation is used by most graders. For width separation, a set of perforated cylinders is used. The smaller and medium-width oats can pass through the first's intermediate perforation, while the larger oats are taken out of the cylinder's end (over-tail). The medium and small oats then flow into a second perforated cylinder, which has smaller pores than the first (Decker et al. 2014).

4.8.3 Dehulling

Since the shell is largely indigestible, it must be removed to get the most nutritional value. A revolving disc with several fins going from the disc's core to the outside is used in modern mills as the initial stage of dehulling. The groat is separated from the hull as the oats fall into the center of this rotating disc and are hurled onto a succession of impact rings on the dehuller's wall. Approximately 85% of the oats should be dehulled by this point because any more dehulling would break the groats and reduce production. The oats are divided into similar weights before dehulling since the efficiency of dehulling depends on oat weight, moisture content, and the throughput of the dehulling machine. The speed at which the disc rotates can be changed to improve dehulling effectiveness without breaking groats. For instance, larger oats require a lower speed of rotation to separate the hull than do little oats. A moisture concentration of 12–13% is desirable because moisture has a significant impact on dehulling performance, with high moisture reducing dehulling and low moisture increasing groat-breakage rates. Dehulling yields a mixture of hulls, groats, unhulled oats, and broken groats, all of which need to be further separated (Girardet and Webster 2011).

After dehulling, the oat stream is aspirated to remove the hulls and fines (minor fragments created during breakage). After passing through a cylinder with a rough interior to remove any sticking hulls, the mainly dehulled oats are aspirated again. The next step is to pass the oat stream through a table or paddy separator, which uses variations in particle density and smoothness to distinguish groats from unhulled oats. Groats pass through the separator more quickly than unhulled oats because they are denser and have a smoother texture. The oat stream enters a table that is tilted and has several bumpers on it, much like a pinball game. The oat stream then bounces off of the bumpers as the table moves back and forth on its longitudinal axis. While the unhulled oats float to the top, the smoother, denser groats sink to the bottom. The dehuller receives the unhulled oats at that point. Several table separators are used in series for end products needing high levels of dehulled oat removal (such as chopped oats). In order to eliminate contaminated grains like wheat and barley that are larger than the groats, a perforated drum separator is used (Decker et al. 2013).

4.8.4 Kiln Drying

The most typical method of kilning involves arranging the groats in extended vertical cylinders and infusing steam and air into the columns. To quickly raise the temperature of the groat, live steam is fed into the top of the column. Because the effectiveness of enzyme inactivation rises with increasing moisture content, the steam causes the groats moisture content to rise. Further down the column, the groats are exposed to radiant heating (dry heat) to evaporate extra moisture because an increase in groat moisture content might affect the quality and storage stability of the finished products. Radiant heating has the added benefit of accelerating the Maillard reaction, which results in delicious nutty flavor and caramel tints. Air is introduced into the groats toward the end of the process to lower temperature and remove moisture, leaving a final water content of 10%. Measuring peroxidase activity allows for the monitoring of the efficiency of enzyme inactivation.

Due to peroxidase's greater heat stability compared to lipase and lipoxygenase, its total inactivation guarantees that lipase and lipoxygenase are also rendered inactive.

The ability to kiln to inactivate bacteria, yeasts, and molds that can shorten shelf life and pose hazards to food safety is another benefit to oat quality. Although kilning will also damage some heat-liable vitamins, such as B vitamins, as with all thermal processing methods, the advantages of being able to extend shelf life greatly outweigh these negative impacts (McClements and Decker 2008).

4.8.5 Converting Groats to Food Ingredients and Products

The finished products after milling include entire groats of various sizes, broken groats, and powdered fines (Fig. 4.2a). These various goods can be used to create oat flakes, steel-cut oats, oat flour, oat bran, and, in some situations, oat components like fiber.

Fig. 4.2 Photographs of oat groats (**a**), steel-cut oats (**b**), rolled oats (**c**), oat flour (**d**), and oat bran (**e**) that show differences in physical appearance due to processing. Available at: http://www.journals.cambridge.org/bjn

4.8.6 Cutting

Simply cutting the full groats into smaller pieces results in steel-cut oats (Fig. 4.2b). A rotary granulator, which consists of a perforated drum with steel knives on the outside, achieves this. The blades cut the groat barrier that extends through the perforation as the drum is rotated. Each groat is typically chopped two to four times. The speed of rotation, the placement of the blades, the size of the perforations, and the beginning size of the groat are only a few of the many variables that impact the size of the sliced groat. This procedure generates very small amounts of fines (2%), which are eliminated by a shifter. Oats that have been sliced into various sizes are separated by an indented cylinder that resembles the cleaning equipment (Decker et al. 2014).

4.8.7 Flaking

Oat flakes (Fig. 4.2c) are easily made by simply rolling two revolving rollers over entire or steel-cut groats. However, due to their low moisture content, the groats coming out of the kiln drying process are extremely vulnerable to being crushed into powder. To prevent this, the groats are agitated and subjected to steam before rolling in an effort to add 3–5% moisture. The groats' temperature will rise during this process, which can last 20–30 min. To limit nutrient degradation, moisture equilibration should be accomplished in an ideal manner with the least amount of temperature increase over the shortest amount of time (Kaur et al. 2012).

The distance between the rollers can be changed to determine the thickness of the flake; the thinner the flake, the quicker it will cook. Generally speaking, whole oat flakes are rolled thicker (0.51–0.76 mm) than quick-cooking flakes (0.36–0.46 mm). Flake thickness is increased even further for items like muesli. For quick-cooking oats, the flakes' thickness might range from 0.7–12 to 0.4 mm. After rolling, the flakes are exposed to an air stream to lower their temperature and moisture content, bringing the water content back down to 10–12%. To break up clumps of flakes and remove fines and small flakes, the flakes are finally passed over a shaking shifter. The starch in the oat flakes may get slightly pre-gelatinized as a result of drying after steaming. Pre-gelatinized starch will more rapidly absorb watered than unprocessed starch, thus decreasing the cooking time (McGee 2004).

4.8.8 Flour

A pin or hammer mill can be used to grind groats or oat flakes into flour. Oat flour (Fig. 4.2d) tends to clump during grinding because of its high fat content; therefore, the air is employed to propel the grain through the mill and reduce heat buildup. The residual large particles are removed from the flour when it exits the mill, and it then goes via a vibrating shifter to be recycled for subsequent milling. Baby food and ready-to-eat cereals are the two main products that employ oat flour (Stevenson and Inglett 2011).

4.8.9 Oat Bran

There are coarse and fine fractions of oat flour. The outer aleurone and sub-aleurone layers of the groat include the coarse fraction, or bran, which is higher in fiber, protein, vitamins, and minerals, as well as somewhat higher in fat. There is a lot of starch in the fine fraction. Shifting the oat flour separates the coarser bran portion. The US Food and Drug Administration and the American Association of Cereal Chemists stipulate that oat bran must have a minimum total dietary fiber level of 16%, with at least a third of that content being soluble, and a minimum b-glucan concentration of 5% (Decker et al. 2014).

4.8.10 Oat Hulls

Utilizing oat hulls, which make up 28–32% of the oat, is difficult because they are a by-product. The hulls include 30–35% fiber, 30–35% pentosans, 10–15% lignans, protein, and ash, which is rich in silicic acid. They also contain 30–35% pentosans. If the hulls are ground to a fine consistency, they can be used as a high-fiber animal feed and food ingredient. Oat hulls have recently been utilized as biomass for power plants; for instance, a nearby biomass power station uses oat hulls from a general mills food-processing facility to generate enough electricity to power over 17,000 homes. The University of Iowa and Quaker Oats have a similar cooperation, with oat hulls replacing coal and providing over 10% of the university's energy requirements. Oat hulls can also be used as a source of cellulosic fiber using alkaline treatments (Decker et al. 2014).

4.9 Value Addition

Because more people are aware of the numerous nutritional advantages of oats, there has been an increase in the demand for products made from them in recent years. Increased consumer health consciousness has placed emphasis on eating a diet high in fiber. A great source of dietary fiber is oats. As a result, its usage is encouraged in oat-based functional food products such as porridge, oatmeal, muesli, granola bars, oat flour, bread, biscuits, and cookies, as well as oat milk, oatrim, probiotic drinks made from oats, morning cereals made from oats, flakes, and infant food. Ice creams can be stabilized with oat β-glucan. Antioxidants in oats are effective for stabilizing dairy and meat products that are susceptible to oxidation of fat during storage. Due to their viscous and emulsifying qualities, oat proteins have been employed in food products, including heat-resistant chocolates. It has been discovered that adding oats to foods increases their general quality (Sanchez-Pardo

Table 4.1 Utilization and applications of oats

Value added products of oats	Component of oat
Bread	Oat flour
	Oat starch
	Oat lecithin
Beverage	Whole oat
Biscuits and cookies	Oat flour
Breakfast cereal	Whole oat
Pasta products	Oat starch
Granola bars and cereals	Whole oat and resistant starch fractions
Infant food	Whole oat
Oat milk	Whole oat extract
Oat-based non-dairy fermented yoghurt	Whole oat
Fat substitute in meat balls, dairy and bakery products	Oat bran
Fat substitute	Oat dextrin
	Soluble β-glucan
Stabilizer in ice creams	β-glucan
Gluten-free diet	Whole oat
Cholesterol-lowering effect	β-glucan
Anticancerous effect	β-glucan
	Short chain fatty acids
Methane production	Biogas Oat husk

et al. 2010). Oats are used in a variety of ways for medical, industrial, and food purposes (Rasane et al. 2015) as shown in Table 4.1.

4.10 Safety Consideration

Consumption of gluten by those who are gluten intolerant causes celiac disease. A complex protein that is alcohol soluble, gluten is mostly found in wheat and other similar grains like barley and rye. Gluten consumption results in an inappropriate small intestinal immune response in genetically susceptible people, characterized by villous atrophy and crypt hyperplasia, which causes malabsorption of protein, fats, carbohydrates, soluble vitamins, folate, and minerals, particularly iron and calcium. Currently, the sole treatment option is to entirely exclude gluten from the patient's diet. Oats may boost a gluten-free diet nutritional content; however, some people disagree with their inclusion. To determine if oats included in a gluten-free diet for patients with celiac disease are safe. However, there have been questions as to whether oats can be safely consumed by celiac disease patients (Arentz-Hansen et al. 2004). This is due to inconsistent clinical findings, where a handful of studies have suggested that some celiac disease patients possess sensitivity to oat proteins. In contrast, other studies have shown that even high amounts of oats are well tolerated by celiac disease subjects. This has led to an unsettled debate about the safety of oats as a dietary component for celiac disease patients, with many clinicians and some regulatory authorities, such as Food Standards Australia New Zealand, recommending against routine oat consumption. Oats are cereal grain with a very high risk of contamination from gluten-containing grains (GCGs). However, it is now widely acknowledged that people with celiac disease can safely consume pure oats that have not been tainted by GCGs. Due to this, a lot of oat suppliers now predicate their capacity to prevent GCG cross-contamination on their ability to make gluten-free labelling claims. Oats and other whole grains need to be identified and given industry guidance regarding their safety for the gluten-free market. Oats and other whole grains can be visually inspected to see if they are gluten-free. This method avoids many issues with antibody-based testing methods, such as sampling restrictions and the uneven distribution of gluten in ground flours (Murray and Makharia 2017).

4.10.1 Future Challenges and Opportunities

Oat compounds offer several options for including oats in food products with beneficial properties. The bioavailability of antioxidants from oats and other food sources must be determined, as well as their varied effects on human and animal health. Antioxidants and β-glucan are just two of the fascinating ingredients found in oats. Oats need more scientific study to support and adjust their nutraceutical status in geriatric and pediatric diets because they are convenience meals that people of all ages consume. To identify novel functional components in oats and extract these components in fractions that can be used in food items, more research and development are required. The optimal utilization of oats and products generated from oats must be ensured in order to help meet the rising nutritional demands, which is why food security is seen as a global concern. Another avenue of study could be the genetic engineering of a particular enzyme or system of enzymes in oats to change the pathways and produce high levels of the desired ingredients (Chen and Wang 2020).

4.11 Conclusions

In conclusion, oats have a fascinating history that spans millennia, from their origins in the ancient Near East to their widespread use in modern cuisine. Their adaptability, nutritional value, and diverse culinary applications have cemented their place as a valuable and enduring crop in the world of agriculture and food production. Successful oat production and cultivation require a combination of careful planning, proper agronomic practices, and effective pest and disease management. By selecting suitable oat varieties, preparing the soil adequately, managing nutrients, and implementing sound weed and pest control strategies, farmers can ensure a bountiful oat harvest. Post-harvest handling practices, including drying and storage, are equally important to preserve grain quality and marketability. Health awareness, changing dietary preferences, and effective marketing strategies have all contributed to this transformation, making oats a ubiquitous and valued component of modern diets worldwide. The chemical composition of oats is a complex and nutritionally valuable blend of macronutrients, micronutrients, dietary fiber, bioactive and functional compounds. Oats are not only a source of energy but also contribute to various aspects of human health, making them a versatile and nutritious addition to the diet. Oats, despite their numerous health benefits, contain antinutrients and potentially toxic compounds that can affect mineral absorption, protein digestion, and, in some cases, gluten-sensitive individuals. Understanding these components and employing appropriate processing and consumption practices can help mitigate their adverse effects and ensure the safe and nutritious incorporation of oats into the diet. Understanding the applications of these components can help promote the consumption of oats as a part of a balanced diet, contributing to overall health and well-being.

References

Anderson JW, Baird P, Davis RH Jr, Ferreri S, Knudtson M, Koraym A, Waters V, Williams CL (2001) Health benefits of dietary fiber. Nutr Rev 59(7):188–205

Anderson GF, Forsido SF, Tola YB, Amare E (2023) Nutritional and phytochemical composition and associated health benefits of oat (*Avena sativa*) grains and oat-based fermented food products. Sci World J 2023:2730175

Andon MB, Anderson JW, Miller ME (2001) The oatmeal-cholesterol connection: 10 years later. Am J Lifestyle Med 14(2):154–159

Arentz-Hansen H, Korner R, Molberg (2004) The oat protein avenin is a major source of T-cell reactivity in patients with celiac disease. Gastroenterology 127(3):574–584

Bora P (2014) Antinutritional factors in foods and their effects. J Acad Ind Res 3(6):285–290

Brennan MA, Cleary LJ (2005) The potential use of cereal (1→3, 1→4)-β-d-glucans as functional food ingredients. J Cereal Sci 42(1):1–13

Butt MS, Nadeem MT, Khan MKI, Shabir R (2008) Oat: unique among the cereals. Eur J Nutr 47:68–79

Chandel S (2013) Oats: a multi-purpose crop. Indian J Agric Res 47(3):203–211

Chen H, Wang Z (2020) Advances in oat research: nutritional benefits and future opportunities. Food Sci Nutr 60(11):1794–1808. https://doi.org/10.1080/10408398.2018.1558605

Chen O, Mah E, Dioum E et al (2021) The role of oat nutrients in the immune system: a narrative review. Nutrients 13(4):1048

Comino I, Moreno MDL, Sousa C (2015) Role of oats in celiac disease. World J Gastroenterol 21(42):11825–11831

De Oliveira Maximino JV, Barros LM, Pereira RM et al (2021) Mineral and fatty acid content variation in white oat genotypes grown in Brazil. Biol Trace Elem Res 199(3):1194–1206

Fellows P (2009) Food processing technology: principles and practice. Woodhead Publishing, Cambridge, UK

Gabrovska D, Fiedlerová V, Holasová M et al (2004) Nutritional changes of common oat (*Avena sativa* L.) and naked oat (*Avena nuda* L.) during germination. Czech J Food Sci 22(5):317–320

Gazal A, Dar ZA, Zaffar G, Lone AA, Abidi I, Shabir A, Khan YN (2014) Trends in breeding oat for nutritional grain quality. J Appl Nat Sci 6:904–912

Girardet N, Webster FH (2011) Oat milling: specifications, storage, and processing. In: Oats: chemistry and technology, 2nd edn. American Association of Cereal Chemists, St Paul, MN, pp 301–316

Godswill AG, Somtochukwu IV, Ikechukwu AO, Kate EC (2020) Health benefits of micronutrients (vitamins and minerals) and their associated deficiency diseases: a systematic review. Int J Food Sci 3:1–32

Graf E, Empson KL, Eaton JW (2018) Phytic acid. A natural antioxidant. J Biol Chem 262(24):11647–11650

Haug W, Lantzsch HJ (1983) Sensitive method for the rapid determination of phytate in cereals and cereal products. J Sci Food Agric 34(11):1423–1426

Hsueh CW, Chia HH, Jeng DH, Mon YY, Shing JW, Chau JW (2011) Inhibitory effect of whole oat on aberrant crypt foci formation and colon tumor growth in ICR and BALB/c mice. J Cereal Sci 53(1):73–77

Huang T, Xu M, Lee A, Cho S, Qi L (2015) Consumption of whole grains and cereal fiber and total and cause-specific mortality: prospective analysis of individuals. BMC Med 13(59):367–442

Jenkins AL, Jenkins DJA, Zdravkovic U, Wursch P, Vuksan V (2002) Depression of the glycemic index by high levels of ß-glucan fiber in two functional foods tested in type 2 diabetes. Eur J Clin Nutr 56:622–628

Kaur KD, Jha A, Sabikhi L, Singh AK (2012) Significance of coarse cereals in health and nutrition: a review. J Food Sci Technol. https://doi.org/10.1007/s11483-012-0290-3

Kaur S, Bhardwaj RD, Kapoor R, Grewal SK (2019) Biochemical characterization of oat (*Avena sativa* L.) genotypes with high nutritional potential. Food Sci Technol 110:32–39

Khalid W, Arshad MS, Jabeen A, Muhammad Anjum F, Qaisrani TB, Suleria HAR (2022) Fiber-enriched botanicals: a therapeutic tool against certain metabolic ailments. Food Sci Nutr 10(10):3203–3218

Kumar L, Sehrawat R, Kong Y (2021) Oat proteins: a perspective on functional properties. Food Sci Technol 152:112307

Lee-Manion AM, Price RK, Strain J, Dimberg LH, Sunnerheim K, Shinde M (2016) Oat β-glucan and serum cholesterol: an analysis. Food Chem 211:531–537

Ludwig DS (2000) Dietary glycemic index and obesity. Am Soc Nutr Sci 130:280–283

Maki KC, Galant R, Samuel P, Tesser J, Witchger MS, Ribaya-Mercado JD, Blumberg JB, Geohas J (2007) Effects of consuming foods containing oat β-glucan on blood pressure, carbohydrate metabolism and biomarkers of oxidative stress in men and women with elevated blood pressure. Eur J Clin Nutr 61:786–795

McClements DJ, Decker EA (2008) Lipids. In: Fennema's food chemistry. CRC Press, Boca Raton, FL, pp 155–216

McGee H (2004) On food and cooking: the science and Lore of the kitchen. Scribner, New York, NY

Menon R, Gonzalez T, Ferruzzi M, Jackson E, Winderl D, Watson J (2016) Oats-from farm to fork. Adv Food Nutr Res 77:1–55

Minatel IO, Borges CV, Ferreira MI, Gomez HA (2014) Lipid profile and oxidative stability of processed oats (*Avena sativa* L.). J Food Sci Technol 51(11):3305–3310

Mostafalou S, Abdollahi M (2014) Pesticides and human chronic diseases: evidences, mechanisms, and perspectives. Toxicol Appl Pharmacol 268(2):157–177

Murray JA, Makharia GK (2017) Oats in the diet of patients with celiac disease. Nutrients 9(2):185

Othman RA, Moghadasian MH, Peter JH (2011) Cholesterol-lowering effects of oat β-glucan. Nutr Rev 69:299–309

Peterson J, Dwyer J, Adlercreutz H, Scalbert A, Jacques P, McCullough ML (2010) Dietary lignans: physiology, potential for cardiovascular disease risk reduction. Nutr Rev 68(10):571–603

Pitt JI, Hocking AD (2009) Fungi and food spoilage. Springer Science and Business Media

Popa A, Duta D, Niculita P (2012) Recent advances in oat-based functional cereal products. Rom Biotechnol Lett 17:7717–7725

Punia S, Sandhu KS, Dhull SB et al (2020) Oat starch: physico-chemical, morphological, rheological characteristics and its applications—a review. Int J Biol Macromol 154:493–498

Rasane P, Jha A, Sabikhi L, Kumar A, Unnikrishnan VS (2015) Nutritional advantages of oats and opportunities for its processing as value added foods—a review. J Food Sci Technol 52(2):662–675

Ratnasari N, Walters M, Tsopmo A (2017) Antioxidant and lipoxygenase activities of polyphenol extracts from oat brans treated with polysaccharide degrading enzymes. Heliyon 3(7):e00351
Rezende ESV, Lima GC, Naves MMV (2021) Dietary fibers as beneficial microbiota modulators: a proposed classification by prebiotic categories. Nutrition 89:111217
Ryan D, Kendall M, Robards K (2007) Bioactivity of oats as it relates to cardiovascular disease. Nutr Res 20(2):147–162
Sanchez-Pardo ME, Jimenez G, Gonzalic-Gracia I (2010) Study about the addition of chemically modified starches (cross-linked corn starches), dextrins, and oat fibers in pound cake. J Biotechnol 150(3):316–319
Sangwan S, Singh R, Tomar SK (2014) Nutritional and functional properties of oats. J Innov Biol 1:3–14
Schwarz PB, Kirkegaard JA, Bithell SL, Krishnamurthy R, Stapper M, Delhaize E (2014) Root morphology and seedling vigour of oat genotypes with and without major QTL for Alp in acid soils. Crop Pasture Sci 65(4):343–353
Singh R, De S, Belkheir A (2013) *Avena sativa* (Oat), A potential nutraceutical and therapeutic agent: an overview. Food Sci Nutr 53(2):126–144
Stevenson DG, Inglett GE (2011) Oat dietary fiber: commercial processes and functional attributes. In: Chemistry and technology, 2nd edn. American Association of Cereal Chemists, St Paul, MN, pp 321–328
Ti H, Zhang R, Zhang M, Li Q, Wei Z, Zhang Y, Jiao Y (2016) Effects of oat consumption on stress, antioxidant status, and heat shock protein response in endurance athletes. Nutrition 32(2):255–261
Tosh SM, Chu Y, McIntosh T (2013) Canadian oats and oat products: a review of processing and impact on glycaemic response. Br J Nutr 110(5):876–886
Tripathi V, Singh A, Ashraf M (2018) Avenanthramides of oats: medicinal importance and future perspectives. Pharmacognosy 12(23):66–71
Van Den Broeck HC, Londono DM, Timmer R, Smulders MJ, Gilissen LJ, Van Der Meer IM (2015) Profiling of nutritional and health-related compounds in oat varieties. Food Secur 5(4):2–11
Wang Y, Harding SV, Eck P, Thandapilly SJ (2015) High-molecular-weight β-glucan decreases serum cholesterol differentially based on the polymorphism in mildly hypercholesterolemic adults. J Nutr 145(4):799–805
Yoo H-U, Ko M-J, Chung M-S (2020) Hydrolysis of beta-glucan in oat flour during subcritical-water extraction. Food Chem 308:125670
Yu J, Pedroso IR (2023) Mycotoxins in cereal-based products and their impacts on the health of humans, livestock animals, and pets. Toxins (Basel) 15(8):480
Zhu F (2017) Structures, properties, modifications, and uses of oat starch. Food Chem 229:329–340

Rye

5

Diksha Kataria, Jaspreet Kaur, Jyoti Singh, Prasad Rasane, Vishesh Bhadariya, and Sawinder Kaur

Abstract

Rye cereal, an ancient grain with a rich history, has recently gained attention in the field of nutritional science for its exceptional nutraceutical potential. This book chapter delves into the multifaceted world of rye, shedding light on its historical roots, nutritional profile, and its remarkable role in promoting health and well-being. It is the second most popular cereal grain worldwide, closely related to wheat. Its unique properties allow it to cross-breed with wheat, to produce triticale. Within the kernel and bran of rye lie some essential nutrients and bioactive compounds. Cereal rye grain is renowned for its abundant fibre content, which has been linked to numerous health benefits, including gallstones and cancer prevention, diabetes management, enhanced metabolic markers, and cardiovascular health maintenance. It provides in-depth knowledge of the nutritional components that make rye a nutraceutical powerhouse. It also provides highlights of various bioactive compounds and phytochemicals such as lignans and alkylresorcinols that contribute to its antioxidant properties. Additionally, this chapter also discusses the applications of rye in the food industry, where it is increasingly utilized in the production of breads, breakfast cereals, and even alcoholic beverages like whiskeys and vodkas. These culinary practices not only showcase ryes versatility but also its potential to enhance the nutritional quality of everyday foods. By

D. Kataria · J. Singh · P. Rasane · S. Kaur
Department of Food Technology and Nutrition, School of Agriculture, Lovely Professional University, Phagwara, Punjab, India

J. Kaur (✉)
Department of Food Technology and Nutrition, School of Agriculture, Lovely Professional University, Phagwara, Punjab, India

V. Bhadariya
School of Chemical Engineering, Oklahoma State University, Stillwater, OK, USA

J. Singh et al. (eds.), *Cereals and Nutraceuticals*,
https://doi.org/10.1007/978-981-97-2542-7_5

exploring its nutraceutical potential, we gain a valuable insight into how this cereal grain contributes to our health and well-being, both now and in the future.

Keywords

Rye · Nutritional profile · Bioactive components · Nutraceutical potential · Processing · Health benefits

5.1 Introduction of Rye (*Secale cereale*)

As the field of nutritional science continues to evolve, there is a growing interest in discovering and utilizing the health benefits of natural food sources. Rye cereal stands out among the many grains and cereals that have been consumed for centuries due to its exceptional collection of nutraceutical compounds. It is the second most popular cereal grain after wheat which is mostly grown in the northern part of the northern hemisphere. It belongs to the grass family Poaceae and the tribe Triticeae with other cereals like wheat, barley, and oats. It is so similar to wheat that it can cross-bred with wheat and form triticale (*Triticosecale wittmack*) (Ayalew et al. 2018). Different nutrients in rye and rye fibre have already been recognized as nutraceuticals for overall health and wellness. Additionally, there is a large number of bioactive compounds and phytochemicals present in rye bran such as lignans, alkylresorcinols, benzoxazinoids, phytosterols, phenolic acids, folates, tocopherols, and tocotrienols that shows its antioxidant properties (Afzal et al. 2013). It also carries abundant amounts of vitamins and minerals such as pantothenic acid, manganese, copper, phosphorous, and magnesium. Rye has several health benefits including preventing gallstones and cancer, controlling diabetes, boosting metabolic performance, and maintaining cardiac health (Chopade et al. 2017). Due to its health-promoting effects, rye has been increasingly used in the baking of breads, breakfast cereals, and beverage whiskey and vodkas in the food industries. This book chapter will summarize the recent advances in the nutritional profile of rye, examine its historical roots, and scrutinize the presence of antinutritional and toxic compounds in its nutraceutical properties. It will also delve into the main molecular family of bioactive and functional components that render rye cereal a nutritional component and a nutritional powerhouse while considering its profound implications for health and well-being. Furthermore, various processing techniques transform rye into a versatile culinary ingredient.

5.2 Origin and Background

Feral rye or cereal rye is the common name of rye in the United States derived from the old English word ryge and from the proto-German word ruig by the Slavic tribe. It is considered an allogamous plant species that shows a high degree of self-incompatibility (Schlegel 2013). It is a diploid species that originated from the Near East (Crespo-Herrera et al. 2017). The primary centre of origin of rye grain was southwestern Asia. It was then distributed to Russia and Scandinavia and from here to Poland and Germany, before further spreading throughout Europe and thus became a European crop. The oldest grown rye seed in central Europe was from the Hallstatt period (1000–500 BC). From 2500– to 2000 BC, rye cultivation was migrated to Poland and Sweden from the northwest. Later during the sixteenth and seventeenth centuries, it was brought to North America and South America by Europeans and thus the production of rye was increased probably due to the winters (Schlegel 2013; Sapirstein and Bushuk 2016).

5.3 Production and Cultivation Practices

The production and cultivation practices of rye involve various agricultural practices to ensure optimal growth, yield, and quality of the crop. Rye is a winter cereal that requires less soil and heat (Chopade et al. 2017). Therefore, it can be cultivated in areas that are not suitable for other cereals. In Germany and the Nordic, Baltic and Eastern European countries, rye grain is considered as an essential part of the cereal food culture (Poutanen 2016).

5.4 Consumption Pattern

Rye grain has lately been referred to as "The overlooked cereal" from the British point of view but the consumption trend is still falling. Whole-grain bread and rye bread are key elements of the nutritious Nordic diet, offering numerous therapeutic benefits. In Finland, 85% of the population regularly consumes rye bread with men consuming an average of 124 g/day and women consuming an average of 78 g/day. Consequently, rye bread serves as a primary contributor of dietary fibre intake in Finland, accounting for a total of 24 g/day among men and 21 g/day among women. Rye crisp bread, the dehydrated form of traditional rye bread, stands out as one of the most well-known rye-based bread varieties as it also possesses positive effects on both glucose metabolism and post-meal satisfaction. It also has a longstanding history of use in dishes like porridge and some breakfast cereals (Poutanen 2016). Some notable Finnish rye dishes include Karelian pastries, composed of thin rye dough shells filled with rice or barley porridge and kalakukko and meat and fish baked in thick rye cover. A large variety of contemporary rye-containing goods, such as pasta, cookies, rice alternatives, frozen dough, and snacks, are also sold throughout the Nordic nations. Scandinavian markets have also provided rye hamburgers, rye baking ingredients, puddings, and flakes (Poutanen et al. 2014; Poutanen 2016).

5.5 Nutritional and Chemical Composition

Rye is considered as second most important grain for bread after wheat. This is due to its high nutritional content of dietary fibre, vitamin B, and several minerals, including Ca, Fe, F, Zn, Mn, Cu, P, and K for which it is considered a suitable ingredient for the production of functional foods (Popović et al. 2022). Bran, the outermost layer of the kernel, contains a higher amount of fibre than any other cereal including arabinoxylan, cellulose, β-glucans, fructans, and lignin (Dapčević-Hadnađev et al. 2018). Arabinoxylan is a non-cellulose polysaccharide that represents a major part of rye dietary fibre and is most abundant in rye compared to other cereals. Arabinoxylan positively and negatively influences the baking quality of rye flour through its water-extractable and non-extractable properties as it is a functional ingredient in baked foods (Chen et al. 2019). The carbohydrate content of the rye is almost equal to wheat which potentiates the prevention of diabetes. Starch is the most evident component found in the rye ranging between 55% and 65%. It is made up of two different types of polymers, i.e., amylose and amylose pectin with amylose ranging between 24% and 26% of the total (Poutanen et al. 2014). A detailed information on the nutritional composition of rye is presented in Table 5.1.

5.5.1 Antinutrient and Toxic Components

Antinutritional factors are the secondary metabolites that hinder the utilization and bioavailability of nutrients and further affect the reproduction, normal growth, and health of humans (Saurabh et al. 2021). The endosperm of the grain is surrounded by the aleurone cells that are rich in phytic acid, which is powerful enough to combine multi-charged metal ions, especially Zn, Fe, and Ca resulting in them being inaccessible to humans. Simple, traditional, and household methods like soaking, heating, roasting, germination, and fermentation are used for the preparation of grain to enhance its nutrition profile (Ikram et al. 2023). Grabiński et al. (2021) reported that anti-nutritive content largely depends on the genotype and weather conditions. During its cultivation season in 2011 and 2013, stress-related factors raised the levels of alkylresorcinols, trypsin inhibitors, and soluble pentosans. Alkylresorcinols are the phenolic lipids found in the outer part of the grain. These are considered specific dietary biomarkers of whole-grain wheat and rye intake. Trypsin inhibitors are the proteins that interfere with the digestion of protein by inhibiting the activity of the enzyme trypsin, and water-soluble pentosans-rich fraction from rye grain has been found to have antinutritional activity.

5.5.2 Bioactive and Functional Components

There are certain bioactive compounds present in the germ and outer layer of the rye such as phenolic acids, lignans, benzoxazinoids, alkylresorcinols, minerals and trace elements, tocotrienols, tocopherols, and folates (Poutanen et al. 2014; Poutanen 2016). These compounds are either found bound to the fibre structure or present close to it so they can be released easily during the digestion process in the small intestine and can be metabolized by the gut microbiota (Jonsson et al. 2018).

5.5.2.1 Alkylresorcinols

Alkylresorcinols are the amphiphilic phenolic lipids in large amounts particularly in rye and wheat among all other cereals. These are particularly present in the outer

Table 5.1 Nutritional composition of rye

Nutrients	Amount	References
Macronutrients		
Energy (kJ/100 g)	1413	Jonnalagadda et al. (2011)
Carbohydrates (%)	60.7–75.9	Petrova and Petrov (2017), Della Pepa et al. (2018)
Sugar (%)	3.7	Frølich et al. (2013)
Starch (%)	68.3	Devi et al. (2014)
Lipids (%)	1.6–2.4	Frølich et al. (2013), Petrova and Petrov (2017), Della Pepa et al. (2018)
Proteins (%)	8.8–13.4	
Fibre (%)	13.2–16.1	
Crude fibre (%)	2.1	Devi et al. (2014)
Arabinoxylan (%)	6–11.5	Frølich et al. (2013), Németh and Tömösközi (2021)
Cellulose (%)	2.9	
β-glucan (%)	0.7–3	
Fructan (%)	3.6–6.6	
Lignin (%)	1.1	
Ash (%)	1.6–2	Devi et al. (2014), Poutanen (2016)
Vitamins (mg/100 g)		
Thiamine	0.316–0.69	Della Pepa et al. (2018), Devi et al. (2014)
Riboflavin	0.22–0.26	Della Pepa et al. (2018), Devi et al. (2014)
Niacin	1.2–4.270	Della Pepa et al. (2018)
Pyridoxine	0.294–0.35	
Folate	0.038–78	
Vitamin-E	0.850–1.60	
Vitamin-K	0.060	
Minerals (mg/100 g)		
Calcium	24–32	Frølich et al. (2013), Della Pepa et al. (2018)
Iron	2.63–3.4	
Magnesium	110–96	
Phosphorous	332–348	
Potassium	510–410	
Sodium	2	
Zinc	2.65–3.0	
Phytochemicals		
Phenolic acids (μg/g)	491–1082	Frølich et al. (2013)
Ferulic acid (mg/100 g)	4	Della Pepa et al. (2018)
Flavonoids (mg/100 g)	7	
Alkylresorcinol (μg/g)	797–1231	Frølich et al. (2013)
Total phenols (mg/100 g)	13.2	Della Pepa et al. (2018)
Phytic acid (μg/g)	540–1460	Frølich et al. (2013)
Tocols (μg/g)	44–67	
Phytosterol (μg/g)	1098–1420	

layer of the grain. Alkylresorcinols are absorbed in the small intestine, transported in lipoproteins, and distributed in red blood cell membranes and adipose tissue, possibly other parts of the body (Jonsson et al. 2018). It has a cholesterol-lowering effect as it decreases cholesterol absorption by interfering with the micelle solubilization of cholesterol in the digestive tract. Its derivatives with unsaturated that account for 20% of total alkyresorcinols in rye and oxygenated side chains also exist. Therefore, dietary alkylresorcinol is bioavailable in the human body and may have in vivo bioactivities. Its daily intake is high in countries like Germany, Scandinavia, Poland, and the Baltic states (Andersson et al. 2014).

5.5.2.2 Phenolic Acids

It is the most important phenolic compound present in the outer layer of the rye grain. It includes two groups, viz*:* *p*-hydroxybenzoic acid (its derivatives vanillic and syringic) and *p*-coumaric acid (its derivatives ferulic acid and caffeic acid). Although free phenolic acids make up less than 3% of the overall phenolic content in rye, their presence plays a substantial role in enhancing the antioxidant properties of rye flour. Moreover, during the fermentation process in the colon by gut microbiota, bound phenolic compounds are released from the food material leading to numerous health benefits. It contains phenolic acid ranging between 65 and 300 mg/100 g of grain. It also contains several other predominant phenolic compounds such as ferulic (85–90%), *p*-coumaric (3–5%), and sinapic (9–10%). These compounds help in the prevention of certain diseases such as cardiovascular diseases, diabetes mellitus, and cancers and also potentiate anti-inflammatory, antioxidant, anti-cancer, anti-mutagenic, anti-glycaemic and neuroprotective properties. Additionally, in comparison to other cereals, it is also a good source of *p*-hydroxybenzoic, gallic, vanillic, and caffeic acid (Dziki 2022).

5.5.2.3 Flavonoids

Flavonoids are a type of phytonutrient with antioxidant properties present in a minor quantity in the bran and the outer layer of the grain. Common flavonoids found in rye include quercetin, rutin, and kaempferol. Flavones are the most important flavonoids as their structure is linked with enhanced inflammatory properties, improved insulin sensitivity and interaction with intestinal membrane transporters (Ravisankar et al. 2020). Kulichová et al. (2019) conducted a study in which free phenolic acids, including p-coumaric, ferulic, and sinapic acids, as well as phenolic acids released from soluble esters and glycosides such as vanillic, caffeic, p-coumaric, ferulic, and sinapic acids, were detected in rye cereal. Among these, ferulic acid emerged as the predominant phenolic acid in rye grains.

5.5.2.4 Lignans

Lignans are the phenolic compounds, present in vegetables, fruits, and seeds. Many structures have also been found in different cereal grains. After consumption, these compounds are converted into mammalian lignans enterodiol and enterolactone by gut microbiota and are further absorbed in the small intestine which exerts various health benefits (Poutanen et al. 2014). High intake of lignin and blood concentration of enterolactone is linked to a decreased risk of illness, particularly a better prognosis following breast cancer by inhibiting cell proliferation (Jonsson et al. 2018).

5.6 Health Implications

Intake of high dietary fibre especially from cereals has been found to exert a protective effect toward many chronic diseases including cardiovascular diseases and diabetes mellitus. Some cereal fibre also enhances the conversion of cholesterol to bile acid in the liver which improves the functionality of the large intestine and reduces the levels of certain biomarkers that are associated with an increased risk of colorectal cancer (Sluková et al. 2021). It is also considered a robust and stress-tolerant cereal (Miedaner and Laidig 2019). The various health benefits of rye are summarized in Fig. 5.1.

5.6.1 Effect on Metabolic Risk Markers

Rye being an excellent source of fibre has been reported to improve metabolic risk markers and also contributes to weight reduction. Iversen et al. (2022a) conducted a 12-week trial, where all the individuals were asked to either have refined wheat products or high-fibre rye products with a 500 kcal/day energy deficit. After the completion of the 12-week intervention, all the participants were found to have reduced their weight and fat, and people consuming rye-based products also had reduced CRP levels and acetic acid concentration whereas, increased levels of

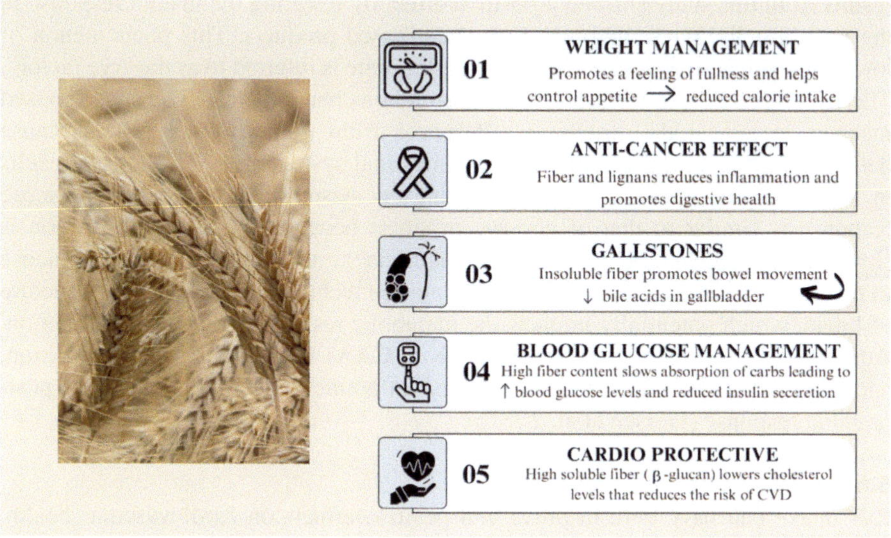

Fig. 5.1 Health benefits of rye

alkylresorcinols and plasma short-chain fatty acid concentration were found to be increased.

5.6.1.1 Anti-cancer Effect of Rye

Rye is a whole-grain cereal that has been studied for its health benefits, including its impact on cancer risk. Colon cancer is one of the most widely spread diseases and the major cause of death and mortality in the Western world. Insoluble dietary fibre in rye affects the intestinal transit time which in turn decreases the risk and relieves the symptoms of constipation (Jonsson et al. 2018). Moreover, proteins and peptides present in rye tend to potentiate the preventive effect against different cancer stages, i.e. initiation, promotion, and progression. It also contains phytochemicals, including lignans and phenolic acids, which have an antioxidant effect that helps protect cells from oxidative stress and may play a role in reducing the risk of cancer (Rodríguez-García et al. 2019).

5.6.1.2 Gallstones

Rye is a type of grain that is often consumed in various forms, including bread, cereal, and pasta. It's a good source of dietary fibre and has several potential health benefits particularly soluble fibre. It binds to cholesterol in bile and prevents it from forming solid particles that can lead to gallstones (Tehrani et al. 2023).

5.6.1.3 Blood Glucose Management

Whole-grain rye contains various beneficial components that also contribute to the management of blood glucose levels. It has a relatively low glycemic index compared to refined grains. Iversen et al. (2022b) also conducted a study to optimize the effect of rye-based food on insulin and glucose in the post-prandial phase. The results from this study showed a positive effect by lowering the insulin response in the post-prandial phase compared to wheat-based products. This phenomenon of lowering the insulin levels in the post-prandial state is referred to as the "rye factor". There are several possible mechanisms behind this phenomenon. The one proposed mechanism is that the structural differences from non-rye control items cause delayed glucose absorption in the gastrointestinal system after intake. This results in lower insulin secretion after consuming rye-based products, but the glucose response is similar to that of non-rye products because if glucose absorption is slower, less insulin would be required to maintain the same concentration of glucose in the blood. Besides this, different fermentation techniques also alter the structure of bread, which potentially impacts the metabolic response to bread consumption. Additionally, branched-chain amino acids (BCAA) may affect insulin secretion, and it has been speculated that cereal product amino acid content alters the post-prandial response (Iversen et al. 2022b).

5.6.1.4 Cardiovascular Health

Rye intake can have both negative and positive effects on cardiovascular health. Afzal et al. (2013) discussed a study in which 22,000 individuals of a certain age range had a negative correlation with the amount of fibre consumed and the prevalence of coronary heart disease. This population group was studied for a considerable amount of time which resulted that compared to the group of people who ate fruit and vegetable fibre, the group that ingested rye as a source of dietary fibre had a greater risk of suffering a myocardial infarction. Along with this, several other studies have also been performed for its protective effect. (Helnæs et al. 2016), performed a study including 54,871 adults aged 50–64 years old, during 13.6 years of follow-up 2329 participants experienced myocardial infarction. A questionnaire was analysed that focused on the consumption of whole-grain items and species. The results from the follow-up showed that rye bread (in both men and women), oatmeal (in men), and whole grain from rye and oats (in men) were significantly linked to lower risk of myocardial infarction.

5.7 Processing of Rye

In various studies, it has been observed that the processing step significantly affects the flavor and texture of rye bread and can also increase its bitterness. The key processing step for transforming rye grain into food for human consumption is milling, which essentially involves two multistep processing steps, namely a break and reduction system. The break system essentially entails crack opening the grain by utilizing corrugated rolls, resulting in the separation of the bran from the endosperm. Furthermore, the endosperm is ground to produce flour. On the other hand, the reduction system's main objective is to refine the middling particles, which consist mostly of pure endosperm, into the final flour (Sapirstein and Bushuk 2016). To ensure high-quality flour, the initial step in the milling process involves the removal of any unwanted contaminants. Specialized cleaning machines are used in a series, with variations in shape, size, and density utilized to enhance cleaning effectiveness. The final stage of the cleaning process is tempering, which requires the addition of water to strengthen the bran while softening the endosperm, making it easier to separate (Campbell et al. 2020). Rye milling necessitates a larger sifting area compared to hard wheat types due to the finer particle size distribution of rye flour, resulting from the endosperm's softer texture, which causes clumping.

5.8 Value Addition

Demand for healthier goods with high dietary fibre and bioactive components is rising with the customers' growing health consciousness. Rye being a good source of nutrition can be used for milling into flour and whole meal for baking into a variety of foods including rye bread, crisp bread, black bread, sourdough bread and pumpernickel and for some beverages like beer, whiskies, and vodkas (Wrigley and Bushuk 2017).

Soft whole-grain rye bread is the most common type of rye bread consumed in Denmark, Russia, Finland, and the Baltic nations. It is made by using a sourdough method that involves mixing the ingredients including whole-grain rye flour, starter culture, and water. The mixed ingredients are fermented for around 8–18 h during that period, lactic acid bacteria and the sourdough yeast grow, and various enzymatic reactions occur leading to the formation of the main components such as lactic acid and acetic acid along with which several flavour compounds and precursors such as amino acids are also formed. After the fermentation process is completed, additional water, rye grain flour, and other ingredients are mixed with the sourdough to prepare the dough. The dough is then left for a short period to rise, post that the breads are shaped and set to rise again, and baked (Poutanen et al. 2014).

In another study by Nasabi et al. (2021), wafers were made by incorporating barley, oats, corn, and rye flour with wheat flour as a control sample. Wheat was replaced by 20% of each selected sample, and thus four different samples were formulated. Each sample batter was prepared using 100 g of flour, 0.4 g of salt, 0.2 g of sodium bicarbonate, 150 g of water, 1 g of palm oil, 0.1 g of ammonium bicarbonate, and 0.5 g of lecithin. The sample with rye flour obtained the lowest hardness score, the highest porosity score and the least acceptability for colour among all the samples. Besides breads, there are several other dietary applications of rye grain such as cookies, snacks, breakfast cereals, beer, whiskeys, vodkas, etc. (Wrigley and Bushuk 2017).

5.9 Future Challenges and Opportunities

Rye like wheat is a multifaceted ingredient that can be used in food product development in many ways. Nowadays, as consumers are becoming more conscious regarding their health, the demand for healthy food products with more bioactive compounds and dietary fibre content is also increasing (Németh and Tömösközi 2021). But the future of rye grain presents both challenges and opportunities. Climate variability and extreme weather events are the first challenges that may affect rye production and quality, making it important to develop climate-resilient varieties (Hackauf et al. 2022). It is a crop that is sensitive to changes in temperature, precipitation, and other climate factors. If there are unexpected shifts in temperature or rainfall patterns, it can disrupt the growth and development of the rye plant. In order to address these challenges, it is crucial to develop rye varieties that can withstand and adapt to changing climate conditions and extreme weather events. These climate-resilient varieties would be better equipped to thrive even in less favourable conditions, ensuring a more reliable and stable rye production. It also encounters a competitive challenge in the food industry. It is a type of grain that can be used for making bread, just like wheat. However, wheat is more widely consumed and preferred for bread-making due to its well-known characteristics. Gluten's unique properties contribute to the formation of a light, airy, and chewy bread texture that many people prefer. Rye bread, on the other hand, tends to be denser and has a different flavour profile.

5.10 Conclusion

This book chapter delves into the multifaceted world of rye grain and its nutraceutical components. Rye often overshadowed by its more prominent cereal counterparts emerges as a nutritional powerhouse with immense potential for promoting human health and well-being. By detailed examination of its chemical composition, we have unveiled a treasure trove of bioactive compounds, including dietary fibres, lignans, phenolic acids, and more, each bearing the promise of various health benefits. These nutraceuticals have been linked to the prevention and management of chronic diseases, including cardiovascular disorders, diabetes, and certain cancers. Furthermore, our exploration of rye as a functional ingredient and its functional properties has highlighted its role in improving glycaemic control, promoting satiety, and aiding in weight management. However, it is imperative to acknowledge the challenges associated with promoting rye's adoption on a broader scale. Consumer preferences, processing techniques, and market dynamics all play pivotal roles in determining its accessibility and acceptance. In light of these findings, it is evident that rye grain and its nutraceuticals offer a compelling avenue for further research, product development, and dietary recommendations. Rye being a functional ingredient can undoubtedly emerge as a vital component of a health-conscious, sustainable diet.

References

Afzal S, Shehzad A, Randhawa MA, Asghar A, Shoaib M, Jahangir MA (2013) Health benefits and importance of utilizing wheat and rye. Pak J Food Sci 23(4):212–222

Andersson AA, Dimberg L, Åman P, Landberg R (2014) Recent findings on certain bioactive components in whole grain wheat and rye. J Cereal Sci 59(3):294–311

Ayalew H, Kumssa TT, Butler TJ, Ma XF (2018) Triticale improvement for forage and cover crop uses in the southern great plains of the United States. Front Plant Sci 9:1130

Campbell GM, Webb C, Owens GW, Scanlon MG (2020) Milling and flour quality. In: Breadmaking. Woodhead Publishing, Duxford, pp 187–219

Chen Z, Li S, Fu Y, Li C, Chen D, Chen H (2019) Arabinoxylan structural characteristics, interaction with gut microbiota and potential health functions. J Funct Foods 54:536–551

Chopade MU, Patil HS, More RB (2017) Cereals and their nutritional value for health benefits. Adv Innov Res 58–61. https://www.researchgate.net/profile/Pravin-Dhakite/publication/326648368_DEVELOPMENT_OF_GREEN_ECOFRIENDLY_PRODUCTS_BASED_ON_NATURAL_VEGETATION/links/5b5acd500f7e9bc79a66b621/DEVELOPMENTOF-GREEN-ECOFRIENDLY-PRODUCTS-BASED-ON-NATURAL-VEGETATION.pdf#page=82

Crespo-Herrera LA, Garkava-Gustavsson L, Åhman I (2017) A systematic review of rye (Secale cereale L.) as a source of resistance to pathogens and pests in wheat (Triticum aestivum L.). Hereditas 154:1–9

Dapčević-Hadnađev T, Hadnađev M, Pojić M (2018) The healthy components of cereal by-products and their functional properties. In: Sustainable recovery and reutilization of cereal processing by-products. Woodhead Publishing, Duxford, pp 27–61

Della Pepa G, Vetrani C, Vitale M, Riccardi G (2018) Wholegrain intake and risk of type 2 diabetes: evidence from epidemiological and intervention studies. Nutrients 10(9):1288

Devi PB, Vijayabharathi R, Sathyabama S, Malleshi NG, Priyadarisini VB (2014) Health benefits of finger millet (Eleusine coracana L.) polyphenols and dietary fiber: a review. J Food Sci Technol 51:1021–1040
Dziki D (2022) Rye flour and rye bran: new perspectives for use. PRO 10(2):293
Frølich W, Åman P, Tetens I (2013) Whole grain foods and health—a Scandinavian perspective. Food Nutr Res 57(1):18503
Grabiński J, Sułek A, Wyzińska M, Stuper-Szablewska K, Cacak-Pietrzak G, Nieróbca A, Dziki D (2021) Impact of genotype, weather conditions and production technology on the quantitative profile of anti-nutritive compounds in rye grains. Agronomy 11(1):151
Hackauf B, Siekmann D, Fromme FJ (2022) Improving yield and yield stability in winter rye by hybrid breeding. Plan Theory 11(19):2666
Helnæs A, Kyrø C, Andersen I, Lacoppidan S, Overvad K, Christensen J et al (2016) Intake of whole grains is associated with lower risk of myocardial infarction: the Danish Diet, Cancer and Health Cohort. Am J Clin Nutr 103(4):999–1007
Ikram A, Saeed F, Noor RA, Imran A, Afzaal M, Rasheed A et al (2023) A comprehensive review on biochemical and technological properties of rye (Secale cereale L.). Int J Food Prop 26(1):2212–2228
Iversen KN, Dicksved J, Zoki C, Fristedt R, Pelve EA, Langton M, Landberg R (2022a) The effects of high fiber rye, compared to refined wheat, on gut microbiota composition, plasma short chain fatty acids, and implications for weight loss and metabolic risk factors (the RyeWeight study). Nutrients 14(8):1669
Iversen KN, Jonsson K, Landberg R (2022b) The effect of rye-based foods on postprandial plasma insulin concentration: the rye factor. Front Nutr 9:868938
Jonnalagadda SS, Harnack L, Hai Liu R, McKeown N, Seal C, Liu S, Fahey GC (2011) Putting the whole grain puzzle together: health benefits associated with whole grains—summary of American Society for Nutrition 2010 Satellite Symposium. J Nutr 141(5):1011S–1022S
Jonsson K, Andersson R, Knudsen KEB, Hallmans G, Hanhineva K, Katina K et al (2018) Rye and health—where do we stand and where do we go? Trends Food Sci Technol 79:78–87
Kulichová K, Sokol J, Nemeček P, Maliarová M, Maliar T, Havrlentová M, Kraic J (2019) Phenolic compounds and biological activities of rye (Secale cereale L.) grains. Open Chem 17(1):988–999
Miedaner T, Laidig F (2019) Hybrid breeding in rye (Secale cereale L.). In: Advances in plant breeding strategies: cereals, vol 5. Springer, Cham, pp 343–372
Nasabi M, Naderi B, Akbari M, Aktar T, Kieliszek M, Amini M (2021) Physical, structural and sensory properties of wafer batter and wafer sheets influenced by various sources of grains. LWT 149:111826
Németh R, Tömösközi S (2021) Rye: current state and future trends in research and applications. Acta Aliment 50(4):620–640
Petrova P, Petrov K (2017) Traditional cereal beverage boza fermentation technology, microbial content and healthy effects. In: Fermented foods, part II. CRC Press, Boca Raton, pp 284–305
Popović DB, Rajičić V, Popović V, Burić M, Filipović V, Gantner V et al (2022) Economically significant production of Secale cereale L. as functional foods. Agric Forest 68(3):133–145
Poutanen K (2016) Rye and rye bread—an important part of the North European bread basket. In: Rye and health. Elsevier, Amsterdam, pp 1–6
Poutanen K, Katina K, Heiniö RL (2014) Rye. In: Zhou W, Hui YH, De Leyn I, Pagani MA, Rosell CM, Selman JD, Therdthai N (eds) Bakery products science and technology. Wiley Blackwell, Hoboken, pp 75–87
Ravisankar S, Queiroz VA, Awika JM (2020) Rye flavonoids—structural profile of the flavones in diverse varieties and effect of fermentation and heat on their structure and antioxidant properties. Food Chem 324:126871
Rodríguez-García C, Sánchez-Quesada C, Toledo E, Delgado-Rodríguez M, Gaforio JJ (2019) Naturally lignan-rich foods: a dietary tool for health promotion? Molecules 24(5):917

Sapirstein HD, Bushuk W (2016) Rye grain: its genetics, production, and utilization. In: Wrigley CW, Corke H, Seetharaman K, Faubion J (eds) Encyclopedia of food grains, vol 1, 2nd edn. Elsevier, Oxford, pp 159–167

Saurabh V, Urhe SB, Upadhyay A, Shankar S (2021) Antinutritional factors in cereals. In: Handbook of cereals, pulses, roots, and tubers. CRC Press, Boca Raton, pp 173–192

Schlegel RH (2013) Rye: genetics, breeding, and cultivation. CRC Press, Boca Raton

Sluková M, Jurkaninová L, Švec I, Skřivan P (2021) Rye—the nutritional and technological evaluation in Czech cereal technology—a review: grain and flours. Czech J Food Sci 39(1):3–8

Tehrani AN, Saadati S, Yari Z, Salehpour A, Sadeghi A, Daftari G et al (2023) Dietary fiber intake and risk of gallstone: a case–control study. BMC Gastroenterol 23(1):119

Wrigley C, Bushuk W (2017) Rye: grain-quality characteristics and management of quality requirements. In: Cereal grains. Woodhead Publishing, Duxford, pp 153–178

Sorghum and Millets

6

Aniket More, Sonia Morya, and Adams Ovie Iyiola

Abstract

This chapter delves into the often overlooked yet pivotal roles of sorghum and millets in the worldwide domains of agriculture and nutrition, paying special attention to their production and consumption trends. India, notably, emerges as a substantial contributor to global sorghum output, displaying commendable yields and agricultural accomplishments. Millets, encompassing pearl and proso millet, stand out as sturdy crops with extraordinary adaptability to climate change, making them indispensable staples in regions grappling with challenging agricultural conditions. Sorghum, originally cultivated in Africa, boasts an intriguing history of evolution and dispersion across continents. Its ability to thrive in diverse environments, withstand droughts, and lack gluten make it an extraordinary crop. Furthermore, sorghum offers notable health benefits due to its rich nutritional content and a variety of phenolic compounds, positioning it as a promising source for functional foods. Millets, comprising a diverse assortment of small-grained forage grasses, flourish under harsh conditions, offering grains brimming with nutrients, including high levels of protein and lipids. These grains play a pivotal role in fortifying processed foods and are gaining recognition as intelligent dietary choices in the twenty-first century, contributing to global nutritional enhancement. In summary, sorghum and millets, ancient grains with modern applications, hold the key to sustainable agriculture and healthier

A. More · S. Morya (✉)
Department of Food Technology & Nutrition, School of Agriculture, Lovely Professional University, Phagwara, Punjab, India

A. O. Iyiola
Department of Fisheries and Aquatic Resources Management, Faculty of Renewable Natural Resources Management, College of Agriculture and Renewable Natural Resources, Osun State University, Osogbo, Nigeria
e-mail: adams.iyiola@uniosun.edu.ng

J. Singh et al. (eds.), *Cereals and Nutraceuticals*,
https://doi.org/10.1007/978-981-97-2542-7_6

diets. Their resilience, adaptability, and nutritional richness establish them as essential components in the constantly evolving landscape of global food security and nutrition.

Keywords

Sorghum · Millet · Consumption pattern · Value addition · Good health · Anti-nutrients · Toxic compounds · Resistant starch · Production · Cereals

6.1 Origin and Background

In the realm of agriculture and nutrition, two unassuming grains have silently shaped the course of history and continue to play a crucial role in diverse cultures worldwide. Sorghum and millets: where ancient traditions merge with modern sustainability, unveiling a wealth of nutritional benefits (Gupta and Morya 2022). Globally, sorghum production exceeds 170 lakh tons, with India being a significant contributor. India's sorghum production accounts for approximately 80% of Asia's total output and constitutes around 20% of the global production. Remarkably, India achieves a sorghum yield of 1239 kg/ha, slightly surpassing the global average yield of 1229 kg/ha. These impressive figures underscore India's crucial role in the global sorghum market and its commendable agricultural accomplishments in sorghum cultivation (FAO 2021). In 2020, the world production of millets is expected to be 89.17 million metric tonnes (MMT) from an area of 74.00 million hectares. Sorghum and pearl millet are the most crucial millet crops, accounting for more than 90% of global millets production, followed by finger millet, foxtail millet, proso millet, barnyard millet, little millet, and kodo millet (ASSOCHAM 2022).

Sorghum, a vital grain in agriculture and nutrition, was first domesticated in Africa. While its exact origin remains uncertain, it likely took place around 3000 BP in the Ethiopia-Sudan regions. The transition from wild to cultivated sorghum was evident in archaeological findings from Kassala, Northeastern Sudan, dating back to 3500–1500 BP (Taylor 2019). Different sorghum races emerged and diffused separately, with bicolor being the most primitive and serving as the origin of other races. Over time, sorghum cultivation expanded globally to Southern Europe, India, China, the Americas, and Australia, with diverse types cultivated for various uses. Today, sorghum continues to be a significant crop worldwide, valued for its adaptability and nutritional benefits (McGinnis and Painter 2020).

Sorghum is renowned for its exceptional agronomic adaptability, thriving in diverse environments. It is drought and heat resistant, flourishing in high altitudes, saline-alkaline and infertile soils. This resilience arises from its robust root system, high root-to-leaf ratio, and adaptive leaves that can respond to external factors. Beyond its agronomic strengths, sorghum offers unique advantages: it is gluten free and rich in resistant starch (Xiong et al. 2019). Sorghum whole grains offer significant health benefits due to their high nutritional value and unique profile of phenolic compounds. Variations in climatic conditions, soil type, crop varieties, fertilizers, and fertilization methods can influence sorghum's chemical and nutritional composition. According to United States Department of Agriculture (USDA 2023), 100 g of sorghum grain contains approximately 72.09 g carbohydrates, 10.6 g proteins, 6.7 g fibers, and 3.46 g lipids and provides 1377 kJ of energy. The abundant and diverse phenolic compounds in sorghum, including phenolic acids, 3-deoxyanthocyanins, flavanones, flavones, and condensed tannins, particularly in sorghum bran, make it a promising cereal for producing healthy functional foods rich in bioactive compounds (Istrati et al. 2019).

'Millets', the umbrella term encompass a diverse group of small-grained forage grasses. These hardy crops predominantly thrive in semi-arid regions of Asia and Africa. As C_4 photosynthetic plants, millets exhibit remarkable resilience to climate change, flourishing under conditions of nutrient-deficient soils, high temperatures, and limited moisture. Their outstanding attributes include low susceptibility to diseases, adaptability to salinity and drought, and high nutritional value. With their short growth cycles, millets represent an ideal staple crop to support growing populations (Muthamilarasan et al. 2016). Millet's species encompasses a variety of types, such as pearl, proso, foxtail, barnyard, little, kodo, browntop, finger, guinea millets, and others like black and white fonio, sorghum, teff, and Job's tears, among others (FAO 2023).

Millets, particularly minor millets, are nutritionally rich, boasting 2–5% lipids and 12–16% proteins. They excel in providing essential micronutrients and offer a superior amino acid profile. Millets are instrumental as nutraceutical components for fortifying processed foods, leveraging their health benefits, including glucose level regulation and cholesterol reduction, thereby lowering the risk of cardiovascular diseases. Their multifaceted nutritional advantages contribute to a healthier diet (Senthamarai Selvi and Malathi 2019). Millets, the overlooked ancient grains, are making a comeback as twenty-first century smart foods, contributing to a healthier global society. These grains are exceptionally nutritious, non-acidic, and non-glutinous, making them easily digestible. Millets are also recognized as among the least allergenic and most digestible grains in the world (Hemamalini et al. 2021).

6.2 Production and Cultivation Practices

In the global sorghum production, the United States of America stood as the dominant force in 2020, contributing a substantial 15.05% share with a production of 9,473,620 metric tons. Nigeria followed closely, representing 10.47% of the total production with 6,590,000 metric tons, while Ethiopia secured the third position with an 8.03% share, producing 5,058,040 metric tons. India, with a 7.58% share, produced 4,772,100 metric tons in the same year. These statistics underscore the significance of these countries in the global sorghum production landscape over the years 2016 to 2020 (FAO 2022).

Sorghum cultivation relies on warm temperatures, well-drained loamy soil with a pH of 5.5 to 8.5, and medium to good rainfall. Ideal temperatures range from 15 to 32 °C, with a minimum germination temperature of 7 to 10 °C. Sorghum requires a short-day photoperiod of 10 to 11 h for flower initiation. Rainfall between 300 and 1200 mm supports optimum yields, with early-maturing varieties needing less water. Land preparation involves fertile, well-drained soil with prior fallowing or legume crops. Farm yard manure incorporation is recommended. Planting dates depend on rainfall, avoiding drought periods. Variety choice considers local climate and disease resistance. Seed rate is 8–10 kg/ha with seed dressing. Planting depth is 5 cm at row spacing of 75 cm and intra-row spacing of 25 to 30 cm (Ajeigbe et al. 2020).

India holds the distinguished title of being the world's foremost millet producer, with millet cultivation spanning across 21 Indian states. Spearheading this significant millet production are states such as Rajasthan, Maharashtra, Karnataka, Andhra Pradesh, Tamil Nadu, Kerala, Telangana, Uttarakhand, Jharkhand, Madhya Pradesh, Haryana, and Gujarat. This extensive cultivation effort results in a bountiful yield of 17.96 MMT, with an impressive average yield of 1247 kg/ha (APEDA 2023). Millets are cultivated across a vast expanse, covering approximately 13.8 million hectares of Indian soil. India's prowess in millet production extends to several millet varieties, including kodo (100%), little millet (100%), barnyard (99.9%), finger (53.3%), and pearl millet (44.5%). Collectively, these varieties contribute to a substantial output of approximately 12.46 MMT emanating from an expansive land area of 8.87 million hectares. According to FAO 2022, India (2020) had 40.51% share, the highest in millet production, followed by Niger (2020) 11.38%.

Millet cultivation necessitates meticulous preparation of the land. It thrives in various soils but is averse to waterlogged conditions. The field should be ploughed once or twice, followed by harrowing to create fine tilth. There are three primary methods of sowing pearl millet: on a flat surface, employing the ridge and furrow system, or utilizing the broad-bed and furrow system. Seeds should be sown at a depth of 2.5 to 3 cm. Regarding the timing of sowing, for kharif pearl millet, it should align with the onset of the monsoon, typically the first fortnight of July in north and central regions. Seed treatment with biopesticides or chemicals is advised to combat soil-borne diseases. The recommended seed rate is 3 kg/ha, with spacing and plant population varying according to regional conditions. For arid regions, rows should be 60 cm apart, maintaining a low plant population of 1.0 to 1.25 lakh/ha. In areas with over 450 mm of rainfall, a spacing of 45 × 10–15 cm between rows, achieving a plant population of 1.75 to 2.0 lakhs/ha, is recommended. Under normal conditions, a population of 1,80,000 plants/ha is suitable, but it may vary based on soil type and management levels (Chapke 2020). The production steps of sorghum and millet have been discussed in Fig. 6.1.

According to (MITCON 2023), ICAR-IIMR (Indian Council of Agricultural Research—Indian Institute of Millets Research), the utilization pattern of sorghum production reveals that approximately three-fourths of the sorghum harvest is directed toward self-consumption, primarily for human consumption. When it comes to the utilization of ragi (finger millet), the pattern mirrors that of sorghum, with the majority, about 90%, being consumed directly by humans. However, there is a notable difference in the remaining 10% of ragi production, which is allocated for the creation of value-added products in the Fast-Moving Consumer Goods

Fig. 6.1 Production of sorghum and millet

(FMCG) sector. Additionally, around 13% of ragi production serves as animal feed. Ragi's presence in the export market is relatively limited, accounting for only 1% of total exports. This cultivation and utilization of millets, including sorghum and ragi, aligns with the principles of Agriculture Revolution 4.0, which underscores the integration of advanced technologies and innovative approaches to enhance agricultural practices and productivity. Global millet exports surged from $400 million in 2020 to $470 million in 2021. India's millet exports also showed growth, reaching $75.46 million in 2022–23, up from $62.95 million in the previous year (2021–2022) (ITC 2023).

6.3 Consumption Pattern

The Lancet Commissions publication highlights the urgent requirement to identify dietary options that prioritize health and sustainability. This entails a substantial shift toward adopting less-utilized plant species like millets, quinoa, sorghum, and teff grains, given their natural resilience to climate challenges and rich nutritional content (Willett et al. 2019). Global millet consumption has been on the rise, with a 14% increase in production between 2013 and 2018, as reported by the FAO. Millets play a vital role in providing nutrition, especially in developing nations (Millet Advisor 2023). Millets and sorghum, being culturally accepted and cost-effective staple foods, play a crucial role in ensuring food security, particularly in the face of climate change and population growth. To promote their consumption, both the government and stakeholders must prioritize millets. Increasing urban and rural consumption can transform the perception of millets from "cheap food" to a "healthy alternative" and a vital component of dietary diversity. To achieve this shift, large-scale awareness campaigns are recommended to expand the market for millets and sorghum, ultimately enhancing access to sustainable food and nutrition security (Liaqat et al. 2024).

A recent study conducted by Senthamarai Selvi and Malathi (2019) on millet consumption patterns revealed intriguing insights among respondents. The most commonly consumed millet emerged as finger millet, with 30% of participants incorporating it into their diet (Singh et al. 2018). Sorghum followed closely behind, with a consumption rate of 28%. Other millets, including pearl millet, proso millet, foxtail millet, little millet, and kodo millet, showed varying levels of consumption, with some not being consumed at all. Moreover, the study shed light on preferred cooking methods, the use of millet products, and storage practices, providing valuable insights into millet consumption behaviours and opportunities for promoting these nutritious grains in diverse culinary contexts.

6.3.1 Sorghum

Consumption patterns of sorghum have changed recently as a result of evolving dietary habits and rising awareness of the food's nutritional benefits. In the past, sorghum was a common ingredient in foods like porridge and flatbreads in areas with arid temperatures, like Africa (Hossain et al. 2022).

According to recent data, sorghum production has expanded globally and reached 65 million metric tonnes in 2020. Since sorghum is naturally gluten-free, this increase is related to the need for gluten-free substitutes (Fu et al. 2021). Sorghum is being utilized in a variety of ways in Western nations where gluten-free diets have grown in popularity, including whole grains in salads and sorghum flour for baking (McGinnis and Painter 2020; Sharanagat et al. 2022).

Furthermore, the creation of sorghum-based goods like syrup and gluten-free snacks has increased consumer interest. Sorghum's consumption habits are changing as people become more aware of its nutritional advantages, establishing it as a versatile grain that can fulfill the world's changing dietary needs (Sirany et al. 2022).

6.3.2 Pearl Millet

Pearl millet's consumption trends have adapted in response to shifting dietary preferences and a heightened awareness of its nutritional value (Rizwana et al. 2023). In the past, pearl millet was a staple food in areas with arid climates, particularly in parts of Africa and Asia, where it was used to make porridge and flatbreads (Tripathi and Vyas 2023).

Recent studies reveal that the output of pearl millet will increase globally and reach 8 to 11 million metric tonnes in 2018. Since pearl millet is naturally gluten-free, the desire for gluten-free foods can be partly blamed for this rise. Pearl millet is used in a variety of ways in Western nations where gluten-free diets have gained popularity, including pearl millet flour for baking and whole grains in salads and accompaniments (Iyer 2021).

Moreover, the creation of pearl millet-based products, such as gluten-free snacks and fortified foods, has expanded its appeal (Selladurai et al. 2022). As awareness of its nutritional benefits continues to expand, pearl millet's consumption pattern is diversifying beyond its traditional uses, positioning it as a versatile grain that caters to evolving dietary preferences worldwide (Noort et al. 2022).

6.3.3 Finger Millet

The consumption pattern of finger millet, commonly known as ragi, has evolved, bridging its historical significance with contemporary recognition as a nutritional powerhouse (Bhargava et al. 2019). Traditionally, a dietary staple crop in Africa and Asia, finger millet has recently seen a significant boost in production, reaching approximately 7 to 10 million metric tons globally in 2020. After sorghum, pearl millet, and foxtail millet, finger millet is the fourth most significant millet in the world (Ashik Somarajan and Morya 2022).

This surge is partly attributed to the heightened awareness of finger millet's nutritional benefits, particularly its richness in calcium and iron. It has gained prominence as a vital component in addressing malnutrition worldwide (Kebede and Bekeko 2020). Beyond its traditional uses in dishes like porridge and flatbreads, finger millet is finding a place in modern dietary trends, especially among those seeking gluten-free options due to its natural absence of gluten (Kala and Nautiyal 2023). Moreover, its high fibre content positions it as a valuable ingredient for health-conscious individuals aiming to increase their dietary fibre intake.

The results of the endline food consumption survey conducted by Pradhan et al. (2019) demonstrated encouraging effects stemming from increased production through recommended farming practices and heightened nutritional awareness initiatives. Notably, there was a 13% uptick in the average finger millet intake compared to the baseline (from 70 g/person/day). Furthermore, the number of households that regularly included finger millet in their daily diets rose from 172 during the baseline to 187. Also, a larger percentage of households (68%) were able to obtain finger millet from their own production, marking a significant improvement compared to the baseline, when only 47% of households had this capability. These findings underscore the favourable impact of agricultural practices and nutritional education efforts, ultimately promoting greater finger millet consumption.

6.3.4 Proso or Common Millet

Proso millet, often overlooked, is a highly nutritious cereal grain. Packed with vitamins, minerals (like calcium, zinc, and iron), essential amino acids, antioxidants, and beta-glucan, it offers a range of health benefits. It can help lower cholesterol levels, support liver health, and boost neural function due to its high lecithin content. This makes it a valuable addition to diets for improved nutrition and well-being (APEDA 2022). Proso millet, also known as broomcorn or common millet, has undergone a transformation with its emerging reputation as a nutritious grain. Traditionally, proso millet served as a dietary cornerstone in regions of Asia and Eastern Europe, commonly featured in dishes like porridge and bread (Ray 2022).

Recent data underscores a substantial increase in proso millet production. This growth can be attributed in part to the growing recognition of proso millet's nutritional value, particularly its richness in protein and essential micronutrients (Wang and Jian 2022).

Moreover, its gluten-free nature has bolstered its appeal among individuals with gluten sensitivities, leading to its inclusion in gluten-free flours and a range of food products. Proso millet is also gaining ground in health-conscious diets due to its potential to enhance dietary diversity and overall nutritional quality (Gupta et al. 2023).

6.3.5 Foxtail Millet

Foxtail millet holds the second position in global millet production. It boasts a nutritional profile comprising 9–14% protein and 70–80% carbohydrates. Additionally, it serves as a significant source of dietary fibre. Foxtail millet stands out for its relatively high chromium content, with approximately 0.030 mg/100 g. The dietary fibre primarily consists of polymers derived from hexoses, pentoses, cellulose, and pertinacious materials. These attributes make foxtail millet a valuable addition to a balanced diet, contributing to its nutritional richness (Geeta et al. 2016). A study on consumer preferences for foxtail millet products revealed interesting findings. The majority of respondents reported consuming foxtail millet rice on a daily basis (62.00%), while foxtail millet holige was consumed as needed by 50.00% of the participants. These insights provide valuable information about the popularity and consumption patterns of foxtail millet-based products among consumers (Amarapurkar and Banakar 2017).

6.3.6 Barnyard Millet

Barnyard millet is a gluten-free grain that can be safely included in the diets of individuals with celiac disease and diabetes. This versatile grain is believed to offer numerous health advantages, including potential benefits for diabetes management, cardiovascular health, weight control, skin health, cancer prevention, and alleviating symptoms of celiac disease. Its gluten-free nature makes it a valuable addition to the dietary options available for those with specific health concerns (Bhatt et al. 2023).

6.3.7 Little Millet

The consumption trends of little millet have undergone significant transformations recently, merging its traditional significance with its emerging reputation as a highly nutritious grain. Traditionally, little millet has held a central role in the diets of regions in India, Africa, and Southeast Asia, where it was frequently featured in local dishes, often serving as a substitute for rice and used in porridge (Anitha et al. 2022; Goswami et al. 2023).

Furthermore, little millet's gluten-free nature has broadened its appeal, particularly among individuals with gluten sensitivities. It is increasingly integrated into gluten-free flours and a variety of food products. Health-conscious consumers are also embracing little millet for its potential to diversify diets and promote balanced nutrition (Jindal and Nikhanj 2023).

Consuming just 100 g of little millet grains can substantially boost your daily nutrient intake. It can provide up to 28% of the recommended dietary allowance (RDA) for iron, crucial for maintaining healthy blood; as much as 37% of the RDA for zinc, important for immune function and overall well-being; and up to 27% of the RDA for protein, vital for muscle and tissue health. Little millet is a nutritious addition to your diet, offering a convenient and natural way to meet these essential nutrient requirements (ICRISAT 2021).

6.3.8 Kodo Millet

Kodo millet's consumption pattern has recently undergone a significant transformation, blending its historical significance with its newfound recognition as a nutritious grain. Traditionally, kodo millet was a dietary staple in South India, prominently featured in regional dishes such as porridge and flatbreads (Hema et al. 2022).

Recent global statistics reveal a substantial surge in kodo millet production, reaching approximately 600,000 to 800,000 metric tons in 2022. This increase is partly due to the growing awareness of kodo millet's nutritional value, including its high fibre content and essential minerals. These qualities position it as a valuable addition to modern diets, catering to health-conscious eating preferences (Dey et al. 2022).

6.4 Chemical Composition

The compositions of sorghum and millet vary significantly due to genetics, classifications, and environmental factors (Table 6.1) (Mawouma et al. 2022). Protein content is notably variable, with sorghum having a fibre-rich pericarp, a germ rich in protein, fat, and minerals (ash), and an endosperm mainly composed of starch and protein (Adebo and Kesa 2023). In contrast, different millet species have distinct compositions. Pearl millet, with a relatively large germ-to-endosperm ratio, contains higher levels of protein and fat (Akanbi et al. 2019). Proso millet also boasts

Table 6.1 Average nutrient composition in sorghum and millets

Nutrients	Sorghum (flour)	Millets (whole grain)
Water (g/100 g)	9.41	10.2
Carbohydrates (g/100 g)	77.4	74.4
Proteins (g/100 g)	11.4	10
Dietary fibre (g/100 g)	8.16	3.33
Fats (g/100 g)	3.3	4.19
Ash (g/100 g)	1.9	1.13
Energy (Kcal/100 g)	375	376
Vitamins and other components		
Thiamin (mg/100 g)	0.27	0.411
Riboflavin (mg/100 g)	0.15	0.136
Niacin (mg/100 g)	3.95	4.86
B-6 (mg/100 g)	0.166	0.192
Biotin (μg/100 g)	24.2	11.4
Minerals		
Calcium (mg/100 g)	11	9
Iron (mg/100 g)	3.72	2.53
Magnesium (mg/100 g)	116	106
Phosphorus (mg/100 g)	262	258
Potassium (mg/100 g)	335	214
Sodium (mg/100 g)	<2.5	<2.5
Zinc (mg/100 g)	1.66	2.26
Copper (mg/100 g)	0.237	0.488
Manganese (mg/100 g)	1.08	0.924
Molybdenum (μg/100 g)	48.2	29.5

Source: USDA (2023)

significant protein and fat content (Karpagapandi et al. 2023). Conversely, fonio, finger millet, and teff have lower protein and fat content, resulting in higher nitrogen-free extract levels (Abdul and Jideani 2019). Kodo and Indian barnyard millets have higher fibre content, likely due to the increased presence of bran in their caryopses (Barbhai and Hymavathi 2022; Kaushik et al. 2022).

6.5 Anti-nutrients and Toxic Compounds

Sorghum is primarily characterized by the presence of two significant anti-nutritional elements. The first is tannin, a polyphenolic compound predominantly situated within the grain. The second is cyanogenic glycoside, known as dhurrin, primarily found in the aerial shoot and germinated seeds (Etuk et al. 2012). Like many other grains and seeds, millets and sorghum contain naturally occurring substances that can either improve or detract from their nutritional value (Rawat et al. 2023). These grains include anti-nutrients such as phytates, tannins, and polyphenols that are natural defences against diseases and pests but can obstruct human mineral absorption (Abbas 2020). Phytates, for example, can bind to important minerals like calcium, iron, and zinc, making them less available for absorption (Castro-Alba et al. 2019). To lessen this, conventional processing methods like soaking, fermenting, and heating are frequently used to break down these anti-nutrients, hence enhancing the nutritious value of the grains (Samtiya et al. 2020).

Furthermore, mycotoxins, harmful substances made by moulds, have the ability to contaminate millets and sorghum (Byakika et al. 2019). These grains frequently contain mycotoxins aflatoxins, and fumonisins, both of which pose health hazards if consumed in excess (Lesuuda 2023). To reduce mycotoxin contamination, proper storage procedures are crucial. These procedures should include moisture management and pest prevention (Hamel et al. 2020). Additionally, facilities for grain production and processing can guarantee that mycotoxin levels stay within safe ranges by conducting routine testing and quality control methods (Kaushik et al. 2022).

It is critical to stress that, despite the potential risks posed by these anti-nutrients and mycotoxins, sorghum and millets nevertheless offer a wealth of health advantages (Ranjan et al. 2023). These grains are a good source of dietary fibre, antioxidants, and important vitamins and minerals like magnesium. They are also gluten-free, making them a good option for people who have celiac disease or gluten sensitivity (Anagha 2023).

Natural chemicals in sorghum and millets may reduce their nutritional value, but with the right processing and preservation methods, these grains can be a component of a balanced diet (Ramakrishnan et al. 2022). By addressing these issues, people can benefit from the nutritional benefits of sorghum and millet while lowering any possible health risks.

6.6 Bioactive and Functional Components

Millets and sorghum are rich sources of bioactive and functional components that have a variety of positive health effects (Qin et al. 2022). Dietary fibre stands out among these ingredients since it comes in both soluble and insoluble forms and helps with regular bowel movements, lowers the likelihood of constipation, and may even lessen the risk of gastrointestinal disorders (Rawat et al. 2023). These grains are also known for having antioxidants including flavonoids and phenolic substances (Adebo and Gabriela Medina-Meza 2020). These bioactive compounds are essential in combating oxidative stress and inflammation, which may lower the risk of developing chronic diseases including heart disease and some types of cancer (Espitia-Hernández et al. 2022). Sorghum in particular has a high antioxidant content, which has earned it the title of nutritional powerhouse (Yahaya and Shimelis 2022).

Additionally, prominent sources of important minerals including iron and magnesium include sorghum and millet (Mawouma et al. 2022). While magnesium promotes bone health, muscle function, and nerve activity, iron plays a key role in avoiding anaemia (Lokhande et al. 2019). They also contain B vitamins including riboflavin, niacin, and thiamin, which are crucial for overall vigour and the metabolism of energy (Abbas and faleh Al-fekeiki 2021). Also, sorghum and millet are suitable for diabetics due to their low glycaemic index (GI) (Sewak et al. 2023). The continuous release of energy is ensured by their slow digestion and carbohydrate absorption, which help to control blood sugar levels (Kraithong et al. 2022). Both grains are gluten-free, making them a great alternative for anyone with celiac disease or gluten intolerance (Khairuddin and Lasekan 2021). This quality increases their marketability in the expanding gluten-free food market, which includes flour, kinds of pasta, snacks, and baked items (Woomer and Adedeji 2021).

Sorghum and millets have a wide variety of bioactive and functional elements that support health and nutritional variety, therefore including them in one's diet can improve overall well-being (Patra et al. 2023).

6.7 Health Implications

Sorghum and millets have garnered recognition as valuable components of a well-rounded diet due to their impressive range of health advantages, supported by research (Rathore et al. 2019). These grains are not only packed with essential nutrients but also provide vital macronutrients, vitamins, and minerals, making them a substantial part of a wholesome dietary regimen (Musundire et al. 2021). One remarkable characteristic of sorghum and millets lies in their abundant soluble and insoluble dietary fibre content. This fibre plays a pivotal role in regulating gastrointestinal function and enhancing digestive health. It is known to lower the risk of gastrointestinal disorders, ensuring a smoother and healthier digestive process (Rawat et al. 2023).

Especially beneficial for individuals with diabetes or those aiming to effectively manage their blood sugar levels, sorghum and millets offer a distinct advantage. These grains possess a low glycaemic index, meaning they induce a gradual and modest increase in blood sugar levels following consumption. This property aids in maintaining stable blood sugar levels, rendering them a suitable choice for individuals managing diabetes (Pradeep 2023). Furthermore, sorghum and millets have gained renown for their antioxidant attributes and dietary fibre content. These qualities jointly combat oxidative stress, reduce inflammation, and regulate cholesterol levels. Consequently, they serve as heart-healthy dietary options, significantly diminishing the risk of cardiovascular disease (Ranjan et al. 2023).

Incorporating sorghum and millets into one's diet not only enriches nutritional diversity but also contributes to overall well-being. These grains offer a multifaceted approach to health, spanning digestive enhancements, blood sugar control, and cardiovascular health. Their culinary versatility and compatibility with various dietary preferences underscore their value as a wise choice for individuals seeking to enhance their dietary habits and foster enduring health and vitality.

6.8 Processing

Sorghum grains have gained worldwide popularity due to their health benefits and gluten-free nature. However, they contain anti-nutritional factors like tannins, phytate, enzyme inhibitors, and protein crosslinkers, which limit nutrient absorption (Palacios et al. 2021). To address these challenges, various processing methods like soaking, germination, fermentation, and thermal treatments are utilized to boost nutritional availability and product quality (Cabrera-Ramírez et al. 2020). These processing techniques are essential for the production and consumption of sorghum-based food products. Traditional methods such as soaking, wooden ash treatment, germination, fermentation, cooking, steaming, and roasting have historically improved the nutritional value of sorghum grains. In recent times, innovative approaches like microwave roasting, nixtamalization, and irradiation have emerged as potential methods to enhance the properties of sorghum grains (Rashwan et al. 2021).

Millets, both in whole and flour forms, are processed using a variety of traditional and modern techniques. Due to their lack of gluten, they don't form dough like wheat or rice flour. Fortification is a common method to address this limitation. Millets can be processed into flour, porridges, popped grains, and more. Key steps include dehulling, milling, and secondary processing, which transform millets into various food products suitable for direct consumption. Special machinery and techniques are used to refine millet grains and produce value-added products (Birania et al. 2020). Millet, with its high nutritional value and resilience, holds promise for food security. However, the short shelf life of millet flour, caused by enzymatic activity, poses a challenge. Recent innovations in both thermal and non-thermal processing techniques aim to extend millet product shelf life while preserving nutrition. These methods aim to stabilize millet products while preserving their nutritional properties. Notably, technologies such as microwave, radiofrequency, infrared, ohmic heating, and non-thermal approaches like pulsed electric field, cold plasma, irradiation, high-pressure processing, ozonation, and ultrasound treatment have gained significant attention (Joshi et al. 2023). The potential applications of processing of sorghum and millets have been discussed in Table 6.2.

6.9 Value Addition

Sorghum is the cereal that is most widely produced in terms of value addition, following wheat, rice, barley, and maize (Rakshit et al. 2023). Sorghum is regarded for its ability to flourish where other grains find it difficult, such as in tropical and arid regions of the world (Vellaichamy Gandhimeyyan et al. 2023). As a result of its relatively short growing season needs, it is also renowned for its tolerance to double cultivating and crop rotation systems (Karlsson Potter et al. 2023). Sorghum use can be broadly divided into three distinct groups: human food, animal food, and industrial use (Lone et al. 2023). The main sorghum-using sectors are animal feed manufacturers, booze distilleries, and starch producers (Stamenković et al. 2020).

Table 6.2 Processing of sorghum and millets with their potential applications

Process stage	Sorghum/millet processing	Applications	References
Cleaning/ sorting	Eliminate foreign substances and contaminants	Customer satisfaction, animal feed	Taylor and Duodu (2017)
Milling	Grind into sorghum/millet flour	Baking, making porridge, flour for products	Banu and Aprodu (2020)
Dehulling	Remove the outer hull if desired (depends on the product)	Animal feed, brewing, health foods	Chikkaballapur Krishnappa et al. (2022)
Fermentation	Optional: For certain products	Brewing, sourdough preparation, enhancing flavour	Mishra and Sabikhi (2020)
Mixing	Mix with other ingredients (e.g. for baked goods)	Baking, snacks, specialty foods	Dayakar Rao and Kiranmai (2020)
Forming	Shape into desired product (e.g. bread, tortillas)	Bread, tortillas, porridge, snacks	Ramakrishnan et al. (2022)
Cooking/ baking	Heat or bake the product	Baking, cooking, frying	Ramakrishnan et al. (2022)
Packaging	Seal in packaging material	Packaging for retail or distribution	Basaran-Akgul (2021)
Labelling	Add labels (nutritional information, etc.)	Customer satisfaction	Barretto et al. (2021)
Quality control	Check for quality and safety	Ensuring product meets standards	Tripathi et al. (2021)
Distribution	Transport to retailers or consumers	Retail sale, export, or direct to consumers	Anerao et al. (2022)

Additionally, it is a significant source of fodder and feed for cattle. It is farmed for animal feed in affluent countries including Australia, the United States, China, and others. In comparison, sorghum grows more quickly and produces significant grain yields and huge amounts of feed (Diao 2017).

Sorghum's value addition entails changing this hardy and nutritious grain into a variety of goods that increase its market appeal and meet various consumer needs (Tan et al. 2020). Sorghum, which is frequently thought of as a traditional staple, is now becoming known for its exceptional attributes and adaptability. Its hidden potential is unlocked by value addition (Hossain et al. 2021).

Sorghum may be processed into a variety of flours, which is one of the main ways to increase its value (Rashwan et al. 2021). Sorghum flour can be used in baking instead of gluten, making it available to people with dietary requirements (Al Majzoub et al. 2023). Sorghum can also be processed into ready-to-eat snacks like popped sorghum or chips made from sorghum, providing quick and nutritious snack options (Gaikwad et al. 2023). Sorghum can also be used to make beverages like sorghum syrup and beer, enhancing the beverage market with distinctive flavours and gluten-free options (Cadenas et al. 2021; Nyoni et al. 2020).

Given that sorghum is drought-resistant and supports sustainable agriculture, adding value to it is important for encouraging nutritional diversity and satisfying a range of dietary preferences (Hossain et al. 2021). Sorghum's economic value can be increased by investigating its value-added potential, and this will also help consumers choose healthier, more sustainable foods (Jindal and Nikhanj 2023).

Millets have emerged as nutritious powerhouses with untapped potential in a time of rising health awareness, sustainability, and dietary diversity (Blesh et al. 2023). These ancient grains are transformed through the process of value addition into a variety of functional and nutrient-dense goods, including gluten-free flour, quick snacks, and dietary supplements (Caeiro et al. 2022). Beyond generating profits, this method promotes dietary diversity, supports resilient crop choices for sustainable agriculture, and honours culinary traditions while encouraging innovation (Wijerathna-Yapa and Pathirana 2022).

Common techniques of value addition include making gluten-free items, cooking classic foods, and grinding millet grains into flour for baking (Hema et al. 2022). Additionally, millet can be used to make snacks that are nutritious and convenient, like toasted millet chips or puffed millet (Bisht 2022). In order to make gluten-free pasta and noodles, millet can be used in place of wheat (Kaur et al. 2022). Baby food, flour mixes, and ready-to-eat breakfast cereals can accommodate various dietary requirements (Dekka et al. 2023). Making millet beer provides gluten-free options while millet oil is used for a variety of industrial and culinary applications (Selladurai et al. 2022). Millet can be used to make a variety of snack foods, including breakfast bars, breakfast crackers, and snack bars (Dharshini and Meera 2023; Sudha et al. 2021).

Millets can also be found in desserts and health products (Kala and Nautiyal 2023). Millet can be used in ethnic cuisine products like African porridge or Indian dosa (Saleem et al. 2023). Instant porridge mixtures offer convenient and wholesome meal options (Hema et al. 2022). Millet-based pasta sauces improve pasta recipes, and exporting millet products can broaden the market (Qin et al. 2022). As millet is well-known for its nutritional advantages and environmental tolerance, these value-added products not only increase the market appeal of millet but also support dietary diversity and sustainability (Girard and Awika 2018; Chera 2017). To maximize the value of millet, entrepreneurs and businesses can select from these options based on market demand and production capacity (Shah et al. 2023).

6.10 Safety Considerations

Ensuring the safe consumption of sorghum and millet is of utmost importance, given their potential health advantages (Rawat et al. 2023). There are some safety issues with these grains that must be taken into account. Phytates, tannins, and polyphenols, for example, are anti-nutritional substances that can prevent the absorption of nutrients (Gowda et al. 2022). This issue is solved by using conventional processing techniques like soaking and fermentation, which also increases the availability of nutrients (Tharifkhan et al. 2021).

Another point that has to be considered is mycotoxin contamination, which comprises poisonous compounds made by moulds such as aflatoxins and fumonisins (Nešić et al. 2023). To lessen this danger, proper storage procedures and moisture management are essential (Hussain et al. 2019). Furthermore, due to the possibility of cross-contamination during processing, people who have grain allergies should proceed with caution (Demirkesen and Ozkaya 2022).

Certain people may experience digestive pain, which is frequently related to the high fibre content of these grains (Takaidza 2023). Such problems can be reduced by gradually incorporating these grains into the diet and making sure you're getting enough water (Zeppieri 2021). The necessity of choosing organic or pesticide-free choices and carefully washing the grains before cooking is highlighted by the presence of pesticide residues (Pandiselvam et al. 2020). To avoid deterioration and the development of mould, proper storage in cool, dry environments is essential. It is important to keep in mind that various sorghum and millet types may have varied nutrient profiles and anti-nutritional chemicals, which emphasizes the need for knowledge about the particular variety being consumed (Tharifkhan et al. 2021).

Sorghum and millet provide nutritional advantages, but it is crucial to carefully take these safety considerations into account (Chandra et al. 2021). These grains' nutritional worth and safety for ingestion are both ensured through thoughtful processing, adequate storage, and sourcing from reliable sources (Mariutti et al. 2021). When introducing sorghum and millets into their diets, people with special dietary issues or allergies should seek advice from medical professionals or dietitians (Rotela et al. 2021).

6.11 Future Challenges and Opportunities

In 2023, the United Nations General Assembly designated it as the International Year of Millets (#IYOM2023). This initiative, supported by the Food and Agriculture Organization, aims to spotlight the nutritional advantages of millets and their ability to thrive in challenging growing conditions. Ancient grains like sorghum and millet have long been used to feed human populations. But in the modern world, they encounter both opportunities and challenges, notably in relation to agriculture, consumer habits, and global food security (Tripathi and Vyas 2023; Teferra and Awika 2019).

The shift in dietary preferences toward more commonplace crops like wheat, rice, and maize presents a huge challenge for sorghum and millet (Satyavathi et al. 2021). Urbanization, globalization, and the perception of status attached to certain crops all contribute to this transformation (Bobiec et al. 2021). Because of this, the production of sorghum and millet has decreased in some areas, endangering local food security and biodiversity (Raheem et al. 2021).

Another pressing issue is climate change. Compared to other grains, sorghum and millets are typically more resistant to harsh weather although they are not immune (Chazovachii et al. 2012). The need to create climate-resilient varieties of these grains is critical since shifting precipitation patterns and rising temperatures can have an impact on yields (Noort et al. 2022). But sorghum and millet have a lot of prospects ahead of them. Their nutritional value is one major advantage (Tonapi et al. 2020). The minerals, dietary fibre, and antioxidant content of these grains are high (Devi et al. 2014). Sorghum and millets may gain popularity as people become more cognizant of their health and nutrition and look for alternative, nutrient-dense cereals (Hema et al. 2022). Another promising avenue is research and development. The field of research and development is another potential one. Sorghum and millets can produce more and compete more favourably with other crops if investment is made in breeding programmes and better farming techniques (Birol et al. 2023). Innovation in food processing methods can also result in the creation of a variety of goods, such as snacks and drinks as well as gluten-free flour, increasing their attractiveness to customers (Woomer and Adedeji 2021).

Sorghum and millet are consumed in different ways around the world (Babé et al. 2020). They continue to be staple ingredients used in a range of traditional recipes in many African and Asian nations (Mundia et al. 2019). They are, however, frequently restricted to specialty health food stores in Western nations (Shah et al. 2023). It is necessary to make an effort to increase awareness of the nutritional advantages and adaptability of these grains in contemporary cooking if we are to completely realize their potential (Wei and Davis 2021).

Sorghum and millets, in conclusion, have great prospects for nutrition, health consciousness, and sustainable agriculture, but they also present issues connected to shifting consumption patterns and climate change. To ensure that these ancient grains have a future in the world's food supply, it will be crucial to strike a balance between these elements.

6.12 Conclusion

In summarization, the production and consumption patterns of sorghum and millets have experienced a resurgence due to their health benefits and versatility. These grains are valuable complements to our diets since they are high in vitamins, minerals, dietary fibre, carbs, and proteins. They do, however, also contain anti-nutrients and potentially harmful substances, which can be reduced by using the right processing methods. Not only are sorghum and millets nutrient-dense, but they also include bioactive substances like antioxidants, phytochemicals, and micronutrients that have a variety of positive health effects, including better digestion and a lower chance of developing chronic diseases. Value-added goods like millet treats and pasta made from sorghum have broadened their market appeal. However, it is still important to make sure they are handled, stored, and processed in a safe manner.

Challenges in the near future include promoting sustainable farming, increasing knowledge of their nutritional worth, and addressing issues associated with climate change. In a world that is constantly expanding, sorghum and millet have the potential to be healthy and sustainable food sources. For a healthier and more sustainable future, it will be crucial for them to realize their potential and overcome obstacles.

Acknowledgements AM, SM, and AOI want to acknowledge Lovely Professional University for their support to provide infrastructure for completing this manuscript.

Conflict of Interest There is no conflict of interest between the authors.

References

Abbas G (2020) An augmented review about anti-nutrients and toxins of feed stuff and their control strategies, a step toward sustainable resource utilization. Int J Agric Innov Res 8(4):361–377

Abbas SF, faleh Al-fekeiki D (2021) Follow-up of the development of water-soluble and fat-soluble minerals and vitamins for germination and non-germination sorghum grains during different germination periods. Nat Volatiles Essent Oils 8(6):2143–2154

Abdul SD, Jideani AI (2019) Fonio (Digitaria spp.) breeding. In: Advances in plant breeding strategies: cereals, vol 5. Springer, Cham, pp 47–81

Adebo OA, Gabriela Medina-Meza I (2020) Impact of fermentation on the phenolic compounds and antioxidant activity of whole cereal grains: a mini review. Molecules 25(4):927

Adebo JA, Kesa H (2023) Evaluation of nutritional and functional properties of anatomical parts of two sorghum (Sorghum bicolour) varieties. Heliyon 9(6):e17296

Ajeigbe HA, Angarawai II, Akinseye F, Inuwa AH, Abdulazeez T, Vabi MB (2020) Handbook on improved agronomic practices for sorghum production in North East Nigeria feed the future Nigeria integrated agriculture activity

Akanbi TO, Timilsena Y, Dhital S (2019) Bioactives from millet: properties and effects of processing on bioavailability. In: Bioactive factors and processing technology for cereal foods. Springer, Singapore, pp 171–183

Al Majzoub R, Khaled S, El Khatib S, Krayem M (2023) A comparison between wheat and sorghum flour in biscuits application: a review. Food Sci Eng 4:89–102

Amarapurkar S, Banakar B (2017) A case study of consumer preference towards foxtail millet products. Int J Com Bus Manage 10(2):267–270. https://doi.org/10.15740/has/ijcbm/10.2/267-270

Anagha KK (2023) Millets: nutritional importance, health benefits, and bioavailability: a review. Energy 329(328):361

Anerao KK, Gadhe KS, Gaikwad SS, Sadaf T, Ghatge PU (2022) Studies on malting characteristics of different sorghum genotypes and its utilization to prepare traditional fermented product i.e., Ambil. Biol Forum Int J 14(4a):232–237

Anitha S, Givens DI, Subramaniam K, Upadhyay S, Kane-Potaka J, Vogtschmidt YD et al (2022) Can feeding a millet-based diet improve the growth of children?—a systematic review and meta-analysis. Nutrients 14(1):225

APEDA (2022). https://apeda.gov.in/milletportal/Proso_Millet.html

APEDA (2023) India's millet production. Agricultural and Processed Food Products Export Development Authority, Government of India. https://apeda.gov.in/milletportal/Production.html

Ashik Somarajan K, Morya S (2022) A review on finger millet properties, processing, health benefits, and applications

ASSOCHAM (2022). https://www.assocham.org/uploads/files/Report_Millets%202022%20(Print%20Version)%20(1).pdf

Babé C, Kidmo DK, Tom A, Mvondo RRN, Boum RBE, Djongyang N (2020) Thermomechanical characterization and durability of adobes reinforced with millet waste fibres (Sorghum bicolour). Case Stud Construct Mater 13:e00422

Banu I, Aprodu I (2020) Assessing the performance of different grains in gluten-free bread applications. Appl Sci 10(24):8772

Barbhai MD, Hymavathi TV (2022) Nutrient, phytonutrient and antioxidant potential of selected underutilized nutri-cereal brans. J Food Meas Charact 16(3):1952–1966

Barretto R, Buenavista RM, Rivera JL, Wang S, Prasad PV, Siliveru K (2021) Teff (Eragrostis tef) processing, utilization and future opportunities: a review. Int J Food Sci Technol 56(7):3125–3137
Basaran-Akgul N (2021) Packaging requirements for non-thermal processed grain-based foods. In: Non-thermal processing technologies for the grain industry. CRC Press, Boca Raton, pp 199–222
Bhargava A, Srivastava S, Bhargava A, Srivastava S (2019) Cereals. In: Participatory plant breeding: concept and applications. Springer, Singapore, pp 129–173
Bhatt D, Rasane P, Singh J et al (2023) Nutritional advantages of barnyard millet and opportunities for its processing as value-added foods. J Food Sci Technol 60:2748–2760. https://doi.org/10.1007/s13197-022-05602-1
Birania S, Rohilla P, Kumar R, Kumar N (2020) Post harvest processing of millets: a review on value added products. Int J Chem Stud 8(1):1824–1829. https://doi.org/10.22271/chemi.2020.v8.i1aa.8528
Birol E, Foley J, Herrington C, Misra R, Mudyahoto B, Pfeiffer W et al (2023) Transforming Nigerian food systems through their backbones: lessons from a decade of staple crop biofortification programming. Food Nutr Bull 44(1_suppl):S14–S26
Bisht A (2022) Small-millet-based traditional and unconventional food products. In: Small millet grains: the superfoods in human diet. Springer, Singapore, pp 95–117
Blesh J, Mehrabi Z, Wittman H, Kerr RB, James D, Madsen S et al (2023) Against the odds: network and institutional pathways enabling agricultural diversification. One Earth 6(5):479–491
Bobiec A, Paderewski J, Gajdek A (2021) Urbanisation and globalised environmental discourse do not help to protect the bio-cultural legacy of rural landscapes. Landsc Urban Plan 208:104038
Byakika S, Mukisa IM, Wacoo AP, Kort R, Byaruhanga YB, Muyanja C (2019) Potential application of lactic acid starters in the reduction of aflatoxin contamination in fermented sorghum-millet beverages. Int J Food Contam 6(1):1–8
Cabrera-Ramírez AH, Luzardo-Ocampo I, Ramírez-Jiménez AK, Morales-Sánchez E, Campos-Vega R, Gaytán-Martínez M (2020) Effect of the nixtamalization process on the protein bioaccessibility of white and red sorghum flours during in vitro gastrointestinal digestion. Food Res Int 134:109234. https://doi.org/10.1016/J.FOODRES.2020.109234
Cadenas R, Caballero I, Nimubona D, Blanco CA (2021) Brewing with starchy adjuncts: its influence on the sensory and nutritional properties of beer. Foods 10(8):1726
Caeiro C, Pragosa C, Cruz MC, Pereira CD, Pereira SG (2022) The role of pseudocereals in celiac disease: reducing nutritional deficiencies to improve well-being and health. J Nutr Metab 2022:8502169
Castro-Alba V, Lazarte CE, Bergenståhl B, Granfeldt Y (2019) Phytate, iron, zinc, and calcium content of common Bolivian foods and their estimated mineral bioavailability. Food Sci Nutr 7(9):2854–2865
Chandra AK, Chandora R, Sood S, Malhotra N (2021) Global production, demand, and supply. In: Millets and pseudo cereals. Woodhead Publishing, Duxford, pp 7–18
Chapke RR (2020) Latest millets production and processing technologies. ICAR-IIMR Publication, Hyderabad
Chazovachii B, Chigwenya A, Mushuku A (2012) Adoption of climate resilient rural livelihoods through growing of small grains in Munyaradzi communal area, Gutu District. Afr J Agric Res 7(8):1335–1345
Chera M (2017) Transforming millets: strategies and struggles in changing taste in Madurai. Food Culture Soc 20(2):303–324
Chikkaballapur Krishnappa S, Saravanan S, Natarajan V (2022) Development and performance evaluation of foxtail millet (Setaria italica L.) dehuller. J Food Process Eng 45(6):e13937
Dayakar Rao B, Kiranmai E (2020) Novel processes, value chains, and products for food, feed, and industrial uses. In: Sorghum in the 21st century: food–fodder–feed–fuel for a rapidly changing world. Springer, Singapore, pp 859–887
Dekka S, Paul A, Vidyalakshmi R, Mahendran R (2023) Potential processing technologies for utilization of millets: an updated comprehensive review. J Food Process Eng 46:e14279

Demirkesen I, Ozkaya B (2022) Recent strategies for tackling the problems in gluten-free diet and products. Crit Rev Food Sci Nutr 62(3):571–597
Devi PB, Vijayabharathi R, Sathyabama S, Malleshi NG, Priyadarisini VB (2014) Health benefits of finger millet (Eleusine coracana L.) polyphenols and dietary fibre: a review. J Food Sci Technol 51:1021–1040
Dey S, Saxena A, Kumar Y, Maity T, Tarafdar A (2022) Understanding the antinutritional factors and bioactive compounds of kodo millet (Paspalum scrobiculatum) and little millet (Panicum sumatrense). J Food Qual 2022:1–19
Dharshini SS, Meera M (2023) Millet bars-healthier alternative to cereal bars: a review. Agric Rev 44(2):155–163
Diao X (2017) Production and genetic improvement of minor cereals in China. Crop J 5(2):103–114
Espitia-Hernández P, Chavez Gonzalez ML, Ascacio-Valdés JA, Dávila-Medina D, Flores-Naveda A, Silva T et al (2022) Sorghum (Sorghum bicolor L.) as a potential source of bioactive substances and their biological properties. Crit Rev Food Sci Nutr 62(8):2269–2280
Etuk EB, Okeudo NJ, Esonu BO, Udedibie ABI (2012) Anti-nutritional factors in sorghum: chemistry, mode of action and effects on livestock and poultry. Online J Animal Feed Res 2(2):113–119
FAO (2021). https://www.fao.org/statistics/en/
FAO (2022) Crop, livestock products. United Nations Food Agriculture Organization, Rome
FAO (2023). https://www.fao.org/millets-2023/about/en
Fu J, Du J, Lin G, Jiang D (2021) Analysis of yield potential and regional distribution for bioethanol in China. Energies 14(15):4554
Gaikwad PS, Reddy NBP, Sandhya K, Yadav BK (2023) Traditional vs. modern usage of nutri-cereals. In: Nutri-cereals. CRC Press, Boca Raton, pp 27–56
Geeta H, Mathad P, Nidoni U, Ramachandra CT (2016) Development of foxtail millet based extruded food product. Int J Food Sci 6:11–22
Girard AL, Awika JM (2018) Sorghum polyphenols and other bioactive components as functional and health-promoting food ingredients. J Cereal Sci 84:112–124
Goswami D, Mandge HM, Rehal J, Rajput H (2023) 7 nutritional advancement in the ethnic and novel foods using the diverse minor millets. In: Nutriomics of millet crops. CRC Press, Boca Raton, pp 141–156
Gowda NN, Siliveru K, Prasad PV, Bhatt Y, Netravati BP, Gurikar C (2022) Modern processing of Indian millets: a perspective on changes in nutritional properties. Foods 11(4):499
Gupta N, Morya S (2022) Bioactive and pharmacological characterization of *Chenopodium quinoa*, *Sorghum bicolor* and *Linum usitassimum*: a review. J Appl Nat Sci 14(3):1067–1084
Gupta R, Singh H, Kumar RP, Meghwal M (2023) Economic, social, and market feasibility of nutri-cereals. In: Nutri-cereals. CRC Press, Boca Raton, pp 345–360
Hamel D, Rozman V, Liška A (2020) Storage of cereals in warehouses with or without pesticides. Insects 11(12):846
Hema V, Ramaprabha M, Saraswathi R, Chakkaravarthy PN, Sinija VR (2022) Millet food products. In: Handbook of millets-processing, quality, and nutrition status. Springer, Singapore, pp 265–299
Hemamalini C, Sam S, Jain S, Sam S, Patro T (2021) Awareness and consumption of small millets. Pharma Innov J 7:34–37
Hossain A, Islam MT, Maitra S, Majumder D, Garai S, Mondal M et al (2021) Neglected and underutilized crop species: are they future smart crops in fighting poverty, hunger and malnutrition under changing climate? In: Neglected and underutilized crops-towards nutritional security and sustainability. Springer, Singapore, pp 1–50
Hossain MS, Islam MN, Rahman MM, Mostofa MG, Khan MAR (2022) Sorghum: a prospective crop for climatic vulnerability, food and nutritional security. J Agric Food Res 8:100300
Hussain MI, Muscolo A, Farooq M, Ahmad W (2019) Sustainable use and management of non-conventional water resources for rehabilitation of marginal lands in arid and semiarid environments. Agric Water Manag 221:462–476

ICRISAT (2021). https://www.icrisat.org/little-millet-that-are-big-on-nutrition-and-yield-identified-at-icrisat/
Istrati DI, Constantin OE, Vizireanu C, Rodica D, Furdui B (2019) Sorghum as source of functional compounds and their importance in human nutrition. Ann Univ Dunarea Jos Galati Fascicle VI Food Technol 43(2):189–205. https://doi.org/10.35219/foodtechnology.2019.2.13
ITC (2023). https://www.trademap.org/Index.aspx
Iyer N (2021) Everyday superfoods. Bloomsbury Publishing, New Delhi
Jindal P, Nikhanj P (2023) A review on processing technologies for value added millet products. J Food Process Eng 46:e14419
Joshi TJ, Singh SM, Rao PS (2023) Novel thermal and non-thermal millet processing technologies: advances and research trends. Eur Food Res Technol 249:1149–1160. https://doi.org/10.1007/s00217-023-04227-8
Kala CP, Nautiyal S (2023) Traditional food knowledge of local people and its sustainability in the mountains of Uttarakhand State of India. J Soc Econ Dev 25(1):32–51
Karlsson Potter H, Yacout DM, Henryson K (2023) Climate assessment of vegetable oil and biodiesel from Camelina grown as an intermediate crop in cereal-based crop rotations in cold climate regions. Sustainability 15(16):12574
Karpagapandi L, Kalaiselvan A, Baskaran V (2023) Nutraceutical and functional attributes of foxtail millet (Setaria italica). In: Nutri-cereals. CRC Press, Boca Raton, pp 141–177
Kaur S, Kumar K, Singh L, Sharanagat VS, Nema PK, Mishra V, Bhushan B (2022) Gluten-free grains: importance, processing and its effect on quality of gluten-free products. Crit Rev Food Sci Nutr 64:1988–2015
Kaushik N, Chauhan K, Aggarwal M, Kumar R (2022) State-of-the-art of knowledge on underutilized millets: Kodo and kutki, grown in tribal areas of India. Int J Agric Sci Res 12:35–56
Kebede E, Bekeko Z (2020) Expounding the production and importance of cowpea (Vigna unguiculata (L.) Walp.) in Ethiopia. Cogent Food Agric 6(1):1769805
Khairuddin MAN, Lasekan O (2021) Gluten-free cereal products and beverages: a review of their health benefits in the last five years. Foods 10(11):2523
Kraithong S, Wang S, Junejo SA, Fu X, Theppawong A, Zhang B, Huang Q (2022) Type 1 resistant starch: nutritional properties and industry applications. Food Hydrocoll 125:107369
Lesuuda L (2023) Knowledge of young children caregivers about Mycotoxin their postharvest handling practices of sorghum and effect of fermentation on aflatoxin and fumonisin levels in sorghum grains and flour in Kerio Valley, Kenya. Doctoral dissertation, Egerton University
Liaqat W, Altaf MT, Barutçular C, Mohamed HI, Ahmad H, Jan MF, Khan EH (2024) Sorghum: a Star Crop to combat abiotic stresses, food insecurity, and hunger under a changing climate: a review. J Soil Sci Plant Nutr:1–28
Lokhande P, Kharche A, Ganpatrao WS, Manithe D, Harke S (2019) Immune boosting super food supplement from natural resources. J Pharm Phytochem 8(5):2108–2113
Lone R, Hassan N, Bashir B, Rohela GK, Malla NA (2023) Role of growth elicitors and microbes in stress management and sustainable production of sorghum. Plant Stress 9:100179
Mariutti LRB, Rebelo KS, Bisconsin-Junior A, de Morais JS, Magnani M, Maldonade IR et al (2021) The use of alternative food sources to improve health and guarantee access and food intake. Food Res Int 149:110709. https://doi.org/10.1016/j.foodres.2021.110709
Mawouma S, Condurache NN, Turturică M, Constantin OE, Croitoru C, Rapeanu G (2022) Chemical composition and antioxidant profile of sorghum (Sorghum bicolour (L.) Moench) and pearl millet (Pennisetum glaucum (L.) R. Br.) grains cultivated in the far-North region of Cameroon. Foods 11(14):2026
McGinnis MJ, Painter JE (2020) Sorghum: history, use, and health benefits. Nutr Today 55(1):38–44
Millet Advisor (2023). https://milletadvisor.com/millet-consumption-trends/
Mishra P, Sabikhi L (2020) Fermented pearl millet weaning food: an innovation of food technology and application in food processing and management. In: Innovations in food technology: current perspectives and future goals. Springer, Singapore, pp 17–27
MITCON (2023). https://www.mitconindia.com/blogs/millets-consumption-in-india/

Mundia CW, Secchi S, Akamani K, Wang G (2019) A regional comparison of factors affecting global sorghum production: the case of North America, Asia and Africa's Sahel. Sustainability 11(7):2135

Musundire R, Dhlakama RB, Serere JH (2021) Physico-chemical and sensory quality evaluation of an extruded nutrient-dense termite (Macrotermes natalensis) and millet (Eleusine coracana) instant porridge. Int J Trop Insect Sci 41:2059–2070

Muthamilarasan M, Dhaka A, Yadav R, Prasad M (2016) Exploration of millet models for developing nutrient rich graminaceous crops. Plant Sci 242:89–97. https://doi.org/10.1016/J.PLANTSCI.2015.08.023

Nešić K, Habschied K, Mastanjević K (2023) Modified mycotoxins and multitoxin contamination of food and feed as major analytical challenges. Toxins 15(8):511

Noort MW, Renzetti S, Linderhof V, du Rand GE, Marx-Pienaar NJ, de Kock HL et al (2022) Towards sustainable shifts to healthy diets and food security in sub-Saharan Africa with climate-resilient crops in bread-type products: a food system analysis. Foods 11(2):135

Nyoni N, Dube M, Bhebhe S, Sibanda B, Maphosa M, Bombom A (2020) Understanding biodiversity in sorghums to support the development of high value bio-based products in sub-Saharan Africa. J Cereals Oilseeds 11(2):37–43

Palacios CE, Nagai A, Torres P, Rodrigues JA, Salatino A (2021) Contents of tannins of cultivars of sorghum cultivated in Brazil, as determined by four quantification methods. Food Chem 337:127970. https://doi.org/10.1016/J.FOODCHEM.2020.127970

Pandiselvam R, Kaavya R, Jayanath Y, Veenuttranon K, Lueprasitsakul P, Divya V et al (2020) Ozone as a novel emerging technology for the dissipation of pesticide residues in foods—a review. Trends Food Sci Technol 97:38–54

Patra M, Bashir O, Amin T, Shams R, Chaudhary KS, Mirza AA, Manzoor S (2023) A comprehensive review on functional beverages from cereal grains-characterization of nutraceutical potential, processing technologies and product types. Heliyon 9(6):e16804

Pradeep M (2023) Footprint of millets on type 2 diabetic mellitus: a short review. World J Biol Pharm Health Sci 14(1):280–284

Pradhan A, Panda AK, Bhavani RV (2019) Finger millet in tribal farming systems contributes to increased availability of nutritious food at household level: insights from India. Agric Res 8(4):540–547. https://doi.org/10.1007/s40003-018-0395-6

Qin H, Wu H, Shen K, Liu Y, Li M, Wang H et al (2022) Fermented minor grain foods: classification, functional components, and probiotic potential. Foods 11(20):3155

Raheem D, Dayoub M, Birech R, Nakiyemba A (2021) The contribution of cereal grains to food security and sustainability in Africa: potential application of UAV in Ghana, Nigeria, Uganda, and Namibia. Urban Sci 5(1):8

Rakshit S, Prabhakar, Kumar P (2023) Maize and millets. In: Trajectory of 75 years of Indian agriculture after Independence. Springer, Singapore, pp 163–187

Ramakrishnan SR, Antony U, Kim SJ (2022) Non-thermal process technologies: Influences on nutritional and storage characteristics of millets. J Food Process Eng 46:e14215

Ranjan R, Singh S, Dhua S, Mishra P, Chauhan AK, Gupta AK (2023) Kodo Millet (Paspalum scrobiculatum): bioactive profile, health benefits and techno-functionality. In: Nutri-cereals: nutraceutical and techno-functional potential. CRC Press, Boca Raton, pp 193–211

Rashwan AK, Yones HA, Karim N, Taha EM, Chen W (2021) Potential processing technologies for developing sorghum-based food products: an update and comprehensive review. Trends Food Sci Technol 110:168–182

Rathore T, Singh R, Kamble DB, Upadhyay A, Thangalakshmi S (2019) Review on finger millet: processing and value addition. Pharma Innov J 8(4):283–291

Rawat M, Varshney A, Rai M, Chikara A, Pohty AL, Joshi A et al (2023) A comprehensive review on nutraceutical potential of underutilized cereals and cereal-based products. J Agric Food Res 12:100619

Ray RC (2022) Lactic acid bacteria in food biotechnology: innovations and functional aspects. Elsevier, London

Rizwana M, Singh P, Ahalya N, Mohanasundaram T (2023) Assessing the awareness of nutritional benefits of millet amongst women in Bangalore. Br Food J 125(6):2002–2018

Rotela S, Borkar S, Borah A (2021) Health benefits of millets and their significance as functional food: a review. J Pharma Innov 10:158–162

Saleem S, Mushtaq NU, Shah WH, Rasool A, Hakeem KR, Seth CS et al (2023) Millets a smart future food with essential phytonutrients for promoting health. J Food Compos Anal 124:105669

Samtiya M, Aluko RE, Dhewa T (2020) Plant food anti-nutritional factors and their reduction strategies: an overview. Food Prod Process Nutr 2:1–14

Satyavathi CT, Ambawat S, Khandelwal V, Srivastava RK (2021) Pearl millet: a climate-resilient nutricereal for mitigating hidden hunger and providing nutritional security. Front Plant Sci 12:659938

Selladurai M, Pulivarthi MK, Raj AS, Iftikhar M, Prasad PV, Siliveru K (2022) Considerations for gluten-free foods-pearl and finger millet processing and market demand. Grain Oil Sci Technol 6:59–70

Senthamarai Selvi L, Malathi D (2019) Consumption pattern and nutritional assessment of minor millets among rural women in Madurai District of Tamil Nadu, India. Int J Curr Microbiol App Sci 8(11):2102–2112. https://doi.org/10.20546/ijcmas.2019.811.244

Sewak A, Singla N, Javed M, Gill GS (2023) Suitability of pearl millet (Pennisetum glaucum (L.) R. Br.) and sorghum (Sorghum bicolor (L.) Moench) based food products for diabetics. Acta Alimentaria 52(10):1–12

Shah P, Dhir A, Joshi R, Tripathy N (2023) Opportunities and challenges in food entrepreneurship: in-depth qualitative investigation of millet entrepreneurs. J Bus Res 155:113372

Sharanagat VS, Singh L, Nema PK (2022) Approaches for development of functional and low gluten bread from sorghum: a review. J Food Process Preserv 46(11):e17089

Singh N, David J, Thompkinson DK, Seelam BS, Rajput H, Morya S (2018) Effect of roasting on functional and phytochemical constituents of finger millet (Eleusine coracana L.). Pharma Innov J 7(4):414–418

Sirany T, Tadele E, Aregahegn H, Wale D (2022) Economic potentials and use dynamics of sorghum food system in Ethiopia: its implications to resolve food deficit. Adv Agric 2022:4580643

Stamenković OS, Siliveru K, Veljković VB, Banković-Ilić IB, Tasić MB, Ciampitti IA et al (2020) Production of biofuels from sorghum. Renew Sust Energ Rev 124:109769

Sudha V, Malleshi NG, Ratnavathi CV, Shobana S, Vetriventhan M, Hariprasanna K et al (2021) Germplasm characterization and novel technologies to unleash the nutritional potential of small millets. In: Orphan crops for sustainable food and nutrition security. Routledge, New York, pp 309–318

Takaidza S (2023) Indigenous South African food: nutrition and health benefits. In: Soto-Hernández M, Aguirre-Hernández E, Palma-Tenango M (eds) Phytochemicals in agriculture and food. IntechOpen. https://doi.org/10.5772/intechopen.110732

Tan XL, Azam-Ali S, Goh EV, Mustafa M, Chai HH, Ho WK et al (2020) Bambara groundnut: an underutilized leguminous crop for global food security and nutrition. Front Nutr 7:601496

Taylor JRN (2019) Sorghum and millets: taxonomy, history, distribution, and production. In: Sorghum and millets: chemistry, technology, and nutritional attributes. Woodhead Publishing, Duxford, pp 1–21. https://doi.org/10.1016/B978-0-12-811527-5.00001-0

Taylor JR, Duodu KG (2017) Sorghum and millets: grain-quality characteristics and management of quality requirements. In: Cereal grains. Woodhead Publishing, Duxford, pp 317–351

Teferra TF, Awika JM (2019) Sorghum as a healthy global food security crop: opportunities and challenges. Cereal Foods World 64(5):1–8

Tharifkhan SA, Perumal AB, Elumalai A, Moses JA, Anandharamakrishnan C (2021) Improvement of nutrient bioavailability in millets: emphasis on the application of enzymes. J Sci Food Agric 101(12):4869–4878

Tonapi VA, Talwar HS, Are AK, Bhat BV, Reddy CR, Dalton TJ (2020) Sorghum in twenty-first century and beyond: perspectives, prospects, strategies and way forward. In: Sorghum in the 21st century: food–fodder–feed–fuel for a rapidly changing world. Springer, Singapore, pp 929–940

Tripathi T, Vyas S (2023) From ancient grains to modern solutions: a history of millets and their significance in agriculture and food security. Int J Home Sci 9(2):72–78

Tripathi MK, Jadam RS, Kumar A (2021) Quality management system in millet and sorghum. Millets Millet Technol 1:363–379

USDA (2023). https://fdc.nal.usda.gov/fdc-app.html#

Vellaichamy Gandhimeyyan R, Chockalingam V, Muthurajan R, Raman R (2023) Accelerating gene discovery in climate-resilient and nutrient-rich major and minor millets through genome-wide association studies: progress and prospects. Plant Breed 142(4):389–406

Wang Y, Jian C (2022) Sustainable plant-based ingredients as wheat flour substitutes in bread making. NPJ Sci Food 6(1):49

Wei D, Davis KF (2021) Culturally appropriate shifts in staple grain consumption can improve multiple sustainability outcomes. Environ Res Lett 16(12):125006

Wijerathna-Yapa A, Pathirana R (2022) Sustainable agro-food systems for addressing climate change and food security. Agriculture 12(10):1554

Willett W, Rockström J, Loken B, Springmann M, Lang T, Vermeulen S, Garnett T, Tilman D, DeClerck F, Wood A, Jonell M, Clark M, Gordon LJ, Fanzo J, Hawkes C, Zurayk R, Rivera JA, De Vries W, Majele Sibanda L et al (2019) Food in the Anthropocene: the EAT–Lancet Commission on healthy diets from sustainable food systems. Lancet 393(10170):447–492. https://doi.org/10.1016/S0140-6736(18)31788-4

Woomer JS, Adedeji AA (2021) Current applications of gluten-free grains—a review. Crit Rev Food Sci Nutr 61(1):14–24

Xiong Y, Zhang P, Warner RD, Fang Z (2019) Sorghum grain: from genotype, nutrition, and phenolic profile to its health benefits and food applications. Compr Rev Food Sci Food Saf 18(6):2025–2046. https://doi.org/10.1111/1541-4337.12506

Yahaya MA, Shimelis H (2022) Drought stress in sorghum: mitigation strategies, breeding methods and technologies—a review. J Agron Crop Sci 208(2):127–142

Zeppieri S (2021) The African vegan in today's modern world: guide to African vegan. Susan Zeppieri

Teff and Fonio

7

Sapna Jarial, Jaspreet Kaur, Jyoti Singh, Rupesh Kaushik,
Sanjeev Kumar, and Ali Abdullahi

Abstract

Teff and Fonio are ancient grains spectacular for their exceptional nutritional attributes and resilience to extreme climatic variations. Often referred to as orphan or underutilised crops, they play a pivotal role in addressing food insecurity resulting from climate change. Teff and Fonio, integral to African diets, vary in consumption across regions. Teff is primarily used in Ethiopian injera, a flatbread, and alcoholic drinks. Fonio, widespread in West Africa, is versatile and suitable for porridge, couscous, and baking. Both grain's global popularity is rising, with Teff's rising demand in the UK and the USA due to its health benefits. Being gluten-free, they are replete with bioactive elements such as dietary fibre, phenolic compounds, resistant starch, and essential amino acids. Teff offers essential amino acids, including cysteine and methionine, while Fonio is abundant in iron. Containing antinutrients like phytic acid and tannins, their absorption can be hindered, but traditional processing like soaking reduces these effects. Both grains are rich in bioactive components, providing health benefits, including aiding digestion, weight management, and providing antioxidant properties. However, they face challenges like complex post-harvest processing and higher

S. Jarial (✉) · R. Kaushik
Department of Agricultural Economics and Extension, School of Agriculture, Lovely Professional University, Phagwara, Punjab, India

J. Kaur · J. Singh
Department of Food Technology and Nutrition, School of Agriculture, Lovely Professional University, Phagwara, Punjab, India

S. Kumar
Department of Horticulture, School of Agriculture, Lovely Professional University, Phagwara, Punjab, India

A. Abdullahi
Department of Agricultural Economics and Extension, Bayero University, Kano, Nigeria

J. Singh et al. (eds.), *Cereals and Nutraceuticals*,
https://doi.org/10.1007/978-981-97-2542-7_7

costs than other cereals. Technical, economic, and policy efforts are needed to support their cultivation and marketability. Their inclusion in diets can enhance digestion, provide antioxidant benefits, and fortify gut and bone health. This chapter explores Teff and Fonio's origin, cultivation, nutritional composition, processing methods, health implications, and safety considerations. Furthermore, it highlights the significance of these crops in confronting climate-induced food security challenges and explores prospects and challenges.

Keywords

Teff · Fonio · Processing · Climate · Antioxidant · Cultivation

7.1 Introduction

Orphan crops are not globally dominant, but they help in food security in food-insecure regions. The term "orphan" highlights the neglect of these crops by the global research community. They are also overlooked underprivileged, forsaken, and underutilised crops. Underutilised crops are crucial for climate change and the need to feed a growing population. Such crops are stress-resilient and are complemented by distinct and nutritionally rich profiles (Sousa and Raizada 2020). Teff (*Eragrostis tef*) and Fonio (*Digitaria exilis)* are novel yet underutilised crops. They are orphan African cereal crops. Millets are small-grained cereals. Ancient grains like these have become popular due to their unique nutritional qualities (Gebremariam et al. 2013). Small millets thrive in arid and semi-arid regions (Choudhary et al. 2023). Food security in drylands depends on Teff and Fonio's development and cultivation (Sousa and Raizada 2020). Teff is known for its high adaptability to extreme climates, making it a "climate-smart crop" (Sharma et al. 2022). Teff straw has silica content comparable to rice husk and wheat straw. Teff straw can produce silica used in construction, healthcare, and wellness (Tessema et al. 2023).

Fonio refers to two species of small-seeded cereals native to Western Africa: white Fonio (*Digitaria exilis*) and Black Fonio *(Digitaria iburua*) (Zhu 2020). Fonio is a drought-resistant, early-maturing cereal with a high yield and nutritional value that promotes food and nutrition security and revenue generation in West Africa. Fonio's nutrient-rich straw is used as a cheap construction material in Sahelian zones, contributing to thermal comfort (Ouedraogo et al. 2019). Regarding human nutrition, Fonio grain is an excellent source of all 20 amino acids, including cysteine and methionine. Thus, it is utilised for its nutritional value (Sharma et al. 2022). It is gluten-free. Therefore, the demand for Fonio is increasing globally (Yisak et al. 2022). Fonio has economic, dietary, and cultural significance (Dansi et al. 2010). Fonio has essential religious and sociocultural significance in West African countries (Togo, Senegal, Mali, Benin, Guinea, Niger, Nigeria, Burkina Faso, Côte d'Ivoire). It is reserved for chiefs and nobility and important events like Ramadan, Christmas, New Year, and other festivals. Women primarily handle Fonio's processing and selling, providing a financial source and ensuring good nutrition for their families (Diop et al. 2018). In Togo, it is used for weddings, baptisms, and for women's initiation rituals (Fanou-Fogny et al. 2011). Nonetheless, Fonio is largely neglected as "minor" millet (Animasaun et al. 2023). Their macro- and micronutrient content is comparable to that of the most common cereals (Taylor and Kruger 2016), but they are still undervalued by conventional research.

7.2 Origin and Background

"Teff, a diminutive grain", originated in Ethiopia and is indigenous to Eritrea, serving as a fundamental food source for millions. In Eastern Africa, it has been growing for nearly 3000 years (Yisak et al. 2022). Teff is an annual cereal grass (family *Poaceae*). "The term 'teff' originates from the Amharic word *'teffa'*, signifying 'small' and referring to the minute seeds" (Petruzzello 2023). "Teff is a clustered grass featuring slender stems and a wide crown. Its shallow, fibrous roots create an extensive root structure that effectively withstands drought and excess water". C4 carbon fixation prevents photorespiration and makes teff drought-tolerant. This grain thrives in diverse climates (Khairuddin and Lasekan 2021). This millet's short growth cycle and high nutrients make it advantageous. Its small size and lack of mechanical processing have kept it a lesser-known grain despite its high nutritious content (Adisa and Adeloye 2012; Chadalavada et al. 2022; Deriu et al. 2022b; Sousa and Raizada 2020) Teff grain is 1 mm long and 1 mm wide, the smallest in the entire flour grain. Colour can range from extremely light ivory to very dark reddish brown (Howell et al. 2022). Fonio and Teff share structural and physicochemical features (starch structure, gelatinisation, water absorption, swelling).

Fonio (*Digitaria exilis*) is regarded as one of the ancient (over 5000 years) West African indigenous cereals, called the "grain of life". Fonio was cultivated in central Mali's Inner Niger Delta region (Anitha et al. 2021). The French word fonio has its origins in the Wolof word foño (Seignobos and Tourneux 2002). West Africa has great potential for Fonio production because it outperforms other grains in drought and poor soil fertility and is less susceptible to pests when stored. Fonio is a staple cuisine in Burkina Faso, Guinea, Senegal, Mali, and Nigeria. Fonio has C4 fixation of carbon, similar to numerous other grass varieties. "The plant's height is medium". The species are diploid (2*n*), tetraploid (4*n*), or hexaploid (6*n*) (Adoukonou-Sagbadja et al. 2007). These cultivated varieties possess a brief growth period and are prepared for harvesting before other grains. It can reduce hunger in the lean season, and white fonio is called "hungry rice" (Petruzzello 2023). Fonio's consumption rises when other grains are scarce due to a poor harvest (Fanou-Fogny et al. 2011). Fonio grain is small, measuring around 1–1.5 mm in length. They resemble small rice grains and are round, elliptical, smooth and glossy, white, black, or brown (Abrouk et al. 2020).

7.3 Production and Cultivation Practices

Production and cultivation practices for Teff and Fonio are crucial for ensuring their continued availability and contribution to food security (Milkias and Beri 2020). Teff and Fonio are produced and grown chiefly using traditional methods, focusing on small-scale farmers. Even though these agronomic, nutritional, and health advantages exist, minor millets are grown in less fertile soils for both grain and fodder, relying on rainwater for cultivation. This often involves experiencing intermittent moisture shortages lasting approximately 25–30 days (Reddy et al. 2021). Teff demands substantial labour and meticulous soil preparation for uniform sowing and appropriate seed depth. Moreover, harvesting, threshing, and winnowing frequently entail manual efforts, and the tiny seeds pose a challenge for handling and transporting without incurring losses (Petruzzello 2023). As Teff grains remain untouched by insect pests, farmers refrain from employing pesticides on them, which stands in contrast to their application on maize and other grains (Demiss et al. 2020; Milkias and Beri 2020). Teff is typically sown using a technique called broadcasting, in which seeds are dispersed over a prepared field either manually or with the aid of a tool or machine. The crop is primarily rain-fed, relying on the rainy season for growth, with minimal use of fertilisers and pesticides. This grain is usually harvested between 2 and 3 months after sown (Yisak et al. 2022). France was a significant player in the teff market, being the top exporter in 2022, and Italy was the leading importer with 123.97 million USD (Tridge 2023b).

Fonio cultivation is similar to Teff cultivation. Fonio is cultivated in tropical monsoon climates in western Guinea and in the Sahel's scorching, arid deserts. Fonio is not likely to thrive as a crop in highly humid regions. Fonio has unfavourable traits like seed shattering, lodging, and lower yields. Small family farms can grow Fonio on land unsuitable for other crops with low water and fertiliser costs (Chadalavada et al. 2022). The growth cycle duration varies between 70 and 130 days, depending upon the specific variety—Guinea is the top producer and primary supplier to the global market. The UK will be the leading exporter of Fonio in 2022. The value of Fonio exported in 2022 is 168.03 K USD (Tridge 2023a). Considerable decline in both export and import suggests a reduced global trade in fonio or disruptions affecting both supply and demand sides of the market. Potential factors could be a drop in consumer demand, competition from other grains, trade barriers, or supply chain challenges. The USA, being the top importer, demonstrates a significant market for Fonio (Tridge 2023a).

7.4 Consumption Pattern

The consumption pattern of Teff and Fonio varies across regions and cultures, but both grains play an important role in local diets and have gained recognition for their nutritional value. Teff and Fonio have been integral to the diets of many African cultures for thousands of years, with consumption patterns often reflecting regional culinary practices and traditions (Khairuddin and Lasekan 2021). Teff and Fonio are used to prepare African cuisines (Fig. 7.1) (Chadalavada et al. 2022). Teff is usually ground into flour and used to prepare traditional Ethiopian flatbread injera (Chadalavada et al. 2022; Gebregewergis 2021). *Injera* acts as a food and an eating utensil, used to scoop up other dishes served on top of it. This consumption pattern is unique to Ethiopian culture and cuisine (Gebregewergis 2021).

Fonio is a typical food in Niger, Nigeria, Mali, Burkina Faso, Benin, Côte d'Ivoire, Senegal, and countries of West Africa (Jocelyne et al. 2020). It is consumed in various forms due to its versatile nature. Its quick cooking time and digestibility make it often used in porridges, especially for the elderly, infants, and the sick. It is also prepared as couscous or is ground into flour for baking (Deriu et al. 2022b; Lydia Pramitha et al. 2023; Petrova and Petrov 2020). Teff is sometimes used as a wheat replacement in bread and pasta products (Hozyasz 2013). Teff is also utilised in producing a regional alcoholic beverage known as '*arak'e*' or '*katikalla*', and a traditional beer called '*t'ella*' or '*fersso*' (FAO n.d.). So, while the traditional consumption of Teff and Fonio is deeply entrenched in specific regional cuisines, their emerging global popularity has led to a wider range of uses and consumption patterns worldwide (Chadalavada et al. 2022). Fonio's gluten-free status has increased interest in Western health food markets (Švec and Hrušková 2015). Teff consumption is increasing in developed countries such as the UK and the USA, a similar increase in Australia is anticipated (Gosine and McSweeney 2019).

7.5 Nutritional and Chemical Composition

Teff and Fonio are cereal crops that have gained attention for their nutritional value (Table 7.1) and potential health benefits (Abewa et al. 2019; Gebregewergis 2021). It is an excellent source of all 20 amino acids, including cysteine methionine (Sharma et al. 2022). Moreover, it is rich in lysine. Fonio is one of the richest cereals in iron, millet, and sorghum. Factors such as soil conditions, climate, and agricultural practices can influence the nutrient content of these grains (Vetriventhan et al. 2020). Teff is abundant in macro- and micronutrients and bioactive, non-essential metabolites such as saponins and phenolic compounds. It was discovered that brown Teff types Fe and Ca contents were higher than white variants. On the other hand, the Cu content of white Teff types is higher than that of brown variants (Aljada et al. 2021).

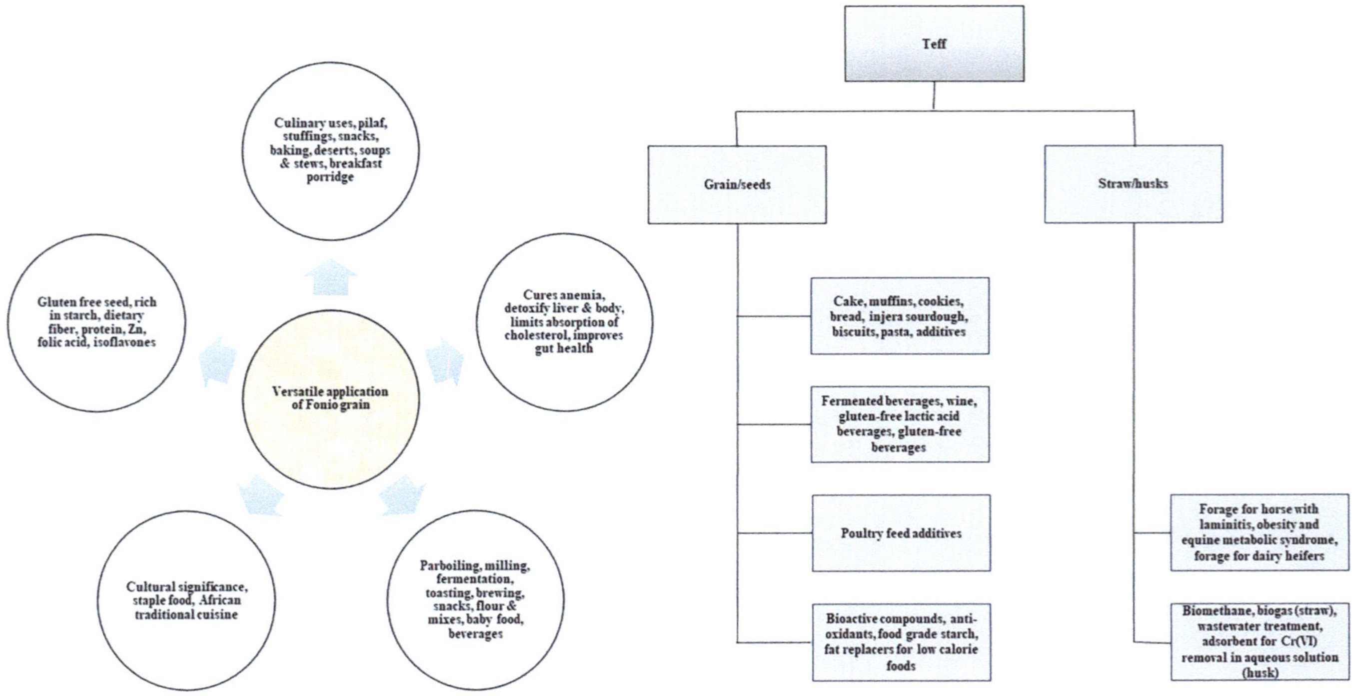

Fig. 7.1 Potentials of Teff and Fonio. (Adapted from source: Barretto et al. 2021)

Table 7.1 Teff and Fonio nutritional composition

Crop	Nutritional components		References
Teff	Essential amino acid level	40.9%	Gebregewergis (2021), Gebru and Sbhatu (2020), Nooshkam et al. (2022)
	Total protein content	9.37–15%	
	Total carbohydrates	85.6%	
	Starch percentage	74–75.5%	
	Amylose concentration	20–30%	
	Amylopectin molecular weight	10.1–16.5 g/mol	
	Amylose molecular weight		
	Calcium	180 mg/100 g	Fukagawa et al. (2022)
	Iron	7.63 mg/100 g	
	Magnesium	184 mg/100 g	
	Phosphorus	429 mg/100 g	
	Potassium	427 mg/100 g	
	Sodium	12 mg/100 g	
	Zinc	3.63 mg/100 g	
Teff grain flour	Lipid content	3.6–4.4%	Ramashia et al. (2018)
	Free form of lipid	80%	Wee and Henry (2020)
Bioactive phytochemicals	Flavones, phenolic compounds: protocatechuic, gentisic, vanillic, syringic 4		Reta et al. (2022)
	Glutelins, globulins, prolamins, albumins		Bartkiene et al. (2019)
Vitamins	Niacin	3.363 mg/100 g	Agriopoulou et al. (2020), Fukagawa et al. (2022)
	B6	0.482 mg/100 g	
	Riboflavin	0.27 mg/100 g	
	K	1.9/100 g	
	A	9 IU	
	Tocopherol	0.08 mg/100 g	
Minerals	P	429 mg/100 g	John et al. (2023)
	K	427 mg/100 g	
	Na	12 mg/100 g	
	Zn	3.63 mg/100 g	
	Ca	180 mg/100 g	
	Fe	7.63 mg/100 g	
	Mg	184 mg/100 g	

(continued)

Table 7.1 (continued)

Crop	Nutritional components		References
Fonio	Carbohydrate	67.1–91%	Nooshkam et al. (2022)
	Lipid	3.25%	Ramashia et al. (2018)
	Albumin	3.5%	Cruz-Garcia et al. (2020)
	Globulin	1.8%	
	Prolamin	5.5%	
	Glutelin	14%	
	Thiamin	0.05–0.1 mg/100 g	Deriu et al. (2022b)
	Riboflavin	0.3–0.48 mg/100 g	
	Nicotinic acid (PP vitamin)	3 mg/100 g	
Acids	Gallic, caffeic, chlorogenic, vanillic, p-coumaric, sinapi, ferulic, protocatechuic, salicylic, syringic		Reta et al. (2022)
Minerals	Magnesium, phosphorus, potassium, copper, manganese, zinc, calcium, sodium		John et al. (2023), Nithiyanantham et al. (2019)

7.6 Antinutrients and Toxic Compounds

Teff and Fonio contain antinutrients—substances that may prevent the absorption of nutrients. These include phytic acid, tannins, and trypsin inhibitors (Molfetta et al. 2022). Tannins are polyphenolic compounds that can inhibit protein digestion, but their content in Teff and Fonio is relatively lower against other grains (Dingeo et al. 2020; Zhang et al. 2014). Trypsin, an enzyme involved in protein digestion, can be inhibited by trypsin inhibitors, thereby interfering with protein absorption. Generally, cooking can inactivate these compounds (Dingeo et al. 2020). Teff and Fonio have no known toxic compounds, making them safe for regular consumption. However, like all foods, they should be included in a balanced diet (Carboni et al. 2020).

Calcium, iron, and zinc can all be bound by phytic acid, which lowers the bioavailability of those minerals. Traditional processing methods for Teff and Fonio, such as soaking, fermentation, and sprouting, can significantly reduce phytic acid levels and increase mineral availability (Deriu et al. 2022b; Dingeo et al. 2020; Petrova and Petrov 2020). Antinutrient content varies according to the grain species, cultivation, and processing techniques.

7.7 Bioactive and Functional Components

An aqueous extract of white Fonio revealed the presence of numerous bioactive compounds, including alkaloids (42 mg/100 g), saponins (1.8 mg/100 g), anthraquinones, terpenoids, phytosterols, cardenolides, and phlobatannins, albeit at relatively low concentrations (Adams and Yakubu 2020).

Different grains include significant bioactive phytochemicals called phytosterols and tocols, two terpenoids. It is yet unknown what these molecules in Fonio are called chemically protocatechuic acid, caffeic acid, syringic acid, vanillic acid, sinapic acid, ferulic acid, chlorogenic acid, gallic acid, and p-coumaric acid were the detected phenolic acids of white Fonio in both free and bound forms. Fonio contains proanthocyanidins, alkaloids, phenolics, saponins, and alkaloids (Adams and Yakubu 2020).

7.8 Health Implications

Teff and Fonio contain bioactive and functional components contributing to their health benefits (Table 7.2). Teff and Fonio can be a part of diets for those sensitive to gluten but do not have celiac disease (Aljada et al. 2021; Bascuñán et al. 2016). Teff and Fonio both contain much fibre, which helps with digestion, prevents constipation, and can contribute to satiety, which may aid in weight management (Dingeo et al. 2020). Resistant starch is a dietary fibre that resists digestion in the small intestine and, as a result, reaches the colon, where it promotes the growth of beneficial bacteria and promotes gut health (Huang et al. 2022; Yaskolka Meir et al. 2021). The phenolic compounds in these grains include antioxidant characteristics that help prevent several ailments, including cancer and heart disease (Zhu 2020). Researchers have discovered that the phenolics in Teff and Fonio have antioxidant, anti-inflammatory, and antidiabetic properties (Yaskolka Meir et al. 2021). Essential amino acids like lysine and methionine are particularly abundant in both cereals. These chemicals must be taken from food in order to synthesise proteins (Al-Khayri et al. 2023).

This is particularly beneficial for vegetarians or those with low meat consumption (Laugesen et al. 2022). Calcium, iron, zinc, and magnesium are among the minerals abundant in Teff and Fonio and essential for numerous bodily processes, such as regulating metabolism and supporting bone health, thus preventing osteoporosis (Reta et al. 2022).

Table 7.2 Health benefits of Teff and Fonio

Crop	Health benefit	References
Teff	High nutrient dense	Barretto et al. (2021)
	Gluten-free	Chochkov et al. (2022)
	Blood sugar regulation	Habte et al. (2022)
	Digestive health	Deriu et al. (2022b), Dingeo et al. (2020), Gebregewergis (2021)
	Bone health	Yisak et al. (2022)
Fonio	Gluten-free	Babarinde et al. (2020a)
	Low glycaemic index	Zhu (2020)
	Digestive health	Bassey et al. (2023)
	Antidiabetic and glycaemic control	Adams and Yakubu (2020)

7.9 Processing

Teff and Fonio are grains that undergo various processing treatments to make them suitable for consumption. Traditional methods for Teff include fermentation, cooking, and baking. These methods can impact the technological and nutritional qualities of Teff products in both positive and negative ways (Sousa and Raizada 2020). However, high levels of organochlorine pesticides were found, posing a consumer risk. Teff processing, like doughing and baking, reduces these pesticides (Mekonen et al. 2019).

Teff, known for its balanced essential amino acids, is suitable for malting and brewing. Often turned into whole-grain flour, Teff retains its bran and germ, enhancing fibre and nutrient content. With its rich nutrient composition, Teff is a promising ingredient for diverse food and beverage products. Roasting Teff grains can boost their nutritional value by increasing bioactive substances like phenols and fatty acids (Gebremariam et al. 2013). A study on microwave (MW) effects on tef flour's properties revealed that MW treatment impacted tef's morphological structure, increasing particle size and hydration. MW-treated samples produced more stable gels with higher viscoelastic properties. The moisture content during microwave (MW) processing affected the flour's properties. The results indicate that microwaving tef flour could improve the quality of food products (Calix-Rivera et al. 2023).

Fonio, however, can be processed into flour for gluten-free beverages and foods (Khairuddin and Lasekan 2021). Value addition processes for these grains can include milling, fermentation, roasting, and cooking. Fonio is traditionally pounded in a mortar with sand and separated into grains, By "popping" it over a flame and pounding it, the grain acquires a toasted hue (CIRAD n.d.).

Fonio flour has higher water absorption and swelling power but lower solubility than other flours. Fonio gels have strong structures, enduring 579% more stress than reference flours at 15% concentration. It also has the highest gelatinisation energy (11.45 J/g) and a compact starch structure, evidenced by a narrow gelatinisation temperature range (9.96 °C). Fonio gel textures remain firmer over time. These traits make Fonio suitable for gel-like food recipes (Deriu et al. 2022a).

The safety of processing methods for Teff and Fonio is crucial.

7.10 Value Addition

Studies have shown that injera made from teff resists staling better than those made from other grains like sorghum. *Wot*, a sauce made from various pulses, is traditionally served with injera to create a nutritionally balanced meal. The grain also has potential in the baby food sector and Western recipes, serving as a thickening agent and in baked products. In Ethiopia, teff is used to produce traditional alcoholic beverages such as beer known as *tella* and *arake* (Gebremariam et al. 2014). The value-addition practice of teff currently needs to be improved and limited to conventional practices (Tadele et al. 2021). To combat malnutrition and ensure the availability of low-cost, nutrient-rich food, novel breakfast foods made from blends of Fonio have been developed (Babarinde et al. 2020b).

7.11 Safety Consideration

A study assessed the safety of foods derived from fonio and ricebean. Formulations included blends of treated fonio, sprouted rice beans, and other ingredients. Findings revealed low levels of oligosaccharides (reducing flatulence risk), minimal phytochemicals (decreasing allergy risk), and dietary fibre content below 5%, ensuring infant digestibility. Despite a low bacterial and fungal count suggesting good hygiene, the near-neutral pH requires careful handling to prevent microbial growth. Overall, the foods are deemed safe for consumption (Ijeomah et al. 2023). In the traditional processing of Fonio, consumers are looking for a well-prepared product, without sand, without grass, husked, and clean.

According to a study, Teff grains are a rich and safe source of minerals and trace elements, with the levels of toxic elements such as aluminium, cadmium, tin, and mercury being below the limits set by the World Health Organization (Koubová et al. 2018). Another study focused on the safety evaluation of Teff for human consumption. Using in vitro and in vivo assays, the study evaluated the potential genotoxic and cytotoxic effects of Teff. The results showed that Teff did not exhibit genotoxic effects and did not induce cytotoxicity (Gebremariam et al. 2014) These findings support the safety of Teff grains for human consumption.

7.12 Future Challenges and Opportunities

Obstacles encompass challenging post-harvest processing, an extended cooking duration, and elevated product expenses relative to alternative grains (Fanou-Fogny et al. 2011). The industry's size and fragmentation prevent a comprehensive grasp of the economic and production landscape. Innovating food products with these grains can tap into new markets. Antinutrients in these grains, influenced by growth and processing conditions, need further study, especially processing's effect on nutrient bioavailability and nutritional value. The impact of roasting on Teff grains needs further research (Ahmed and Ali 2013) (old ref.). Locally successful traditional cultivation methods are laborious and yield low yields, making scaling up difficult. To increase yield, mechanised methods are needed (Yimer 2022; Zeweld et al. 2018). It is crucial to understand how to make these products, their sensory and nutritional profiles, and consumer acceptance. Investigation into enrichment and fortification methods is required (Demirkesen and Ozkaya 2022; Howell et al. 2022).

Understanding their complete nutritional profiles and health impacts necessitates more research (Vetriventhan et al. 2020). The market for Teff and Fonio is potentially viable when supported by advanced production strategies, like CRISPR/Cas-based genome editing, to enhance their productivity (Sharma et al. 2022). To encourage adoption among growers, conducting more comprehensive examinations across the industry regarding harvesting and processing is essential. Exploring the creation of suitable mechanical harvesting and threshing mechanisms holds promise for boosting harvest efficiency and minimising grain loss during this phase. Supply chain and market research on Teff and Fonio will help the industry reach its potential. Teff and Fonio need help meeting increased demand due to their underutilisation and confined cultivation to particular African regions (Sousa and Raizada 2020). Research is needed to promote these grains widely.

7.13 Conclusion

Though not globally dominant, orphan crops like Teff and Fonio are vital for food security in developing regions. Despite being labelled "underutilised", both grains display resilience to environmental stresses and provide unique nutritional benefits. Teff, originating from Eastern Africa, particularly Ethiopia and Eritrea, is known for its tiny size and drought tolerance. Its cultivation is labour-intensive, requiring careful soil preparation and primarily manual post-harvest techniques. In contrast, Fonio, known as the "grain of life" in West Africa, thrives despite droughts and marginal soils. Small-scale farmers cultivate both crops using traditional methods. According to 2022 trade data, France is the leading exporter of teff while Italy is the leading importer. Guinea emerges as the primary producer of Fonio, with the UK as its chief exporter. Fonio and Teff face limited financing, outdated farming methods, and scant research focus. To ensure their sustainable production and enhanced market presence, it is imperative to offer financial backing to farmers, advocate advanced farming methods, conserve diversity, and boost research investments for these grains.

The fragmented nature of their industry clouds a holistic understanding, emphasising the need for in-depth research across various facets, from nutritional profiling to mechanised harvesting. Scalability demands modern biotechnological methods and advanced techniques. As these grains hold promise for innovative food products, comprehensive supply chain analysis and market research are essential to harness their potential fully. Addressing these challenges is not just a business imperative but a necessary step to champion Teff and Fonio's wider acceptance and global recognition, magnifying their role in global food security.

References

Abewa A, Adgo E, Yitaferu B, Alemayehu G, Assefa K, Solomon JKQ, Payne W (2019) Teff grain physical and chemical quality responses to soil physicochemical properties and the environment. Agronomy 9(6):283. https://doi.org/10.3390/AGRONOMY9060283

Abrouk M, Ahmed HI, Cubry P, Šimoníková D, Cauet S, Pailles Y, Bettgenhaeuser J, Gapa L, Scarcelli N, Couderc M, Zekraoui L, Kathiresan N, Čížková J, Hřibová E, Doležel J, Arribat S, Bergès H, Wieringa JJ, Gueye M et al (2020) Fonio millet genome unlocks African orphan crop diversity for agriculture in a changing climate. Nat Commun 11(1):4488. https://doi.org/10.1038/S41467-020-18329-4

Adams DM, Yakubu MT (2020) Aqueous extract of Digitaria exilis grains ameliorate diabetes in streptozotocin-induced diabetic male Wistar rats. J Ethnopharmacol 249:112383. https://doi.org/10.1016/J.JEP.2019.112383

Adisa BO, Adeloye KA (2012) Factors associated with performance of farmer field school as extension approach to Cocoa production in Osun State, Nigeria. Asian J Agric Rural Dev 2(2):162–170

Adoukonou-Sagbadja H, Schubert V, Dansi A, Jovtchev G, Meister A, Pistrick K, Akpagana K, Friedt W (2007) Flow cytometric analysis reveals different nuclear DNA contents in cultivated Fonio (Digitaria spp.) and some wild relatives from West-Africa. Plant Syst Evol 267(1–4):163–176. https://doi.org/10.1007/S00606-007-0552-Z/METRICS

Agriopoulou S, Stamatelopoulou E, Varzakas T (2020) Advances in occurrence, importance, and mycotoxin control strategies: prevention and detoxification in foods. Foods (Basel, Switzerland) 9(2):137. https://doi.org/10.3390/FOODS9020137

Ahmed FA, Ali RFM (2013) Bioactive compounds and antioxidant activity of fresh and processed white cauliflower. Biomed Res Int 2013:367819. https://doi.org/10.1155/2013/367819

Aljada B, Zohni A, El-Matary W (2021) The gluten-free diet for celiac disease and beyond. Nutrients 13(11):3993. https://doi.org/10.3390/NU13113993

Al-Khayri JM, Alshegaihi RM, Mahgoub I, Mansour E, Atallah OO, Sattar MN, Al-Mssallem MQ, Alessa FM, Aldaej MI, Hassanin AA (2023) Association of high and low molecular weight glutenin subunits with gluten strength in tetraploid durum wheat (Triticum turgidum spp. Durum L.). Plants 12(6):1416. https://doi.org/10.3390/PLANTS12061416/S1

Animasaun DA, Adedibu PA, Olawepo GK, Oyedeji S (2023) Fonio millets: an underutilised crop with potential as a future smart cereal. In: Farooq M, Siddique KHM (eds) Neglected and underutilised crops: future smart food. Academic Press, Oxford, pp 201–219. https://doi.org/10.1016/B978-0-323-90537-4.00028-4

Anitha S, Kane-Potaka J, Tsusaka TW, Botha R, Rajendran A, Givens DI, Parasannanavar DJ, Subramaniam K, Prasad KDV, Vetriventhan M, Bhandari RK (2021) A systematic review and meta-analysis of the potential of millets for managing and reducing the risk of developing diabetes mellitus. Front Nutr 8:687428. https://doi.org/10.3389/FNUT.2021.687428/BIBTEX

Babarinde GO, Adeyanju JA, Ogunleye KY, Adegbola GM, Ebun AA, Wadele D (2020a) Nutritional composition of gluten-free flour from blend of fonio (Digitaria iburua) and pigeon pea (Cajanus cajan) and its suitability for breakfast food. J Food Sci Technol 57(10):3611–3620. https://doi.org/10.1007/S13197-020-04393-7

Babarinde GO, Ebun AA, Adegbola PI (2020b) Hepatotoxicity and biochemical evaluation of a novel breakfast food produced from the blend of fonio (Digitaria iburua Stapf) and pigeon pea (Cajanus cajan (L.) Millspaugh) in albino rats. Bull Natl Res Cent 44(1):1–10. https://doi.org/10.1186/S42269-020-00409-6

Barretto R, Buenavista RM, Rivera JL, Wang S, Prasad PVV, Siliveru K (2021) Teff (Eragrostis tef) processing, utilisation and future opportunities: a review. Int J Food Sci Technol 56(7):3125–3137. https://doi.org/10.1111/IJFS.14872

Bartkiene E, Lele V, Ruzauskas M, Domig KJ, Starkute V, Zavistanaviciute P, Bartkevics V, Pugajeva I, Klupsaite D, Juodeikiene G, Mickiene R, Rocha JM (2019) Lactic acid bacteria isolation from spontaneous sourdough and their characterisation including antimicrobial and antifungal properties evaluation. Microorganisms 8(1):64. https://doi.org/10.3390/MICROORGANISMS8010064

Bascuñán KA, Vespa MC, Araya M (2016) Celiac disease: understanding the gluten-free diet. Eur J Nutr 56(2):449–459. https://doi.org/10.1007/S00394-016-1238-5

Bassey SO, Chinma CE, Ezeocha VC, Adedeji OE, Jolayemi OS, Alozie-Uwa UC, Adie IE, Ofem SI, Adebo JA, Adebo OA (2023) Nutritional and physicochemical changes in two varieties of fonio (Digitaria exilis and Digitaria iburua) during germination. Heliyon 9(6):e17452. https://doi.org/10.1016/J.HELIYON.2023.E17452

Calix-Rivera CS, Villanueva M, Náthia-Neves G, Ronda F (2023) Changes on techno-functional, thermal, rheological, and microstructural properties of Tef flours induced by microwave radiation-development of new improved gluten-free ingredients. Foods (Basel, Switzerland) 12(6):1345. https://doi.org/10.3390/FOODS12061345

Carboni J, Reed S, Kolba N, Eshel A, Koren O, Tako E (2020) Alterations in the intestinal morphology, gut microbiota, and trace mineral status following intra-amniotic administration (Gallus gallus) of Teff (Eragrostis tef) seed extracts. Nutrients 12(10):3020. https://doi.org/10.3390/NU12103020

Chadalavada K, Anbazhagan K, Ndour A, Choudhary S, Palmer W, Flynn JR, Mallayee S, Pothu S, Prasad KVSV, Varijakshapanikar P, Jones CS, Kholová J (2022) NIR instruments and prediction methods for rapid access to grain protein content in multiple cereals. Sensors 22(10):3710. https://doi.org/10.3390/S22103710/S1

Chochkov R, Savova-Stoyanova D, Papageorgiou M, Rocha JM, Gotcheva V, Angelov A (2022) Effects of Teff-based sourdoughs on dough rheology and gluten-free bread quality. Foods (Basel, Switzerland) 11(7):1012. https://doi.org/10.3390/FOODS11071012

Choudhary P, Shukla P, Muthamilarasan M (2023) Genetic enhancement of climate-resilient traits in small millets: a review. Heliyon 9(4):e14502. https://doi.org/10.1016/J.HELIYON.2023.E14502

CIRAD (n.d.) Processing—Fonio (Digitaria exilis). https://fonio.cirad.fr/en/processing. Accessed 28 Aug 2023

Cruz-Garcia G, Visser B, Nera E, Salazar R (2020) General framework for FFS curriculum neglected and underutilised species. Women and Nutrition. www.sdhsprogram.org

Dansi A, Adoukonou-Sagbadja H, Vodouhè R (2010) Diversity, conservation and related wild species of Fonio millet (Digitaria spp.) in the northwest of Benin. Genet Resour Crop Evol 57(6):827–839. https://doi.org/10.1007/S10722-009-9522-3/TABLES/6

Demirkesen I, Ozkaya B (2022) Recent strategies for tackling the problems in gluten-free diet and products. Crit Rev Food Sci Nutr 62(3):571–597. https://doi.org/10.1080/10408398.2020.1823814

Demiss M, Mamo T, Beyene S, Kidanu S (2020) Effect of potassium levels on teff (Eragrostis tef (Zucc.) Trotter) growth and yield in Central Highland Vertisols of Ethiopia. Eurasian J Soil Sci 9(2):105–118. https://doi.org/10.18393/EJSS.663486

Deriu AG, Vela AJ, Ronda F (2022a) Techno-functional and gelling properties of Acha (Fonio) (Digitaria exilis stapf) flour: a study of its potential as a new gluten-free starch source in industrial applications. Foods (Basel, Switzerland) 11(2):183. https://doi.org/10.3390/FOODS11020183

Deriu AG, Vela AJ, Ronda F (2022b) Techno-functional and gelling properties of Acha (Fonio) (Digitaria exilis stapf) flour: a study of its potential as a new gluten-free starch source in industrial applications. Foods 11(2):183. https://doi.org/10.3390/FOODS11020183/S1

Dingeo C, Difonzo G, Paradiso VM, Rizzello CG, Pontonio E (2020) Teff type-I sourdough to produce gluten-free muffin. Microorganisms 8(8):1149. https://doi.org/10.3390/MICROORGANISMS8081149

Diop BM, Gueye MC, Agbangba CE, Cissé N, Deu M, Diack O, Fofana A, Kane NA, Ndir KN, Ndoye I, Ngom A, Leclerc C, Piquet M, Vigouroux Y, Zekraoui L, Billot C, Barnaud A (2018) Fonio (Digitaria exilis (Kippist) Stapf): a socially embedded cereal for food and nutrition security in Senegal. Ethnobiol Lett 9(2):150–165. https://doi.org/10.14237/EBL.9.2.2018.1072

Fanou-Fogny N, van Dam B, Koreissi Y, Dossa RAM, Brouwer ID (2011) Factors predicting consumption of Fonio grain (Digitaria exilis) among urban Malian women of reproductive age. J Nutr Educ Behav 43(4):219–228. https://doi.org/10.1016/J.JNEB.2010.03.008

FAO (n.d.) Teff. https://www.fao.org/traditional-crops/teff/en/. Accessed 27 Aug 2023

Fukagawa NK, McKillop K, Pehrsson PR, Moshfegh A, Harnly J, Finley J (2022) USDA's FoodData Central: what is it and why is it needed today? Am J Clin Nutr 115(3):619–624. https://doi.org/10.1093/AJCN/NQAB397

Gebregewergis A (2021) Comparison of the concentrations of metals found in three types of Teff samples. Am J Agric Biol Sci 16(1):33–37. https://doi.org/10.3844/AJABSSP.2021.33.37

Gebremariam MM, Zarnkow M, Becker T (2013) Effect of teff (Eragrostis tef) variety and storage on malt quality attributes. J Inst Brew 119(1–2):64–70. https://doi.org/10.1002/JIB.65

Gebremariam MM, Zarnkow M, Becker T (2014) Teff (Eragrostis tef) as a raw material for malting, brewing and manufacturing of gluten-free foods and beverages: a review. J Food Sci Technol 51(11):2881–2895. https://doi.org/10.1007/S13197-012-0745-5/FIGURES/5

Gebru YA, Sbhatu DB (2020) Isolation and characterisation of probiotic LAB from Kimchi and spontaneously fermented Teff (Eragrostis tef (Zucc.) Trotter) batter: their effects on phenolic content of Teff during fermentation. Biomed Res Int 2020:4014969. https://doi.org/10.1155/2020/4014969

Gosine L, McSweeney MB (2019) Consumers' attitudes towards alternative grains: a conjoint analysis study. Int J Food Sci Technol 54(5):1588–1596. https://doi.org/10.1111/IJFS.14126

Habte ML, Beyene EA, Tilahun TOFA, Diribsa GC, Admasu FT (2022) Nutritional values of Teff (Eragrostis tef) in diabetic patients: narrative review. Diabetes Metab Syndr Obes 15:2599–2606. https://doi.org/10.2147/DMSO.S366958

Howell K, Tavan M, Biju S (2022) An overview of the Australian teff industry a feasibility report. https://agrifutures.com.au/product/an-overview-of-the-australian-teff-industry-a-feasibility-report/

Hozyasz KK (2013) Letter to the editor re: Comino, I., et al. Nutrients 2013, 5, 4250–4268. Nutrients 5(12):4964–4965. https://doi.org/10.3390/NU5124964

Huang L, Liu Z, Wu P, Yue X, Lian Z, He P, Liu Y, Zhou R, Zhao J (2022) Puerariae lobatae radix alleviates pre-eclampsia by remodeling gut microbiota and protecting the gut and placental barriers. Nutrients 14(23):5025. https://doi.org/10.3390/NU14235025/S1

Ijeomah OC, Ndukwu CLC, Azubuike CU (2023) Safety evaluation of FONIO (Digitaria exilis)/Ricebean (Vigna Umbellata) based complementary food. Eur J Nutr Food Saf 15(4):16–22. https://doi.org/10.9734/EJNFS/2023/V15I41303

Jocelyne RE, Béhiblo K, Ernest AK (2020) Comparative study of nutritional value of wheat, maize, sorghum, millet, and fonio: some cereals commonly consumed in Côte d'Ivoire. Eur Sci J 16(21):118. https://doi.org/10.19044/ESJ.2020.V16N21P118

John JE, Tytler AB, Habila J, Apeji Y, Isimi CY, Dada A (2023) Effect of citric acid concentration on the physicochemical properties and release profile of ibuprofen-loaded Digitaria exilis starch nanoparticles. Br J Pharm 8(1):1010. https://doi.org/10.5920/BJPHARM.1010

Khairuddin MAN, Lasekan O (2021) Gluten-free cereal products and beverages: a review of their health benefits in the last five years. Foods 10(11):2523. https://doi.org/10.3390/FOODS10112523

Koubová E, Sumczynski D, Šenkárová L, Orsavová J, Fišera M (2018) Dietary intakes of minerals, essential and toxic trace elements for adults from Eragrostis tef L.: a nutritional assessment. Nutrients 10(4):479. https://doi.org/10.3390/NU10040479

Laugesen SB, Dethlefsen SL, Petersen IL, Aaslyng MD (2022) Texturised vegetable protein as a source of protein fortification of wheat buns. Foods 11(22):3647. https://doi.org/10.3390/FOODS11223647/S1

Lydia Pramitha J, Ganesan J, Francis N, Rajasekharan R, Thinakaran J (2023) Revitalisation of small millets for nutritional and food security by advanced genetics and genomics approaches. Front Genet 13:1007552. https://doi.org/10.3389/FGENE.2022.1007552/BIBTEX

Mekonen S, Ambelu A, Spanoghe P (2019) Reduction of pesticide residues from teff (*Eragrostis tef*) flour spiked with selected pesticides using household food processing steps. Heliyon 5:e01740. https://doi.org/10.1016/j.heliyon.2019.e01740

Milkias D, Beri G (2020) Assessing farmer's perception towards improved Quncho Teff variety in Gindeberet District, West Showa Zone, Oromia Region Ethiopia. J Plant Sci 8(5):106. https://doi.org/10.11648/J.JPS.20200805.11

Molfetta M, Morais EG, Barreira L, Bruno GL, Porcelli F, Dugat-Bony E, Bonnarme P, Minervini F (2022) Protein sources alternative to meat: state of the art and involvement of fermentation. Foods 11(14):2065. https://doi.org/10.3390/FOODS11142065/S1

Nithiyanantham S, Kalaiselvi P, Mahomoodally MF, Zengin G, Abirami A, Srinivasan G (2019) Nutritional and functional roles of millets-a review. J Food Biochem 43(7):e12859. https://doi.org/10.1111/JFBC.12859

Nooshkam M, Varidi M, Alkobeisi F (2022) Bioactive food foams stabilised by licorice extract/whey protein isolate/sodium alginate ternary complexes. Food Hydrocoll 126:107488. https://doi.org/10.1016/J.FOODHYD.2022.107488

Ouedraogo M, Dao K, Millogo Y, Aubert JE, Messan A, Seynou M, Zerbo L, Gomina M (2019) Physical, thermal and mechanical properties of adobes stabilised with fonio (Digitaria exilis) straw. J Build Eng 23:250–258. https://doi.org/10.1016/J.JOBE.2019.02.005

Petrova P, Petrov K (2020) Lactic acid fermentation of cereals and pseudocereals: ancient nutritional biotechnologies with modern applications. Nutrients 12(4):1118. https://doi.org/10.3390/NU12041118

Petruzzello M (2023) Teff | Gluten-Free, Ethiopian Cuisine | Britannica. https://www.britannica.com/plant/teff

Ramashia SE, Gwata ET, Meddows-Taylor S, Anyasi TA, Jideani AIO (2018) Some physical and functional properties of finger millet (Eleusine coracana) obtained in sub-Saharan Africa. Food Res Int 104:110–118. https://doi.org/10.1016/J.FOODRES.2017.09.065

Reddy YAN, Reddy YNP, Ramya V, Suma LS, Reddy ABN, Krishna SS (2021) Drought adaptation: approaches for crop improvement. In: Millets and pseudo cereals: genetic resources and breeding advancements. Woodhead Publishing, Duxford, pp 143–158. https://doi.org/10.1016/B978-0-12-820089-6.00008-2

Reta C, Atlabachew M, Asmellash T, Hilluf W, Yayinie M, Wubieneh TA (2022) Polyphenol and flavonoid content in major Teff [Eragrostis tef (Zuccagni) Trotter] varieties in Amhara Region, Ethiopia. PLoS One 17(8):e0272010. https://doi.org/10.1371/JOURNAL.PONE.0272010

Seignobos C, Tourneux H (2002) Le Nord-Cameroun à travers ses mots: Dictionnaire de termes anciens et modernes: Province de l'extrême-nord (in French). Karthala, Paris

Sharma KK, Palakolanu SR, Bhattacharya J, Shankhapal AR, Bhatnagar-Mathur P (2022) CRISPR for accelerating genetic gains in under-utilised crops of the drylands: progress and prospects. Front Genet 13:999207. https://doi.org/10.3389/FGENE.2022.999207/BIBTEX

Sousa EC, Raizada MN (2020) Contributions of African crops to American culture and beyond: the slave trade and other journeys of resilient peoples and crops. Front Sustain Food Syst 4:1–24. https://doi.org/10.3389/FSUFS.2020.586340/FULL

Švec I, Hrušková M (2015) The Mixolab parameters of composite wheat/hemp flour and their relation to quality features. LWT Food Sci Technol 60(1):623–629. https://doi.org/10.1016/J.LWT.2014.07.034

Tadele E, Hibistu T, Ercolano S (2022) Spatial production distribution, economic viability and value chain features of teff in Ethiopia: systematic review. Cogent Econ Financ 10(1). https://doi.org/10.1080/23322039.2021.2020484

Taylor JRN, Kruger J (2016) Millets. In: Encyclopedia of food and health. Academic Press, Oxford, pp 748–757. https://doi.org/10.1016/B978-0-12-384947-2.00466-9

Tessema B, Gonfa G, Hailegiorgis SM, Sundramurthy VP (2023) Characterisation of teff straw from selected teff varieties from Ethiopia. Heliyon 9(6):e17422. https://doi.org/10.1016/J.HELIYON.2023.E17422

Tridge (2023a) Fonio global market overview today. https://www.tridge.com/intelligences/fonio

Tridge (2023b) Teff production and top producing countries—Tridge. https://www.tridge.com/intelligences/teff/production

Vetriventhan M, Azevedo VCR, Upadhyaya HD, Nirmalakumari A, Kane-Potaka J, Anitha S, Ceasar SA, Muthamilarasan M, Bhat BV, Hariprasanna K, Bellundagi A, Cheruku D, Backiyalakshmi C, Santra D, Vanniarajan C, Tonapi VA (2020) Genetic and genomic resources, and breeding for accelerating improvement of small millets: current status and future interventions. Nucleus (India) 63(3):217–239. https://doi.org/10.1007/S13237-020-00322-3/TABLES/7

Wee MSM, Henry CJ (2020) Reducing the glycemic impact of carbohydrates on foods and meals: strategies for the food industry and consumers with special focus on Asia. Compr Rev Food Sci Food Saf 19(2):670–702. https://doi.org/10.1111/1541-4337.12525

Yaskolka Meir A, Tuohy K, von Bergen M, Krajmalnik-Brown R, Heinig U, Zelicha H, Tsaban G, Rinott E, Kaplan A, Aharoni A, Zeibich L, Chang D, Dirks B, Diotallevi C, Arapitsas P, Vrhovsek U, Ceglarek U, Haange SB, Rolle-Kampczyk U et al (2021) The metabolomic-gut-clinical axis of mankai plant-derived dietary polyphenols. Nutrients 13(6):1866. https://doi.org/10.3390/NU13061866/S1

Yimer AH (2022) Postharvest management practices of maize in Ethiopia: a review. Agric Sci Dig 42(5):521–527. https://doi.org/10.18805/AG.DF-414

Yisak H, Yaya EE, Chandravanshi BS, Redi-Abshiro M (2022) GC-MS profiling of fatty acids and nutritional properties of the white and brown teff [Eragrostis tef (Zuccagni) Trotter] vari-

eties cultivated in different parts of Ethiopia. J Food Compos Anal 107:104405. https://doi.org/10.1016/J.JFCA.2022.104405
Zeweld W, Van Huylenbroeck G, Tesfay G, Azadi H, Speelman S (2018) Impacts of socio-psychological factors on actual adoption of sustainable land management practices in Dryland and water stressed areas. Sustainability 10(9):2963. https://doi.org/10.3390/SU10092963
Zhang L, Liu R, Niu W (2014) Phytochemical and antiproliferative activity of proso millet. PLoS One 9(8):e104058. https://doi.org/10.1371/JOURNAL.PONE.0104058
Zhu F (2020) Fonio grains: physicochemical properties, nutritional potential, and food applications. Compr Rev Food Sci Food Saf 19(6):3365–3389. https://doi.org/10.1111/1541-4337.12608

Triticale

8

Hamid, Deepika Kathuria, Sunakshi Gautam, Shweta Suri, and Amit K. Jaiswal

Abstract

Triticale is a man-made cross between rye and wheat, which was developed in the late nineteenth century. Europe had the largest area and production under triticale cultivation followed by Asia and America in the year 2021. On a global level, Poland is the leading country covering the largest area under triticale cultivation (33.45%) with a production share of 36.03%. Triticale is a highly useful crop with multiple uses for fodder, food grains, and higher adoption to adverse environmental conditions. Triticale resembles more similar to wheat in composition than rye being a good source of carbohydrates, protein (lysine rich), and dietary fiber (insoluble and soluble) with antioxidant potential (alkylresorcinols, proanthocyanidin). It also includes good amounts of phenolic acids, flavonoids, vitamins with bioactive qualities besides the presence of various polysaccharides

Hamid (✉)
Food Technology and Nutrition, Lovely Professional University, Phagwara, Punjab, India

D. Kathuria
Department of Food Science and Technology, Graphic Era University, Dehradun, Uttarakhand, India

S. Gautam
School of Bioengineering and Food Technology, Shoolini University, Solan, Himachal Pradesh, India

S. Suri
Amity Institute of Food Technology, Amity University, Noida, Uttar Pradesh, India

A. K. Jaiswal
School of Food Science and Environmental Health, Technological University Dublin-City Campus, Dublin, Ireland

Environmental Sustainability and Health Institute (ESHI), Technological University Dublin-City Campus, Dublin, Ireland

J. Singh et al. (eds.), *Cereals and Nutraceuticals*,
https://doi.org/10.1007/978-981-97-2542-7_8

(arabinoxylans, fructan, β-glucan, and cellulose) which are exhibiting many health benefits. Its elevated lysine concentration, improved protein digestibility, and mineral balance make it a suitable replacement or supplement in human and animal diets. Triticale products like whole berries, flakes, and flour are available for human consumption, often in specialty health stores. Despite all, triticale grains are still underutilized and consumed very less than other major grains, which may be due to lack of detailed information on their physicochemical and nutritional aspects or less awareness related to potential benefits which are highlighted in this chapter in detail.

Keywords

Bioactive composition · Health benefits · Nutritional value · Production · Triticale

8.1 Introduction, Origin, and Background

Triticale (X Triticosecale Wittmack) is a man-made cross between rye (*Secale* sp.) and wheat (*Triticum* sp.). It was developed in the late nineteenth century with the goal of combining rye's resistance to abiotic and biotic stresses with wheat higher yield or production potential beside good grain qualities (Leon et al. 2008). The term triticale, which is used to refer to a variety of crops worldwide, is a combination of the scientific names for the two different genera involved—rye (*Secale*) and wheat (*Triticum*). Triticale history dates way to the later 1870s, as the very first crosses were made in Scotland even though it is a relatively recent crop. In 1875, Wilson (1875) published the first account of the development of sterile wheat-rye hybrid plants. From crossings of *Triticum aestivum* and rye, Rimpau created the first viable, "true" triticale in 1888, which underwent spontaneous chromosomal duplication (Rimpau 1891). Europe had sporadic testing and study during the subsequent 50 years.

The first commercial release was not made accessible to producers until the 1960s. Triticale cultivars may be divided into three fundamental types: winter, spring, and intermediate (facultative). These cultivars are produced for both forage and grain. Early in their existence, spring varieties generate a lot of fodder and grow erect. They have minimal tillering and are often photoperiod insensitive. Winter varieties can occasionally be planted in the spring but are typically planted in the autumn. Early on in their development, winter kinds grow in a prostrate manner. Winter varieties often provide more fodder than spring varieties, mostly because of their lengthy growing cycle. As their name suggests, intermediate (facultative) types are in between winter and spring kinds (Salmon et al. 2004; Mergoum et al. 2004, 2009).

Triticale production on a global scale has increased gradually over the last past 20 years, approaching 17 million tonnes in 2014. Poland, Belarus, Germany, Russia, and France are the largest producers. Outside of Europe, China is the biggest producer. According to Hosseinian and Mazza (2009), Makowska et al. (2014), and

Straumīte et al. (2017), triticale is a source of carbohydrates, protein, and dietary fiber (insoluble and soluble) with antioxidant potential (alkylresorcinols, proanthocyanidin). It also includes good amounts of phenolic acids, flavonoids, vitamins with bioactive qualities besides the presence of various polysaccharides (arabinoxylans, β-glucan, fructan, cellulose) (Hosseinian and Mazza 2009; Jonnala et al. 2010; Rakha et al. 2012; Knudsen 2014). Triticale appears to be more chemically related to wheat than rye overall. Lysine, the limiting amino acid of cereals, is present in unusually high concentrations in several triticale genotypes (McGoverin et al. 2011). Zhu (2018) has reported that triticale is a superior source of protein (13.18 g), carbohydrates (73.14 g), total dietary fibers (14.6 g), besides variable quantity of vitamins like thiamin (0.378 mg), riboflavin (0.132 mg), niacin (2.86 mg), vitamin B-6 (0.403 mg), folate (74 μg DFE), vitamin E (alpha-tocopherol) (0.9 mg), and variable quantity of various minerals like calcium (35 mg), iron (2.59 mg), magnesium (153 mg), phosphorus (321 mg), potassium (466 mg), sodium (2 mg), zinc (2.66 mg), and lower amount of total lipids (1.81 g), having higher fatty acids such as total polyunsaturated (0.794 g) followed by total saturated (0.318 g) fatty acids and total monounsaturated (0.183 g) fatty acids. Non-starch polysaccharides (like arabinoxylans), starch, polyphenols (like phenolic acids), alkylresorcinols, and vitamins (like vitamin B_1) are triticale nutrients that are receiving more attention in recent studies (Rakha et al. 2012; Buchholz et al. 2012). Triticale has been linked to certain positive health benefits, including in vitro antioxidant beside anticholinesterase properties (Senol et al. 2012). Thus triticale may therefore have an impact on the development of new cereal functional products as well as the expanding demand for healthy foods. Triticale contains gluten and should not be consumed by those who have celiac disease.

Alkylresorcinols, phytoestrogens, vitamins, and microelements are also abundant in triticale, along with several phenolics having antioxidant activity (Jonnala et al. 2010). Triticale's inferior quality compared to wheat and strong alpha-amylase activity make it difficult to use in baking due to the dough's poor rheological characteristics and low gluten concentration. Triticale has been tried to make bread, cookies, extrudates, tortillas, and pasta over the years, but its application in food production is still minimal. The primary causes of this include triticale grain and flour's strong amylolytic activity, fragile gluten system, and low technical stability (Woś and Brzeziński 2015; Zhu 2018). It has been said that foods fortified with arabinoxylans or β-glucan can satisfy the typical daily requirements of dietary fibers for humans.

Triticale grains are usually helpful in controlling diabetes (Havensone et al. 2017), beside it helps in improving digestion, elevates cell production, helps in circulation and boosting the bone growth. Various studies have reported that flour of triticale was better fit to produce products like crackers, cookies, waffles, cakes, tortillas, and supplementary unleavened food products than for the preparation of bread (Leon et al. 2008). Despite all triticale grains are still underutilized and consumed very less than other major grain crops, which may be due to lack of detailed information on their physicochemical and nutritional aspects or less awareness related to potential benefits which are highlighted under the following heads.

8.2 Production and Cultivation Practices

Triticale is a pioneering crop of grain that was generated artificially and has some traits resembles common with crops like rye and wheat. Considering its high potential for productivity and consistently consistent yields over the course of a year, this crop can hold an enviable position among farmed crops (Zhukov et al. 2020). Currently, European nations account for a significant percentage of the area planted under triticale for grain production. According to FAO, the global area under triticale cultivation was four million hectares in the year 2018, with a total grain production of 13.5 million tons in the same year. However, the world gross production value (current thousand US$) of triticale in the year 2021 was recorded 2,910,752 thousand USD globally according to FAO which was contributed largely by Europe as 2,787,327 thousand USD followed by Asia as 56,731 thousand USD, while the remaining by Oceania (Australia and New Zealand) as 40,197 thousand USD and by African countries as 3580 thousand USD, respectively (FAO 2022). According to FAO, this world gross production value (current thousand US$) of triticale was continuously increasing starting from the year 1991 as 514,709 thousand USD to the year 2021 as 2,910,752 thousand USD which was reported due to increased area under the triticale cultivation among the different countries. According to FAO, the total area under triticale in the world was 3,830,794 ha with production of 14,846,162.66 t in the year 2021 (Table 8.1). However, among the different countries Europe has the higher area under triticale cultivation followed by Asia, Oceania, America, and Africa. Similarly among the different countries, Europe has the highest production under triticale cultivation followed by Asia, America, Oceania, and Africa (FAO 2021). However, Poland, Germany, France, Belarus, Spain, China, Russia, Hungary, Turkey, and Romania were the top 10 leading countries covering the largest area (Fig. 8.1) under triticale cultivation and having the highest production on global level in 2021. On Global level Poland is the leading country covering largest area under triticale cultivation (33.45%) with a production share of 36.03% (Fig. 8.2).

Agricultural producers started actively using chemicals in the production, including plant growth regulators, as a result of agricultural firms' inclination to get a high output (Trofimova et al. 2018). The behavior of processes including seed germination, photosynthesis, protection from pathogenic microflora, and the development of vegetative and generative organs of plants can all be stimulated by growth regulators (Yakushev et al. 2015). The use of bacterial fertilizers and growth regulators in crop production techniques is currently more advantageous from an economic and environmental standpoint. Farmers were compelled to use fewer pesticides and mineral fertilizers as a result of the biologization of grain production techniques and consumer demand for organic goods. They also had to make use of new technological advancements to raise product quality while lowering costs. Winter triticale seeds treated with growth regulators prior to sowing result in better sprouting, increased field germination, and increased disease resistance in the plant. These processes are believed to affect the development of crop structure components, seedling density, the number of plants that survive the growing season, productive

Table 8.1 Area and production statistics with percentage share to global total area under triticale and production of year 2021 by FAO

Countries	Area (ha)	% Share in total area	Production (t)	% Share in total production
Poland	1,281,600	33.45	5,349,420	36.03
Germany	328,300	8.57	1,908,600	12.85
France	338,600	8.83	1,736,910	11.69
Belarus	513,270	13.39	1,667,284	11.23
Spain	267,510	6.98	787,820	5.30
China	194,354	5.07	367,123.6	2.47
Russia	119,480	3.11	288,488.1	1.94
Hungary	61,880	1.61	277,440	1.86
Romania	64,550	1.68	259,210	1.74
Turkey	93,053	2.42	228,000	1.53
Algeria	379	0.009	615	0.004
Australia	96,207	2.51	170,812	1.15
Belgium	5050	0.13	32,040	0.21
Bosnia and Herzegovina	14,533	0.37	61,781	0.41
Brazil	38,226	0.99	100,545	0.67
Bulgaria	15,940	0.41	54,210	0.36
Canada	19,500	0.50	39,148	0.26
Chile	10,913	0.28	55,182.83	0.37
Croatia	9390	0.24	42,990	0.28
Cyprus	700	0.018	1140	0.007
Czechia	40,860	1.06	193,450	1.30
Denmark	6900	0.18	49,020	0.33
Estonia	7390	0.19	27,690	0.18
Finland	2500	0.065	9580	0.064
Greece	14,770	0.38	37,950	0.25
Italy	13,210	0.34	63,730	0.42
Kyrgyzstan	8	0.0002	16	0.0001
Latvia	7700	0.20	23,900	0.16
Lithuania	74,510	1.94	206,120	1.38
Luxembourg	4720	0.12	26,730	0.18
Mexico	10,582	0.27	27,587.88	0.18
Netherlands (Kingdom of the)	1200	0.03	5630	0.037
Portugal	13,610	0.35	20,660	0.13
Serbia	24,746	0.64	125,784	0.84
Slovakia	9020	0.23	30,810	0.20
Slovenia	5150	0.13	26,390	0.17
Sweden	29,710	0.77	153,300	1.03
Switzerland	7429	0.19	37,747	0.25
Tunisia	10,000	0.26	23,300	0.15
Ukraine	8001	0.20	2746.08	0.018
United Kingdom of Great Britain and Northern Ireland	15,393	0.40	59,662	0.40

(continued)

Table 8.1 (continued)

Countries	Area (ha)	% Share in total area	Production (t)	% Share in total production
Total world production (2021)	**3,830,794**	–	**14,846,163**	–
Asia	288,115	7.52	596,279.6	4.01
Africa	10,379	0.27	23,915	0.16
America	79,221	2.06	222,463.7	1.49
Europe	3,356,872	87.62	13,832,692	93.17
Oceania	96,207	2.51	170,812	1.15

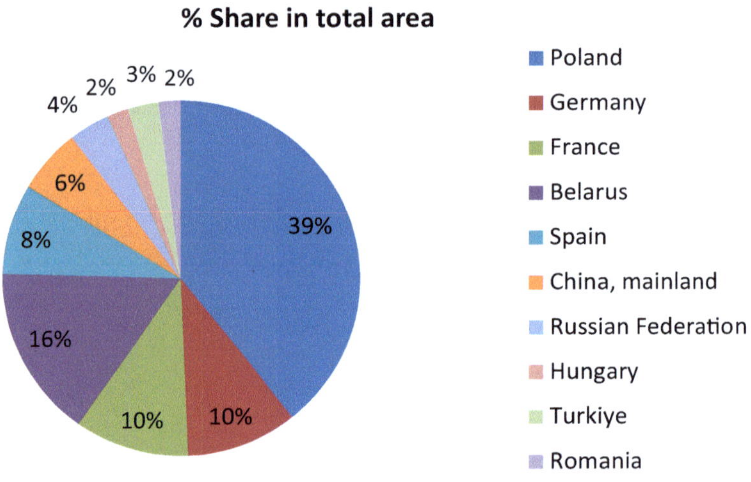

Fig. 8.1 Top ten leading countries with highest area under triticale cultivation

tilling capacity, and other factors, leading to a 6.8–17.4% increase in the yield of the winter triticale varieties as reported by Zhukov et al. (2020).

In order to determine the ideal rate for optimum output, fertilization studies are carried out all over the world. The average grain yield during the study period was 12.4% worldwide. The foliar fertilizers that have been tested according to the elevated values of the structural components of the yield and the intended use of the crop, fertilization should be optimized. According to research by Ivanova and Tsenov (2014a, b), the yield of new triticale types is mostly determined by the time of year and the fertilization level. According to a study by Kolev (2010), the addition of the growth stimulator "Imunotsitofit" during the tillering + ear development phases enhances grain yield by an average of 14.9% across the experiment period compared to the untreated control. Kolev (2014) discovered that the use of the liquid fertilizer "Amalgerol premium" at the tillering stage enhanced the yield by an average of 11.2% across the study period when compared to the unfertilized reference.

Germany, France, China, Australia, Hungary, and the United States are the top producers of triticale, with Poland having the largest area with more than one million hectares (Dobreva 2016). The high quantities of protein in triticale are

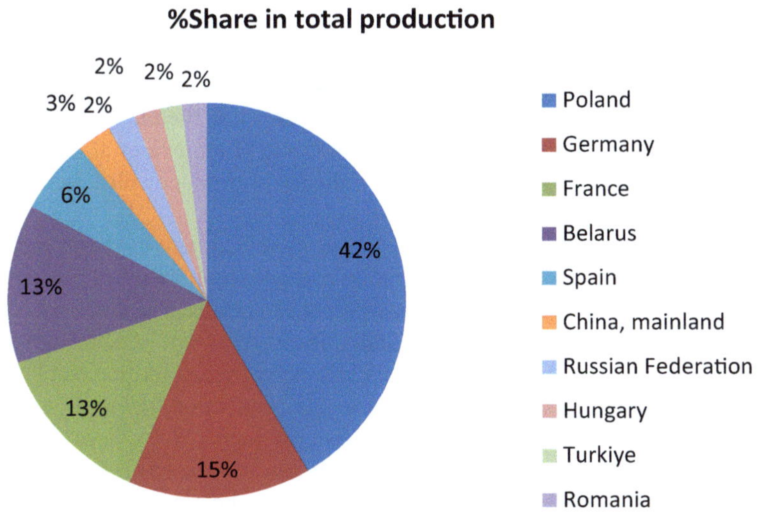

Fig. 8.2 Top ten leading countries with highest triticale production share

primarily responsible for the grain's biological value. According to Kirchev and Popov (2010), fertilization with 120 kg/ha of nitrogen results in the greatest accumulation of amino acids.

Triticale is most likely best suited for use as a feed grain with a similar value to wheat on marginal and lighter soils. It can withstand droughts and the worst winters. Similar conditions for the seedbed and sowing techniques apply to wheat. 270–330 seeds/m^2 (125–160 kg/ha); 150/m^2 plants minimum after the winter. Mid-September to October is the best time to plant. Since winter triticale needs to be vernalized, it shouldn't be planted after the middle of February. Further the nitrogen requirements are between 40 and 120 kg/ha top-dressed at the end of March or beginning of April (Finch et al. 2002).

8.3 Consumption Pattern

A man-made hybrid created by combining the B and A genomes of wheat (*Triticum turgidum* L., *Triticum aestivum* L.) with genome R of rye (*Secale cereal* L.) popularly known as triticale. This cereal grain was developed to combine the higher yield and excellent grain quality of wheat with higher disease resistance beside high environmental tolerances (Zhu 2018). The year of development of triticale was documented as 1875 and in 1968 first cultivar of triticale released by Hungarian breeding programmer was commercialized and made available on farmer's level. The names of these cultivars are Triticale no. 64 and Triticale no. 57 and are developed through octoploid hexaploid crossing (McGoverin et al. 2011). The nutritional profile of triticale is higher than both of its parents. It is an optimum source of protein that contains approximately 11–14% protein. Also this species is having a balanced source of essential amino acid mainly of lysine which is absent in most of the cereal crops (Fraś et al. 2016). The grain structure of triticale is suitable for the production of high-quality flour such as baker's grade flour, thus prominently used in bakery and confectionary industry to develop bread, rusk, diet bread, breakfast cereals, crisp bread, and multigrain bread (Leonova et al. 2022). Other than that, it is also being used in beverage industry to produce spirit, beer, etc. and in pasta industry.

Nutritionally, triticale is better but due to low gluten content its rheological properties are poor (Fraś et al. 2016). The feature triticale flour shows is similar to soft flour which is more suitable for the development of cookies and crackers. The textural properties shown by triticale is generally observed as soft which is not suitable for cookies and cake development, whereas absence of secaloidoline can be used to regenerate allelic variation which can lead to the production of harder endosperm which is a good characteristic in baking of food products (McGoverin et al. 2011). Soft triticale flour with the addition of lecithin has been used to develop layered cakes (Pena 2004). Chapati, flat bread, and noodles can be prepared by triticale flour as grayish color of grain does not impart much color to these. Although this grain is very suitable for human consumption, its application is very limited in food industries. Its functional component needs to be explored thoroughly, especially in the value addition of dairy products. Other than food products this grain is also popular for the production of bioethanol and extraction of alpha amylase enzyme and fungal amylase (Tomic et al. 2017). Its applicability also reported in packaging industry for the production of edible films with goof tensile strength. Salvucci et al. (2019) utilizes triticale flour for the production of biodegradable packaging film and used it for food application with bacteriocin like substances. The film showed good tensile strength (2.7 MPa), elongation at break (190%), and other desirable mechanical properties. Whereas Naguleswaran et al. (2014) utilizes it for the extraction of starch and compared with wheat, corn, and barley starch. They have reported comparable results of amylopectin and amylolysis of triticale and barley with corn and wheat which can be effectively utilized in food industry as sugar derivatives and for the production of bioethanol.

8.4 Chemical Composition

Triticale grain shares chemical composition similarities with both wheat and rye, displaying intermediate values. Research has continued to explore triticale's genetic resources, exhibiting wider variations in nutritional composition. Early triticale cultivars were less favorable for feeding, but genetic improvements have enhanced plumpness while reducing protein content. Notably, triticale has been recognized for its higher protein content (9.8–15.2%) with balanced amount of amino acid that varies on the amount of fertilizer used during growing (Heger and Eggum 1991; Stankowski et al. 2017; Dennett et al. 2013). Triticale possesses lesser amount of protein in the endosperm compared to wheat (Pattison et al. 2014).

On fractionation, triticale protein has higher albumin content (38–45.4%) followed by globulin (19.7–30.2%) then gliadin (11.4–26.1%) and lower value for glutenin (6.8–9.6%) (Navarro-Contreras et al. 2014). Gatel et al. (1985) reported that compared to wheat, triticale is richer in lysine (0.33–0.71%), comparable to wheat in threonine and sulfur amino acid contents except slightly lower in tryptophan. Triticale also surpasses wheat in arginine, aspartic acid, and alanine content (Salmon 1984). Its elevated lysine with improved protein digestibility and mineral balance make it an appropriate supplement or replacement in human and animal diets. Triticale products like whole berries, flakes, and flour are available for human consumption, often in specialty health stores. Nonetheless, the majority of global triticale production serves as animal feed. Carbohydrates constitute over 70% of triticale's dry weight, with genetics and grain development influencing total carbohydrate content. Starch is a major component (ranging between 68.2 and 77.5%), with granules larger than wheat (Berry et al. 1971). Upon milling, the range decreased from 63.3% to 68.8% (Dennett et al. 2013). Within the starch of triticale, amylose constitute of range from 12.8% to 35.1% by Dennett et al. (2009) while 19% to 23.8% by Makowska et al. (2014). While triticale starch typically has amylose content akin to "normal" wheat (20–30%), variations from 13–35% have been observed, offering breeding possibilities (Dennett et al. 2009). Furthermore, non-starch polysaccharides especially dietary fiber content is similar to wheat but has more soluble fraction (Rakha et al. 2011). The total dietary fiber present in triticale is about 13–16% of which 6.8% is arabinoxylan, 2.3% fructan, 1.6% klason lignin, 2.1% cellulose, and 0.7% β-glucan (Rakha et al. 2011). Lignin in grains ranged from 2.1% to 2.8% while 0.6% to 1.2% in flour (Fraś et al. 2016). Furthermore, soluble and insoluble fiber content ranged from 1.5–2.8% and 7.7–9.1%, respectively. On the other hand, crude fiber content of triticale is similar to that of wheat with a value ranging between 3.35% and 4.75% (Dennett et al. 2013). Besides starch, fat content of triticale is comparable to its parent species. The ash content of triticale varies between 1.3% and 1.9% that plays a critical role in affecting the end use of it. Table 8.2 represents the chemical composition of triticale. The vitamin composition goes parallel to wheat but exhibits slightly lower nicotinic acid levels. Triticale, like rye, wheat, barley and oat, lacks thiamine triphosphate or adenosine thiamin triphosphate. The overall thiamine content (9.74 nmol/g db), including non-phosphorylated thiamin (T), thiamin diphosphate (TDP), and thiamin monophosphate (TMP) remains consistent during seed development. The mineral elements in triticale grain samples do not seem to be related to total ash. On average, mineral composition in triticale resembles wheat, with iron potentially being an exception, often found in higher amounts in triticale than wheat (Lorenz 1982). The predominant mineral present in triticale is phosphorus, potassium, and magnesium (Waggle et al. 1967). Table 8.3 represents different mineral elements of triticale reported by different authors. During milling of grain into flour, researchers have reported reduced mineral element compared to grain. About 18%, 67%, and 82% of total magnesium, calcium, and sodium content while 25% to 40% of potassium, phosphorous and iron were present in flour when compared with grain.

Table 8.2 Chemical composition of triticale

Composition (% dw)	Pettersson and Aman (1987)	Serna Saldivar (2010)	Rakha et al. (2011)	Fraś et al. (2016)	Sibian et al. (2017)	Pena (2004)	Dziki et al. (2023)	Rodríguez-Perez et al. (2023)
N (no. of samples)	80	3	18	10	3	–	16	20
Protein	9.4–16.5	12.6–17.2	11.2–15.7	11.8–15.2	10.25	10.2–15.6	9.5–12.77	9.15–15.54
Starch	61.9–69.5	ND	60.5–70.4	60.8–67.6	64.37	56–65	67.64–72.53	54.16–67.39
Ash	1.1–3.0	ND	ND	1.6–2.0	1.27	1.4–2.9	1.74–2.04	1.82–2.24
Lipids	1–2.9	1.6–2.2	1–2.4	1.9–2.4	1.24	1.1–2.4	1.65–2.02	1.23–4.72
Dietary fiber	9.7–18.0	ND	13.2–16	11.7–13.6	6.36	3.0–4.5	14.53–17.43	ND

ND not detected

Table 8.3 Essential amino acid, vitamin, and mineral content of triticale

Parameters	Values
Amino acids (g/100 protein)	
Phenylalanine	3.9–4.6
Histidine	1.87–2.5
Isoleucine	3.3–3.9
Leucine	6.0–6.5
Lysine	3.1–4.0
Methionine	1.6–1.9
Threonine	3.0–3.5
Tryptophan	1.2–3.2
Valine	4.4–4.9
Minerals (ppm)	
Potassium	358–508
Phosphorus	437–534
Magnesium	110–198
Calcium	27–150
Iron	5.1–57
Manganese	1.0–63.2
Sodium	4.5–57
Copper	0.3–13.8
Zinc	1.3–35.9
Vitamins (mg)	
Thiamine	0.34–0.38
Riboflavin	0.09–0.13
Niacin	16.2–17.9
Pantothenic acid	0.68–0.71
Pyridoxine	0.42–0.51
Folate	78
Provitamin A	0.015–0.017
Vitamin E	1.4

Source: Glamočlija et al. (2018), Pena (2004), Lorenz et al. (1974)

8.5 Antinutrients and Toxic Compounds

Triticale, a hybrid grain derived from wheat and rye, contains certain compounds that hinder the absorption of nutrients by humans as well as pose higher health risks. These substances include resorcinols, ergot, phytates beside enzyme inhibitors. Ergot is the most common toxin in triticale that occurs due to infection caused by the fungus *Claviceps purpurea*. Ergot infection reduces the potential grain yields to a reserved extent (5–10%), however, causing major damage of contamination to the harvested grain by toxic alkaloids. *Claviceps purpurea* produces three ergot alkaloids (d-lysergic acid and its derivatives, clavine alkaloids, ergopeptines) with the prospective to cause rigorous health problems in both animals and humans. Gangrenous ergotism is characterized by restricted blood flow to extremities, leading to severe vasoconstriction. This can result in gangrene, where tissues die due to lack of blood supply. It often leads to amputation of affected body parts. Whereas convulsive ergotism is characterized by neurological symptoms, including muscle spasms, convulsions, hallucinations, and other psychological disturbances. It can be particularly severe and is historically associated with outbreaks referred to as "St. Anthony's Fire." Triticale, like other grains, can be susceptible to fungal contamination, which can lead to the production of mycotoxins—harmful compounds that can have negative health effects. Proper storage and handling practices are important to prevent mycotoxin formation. Mycotoxins, being potential contaminants of cereal, are also found in triticale. Perkowski et al. (2012) analyzed the trichothecenes in triticale grown in Poland. Notably, mycotoxins included deoxynivalenol (DON) (ranging from 0 to 102 μg/kg, with a prevalence of 83%), 3-Acetyl-DON (3-Ac-DON) (ranging from 0 to 55 μg/kg, with a prevalence of 94%), and nivalenol (NIV) (ranging from 0 to 19 μg/kg, with a prevalence of 50%) were observed in triticale. Trace amounts (less than 4 μg/kg) of 15-Acetyl-DON, T-2351 triol, HT-2, and T-2 mycotoxins were also detected. They further stated that the triticale's contamination level appeared to be less contaminated than durum wheat and more contaminated than barley. Deoxynivalenol (referred as vomitoxin) is a type B trichothecene that less often occurs in triticale. It may not be highly acute toxin but its prevalence has significance. This mycotoxin thrives in cool and wet environmental conditions. Although processing can reduce its levels in grains, it doesn't completely eliminate its presence. Notably, deoxynivalenol can withstand high temperatures, ranging from 170 °C to 350 °C (Sobrova et al. 2010). Anti-nutritional compound, i.e., phytic acid forms insoluble compounds with minerals, rendering these elements unavailable for absorption. In addition to phytic acid, triticale also contains trypsin and chymotrypsin inhibitors, both of which have been extracted from triticale in quantities falling between those found in wheat and rye. For instance, the concentration of phytic acid in triticale was 1.92 mg/g (db), which observed to be similar to rye, but higher than wheat (0.26 mg/g) and lower than maize (2.3 mg/g). Comparatively, the total phosphorus concentration in triticale (3.48 mg/g, dry basis) was similar to that in rye and wheat, but higher than maize (Mikulski and Kłosowski 2015). Resorcinols are compounds that can interfere with nutrient absorption and digestion while enzyme inhibitors impede the activity of digestive enzymes, potentially affecting nutrient breakdown and absorption. Other than these, the soluble fraction of arabinoxylan (AX) and β-glucan also contribute towards anti-nutritional properties of triticale (Annison and Choct 1991). Arabinoxylans can bind to essential nutrients like minerals (e.g., calcium, iron, zinc) and reduce their bioavailability. AX possesses higher affinity for these minerals, forming insoluble complexes that are resistant to digestion and absorption. AX also increases the viscosity of the contents in the digestive tract that reduces transit time and hinders the absorption of nutrients. It may also affect the activity of digestive enzymes and microbial populations in the gut. Similar to AX, β-glucans also increase the viscosity of the gut contents, thereby slowing down the nutrient absorption and interfering with digestion.

8.6 Bioactive and Functional Components

Triticale is comprised of phenolic acids possessing excellent antioxidant effects that enhance the health fullness of foods (Hosseinian and Mazza 2009; Kandil et al. 2012). Usually, whole cereal grains like triticale are an excellent source of bioactive polyphenols such as benzoic acid, derivatives of cinnamic acid, anthocyanidins, quinine, flavonols, chalcone, flavones, flavanones, and amino-phenolic compounds. Triticale also contains antioxidant tocotrienols, tocopherols, and oryzanols (Kruma et al. 2016). The polyphenolic compounds present in the triticale are illustrated in Fig. 8.3. The phenolic acid content in different cultivars of triticale viz. Borwo, Fredro, and Panteon are 1.37 mg/g, 1.27 mg/g, and 1.20 mg/g, respectively. Ferulic acid is one of the major polyphenols in triticale cultivars. The ferulic acid content of the triticale cultivars varied from 42% to 44% of the total phenolic acid present in the seed. Besides, sinapic acid and diferulic acid are the other polyphenolic compounds present in triticale (Kaszuba et al. 2021). Furthermore, triticale *KserSbah,* a variety of triticale, showed a total phenol content of 0.027 mg EAG/g while the total

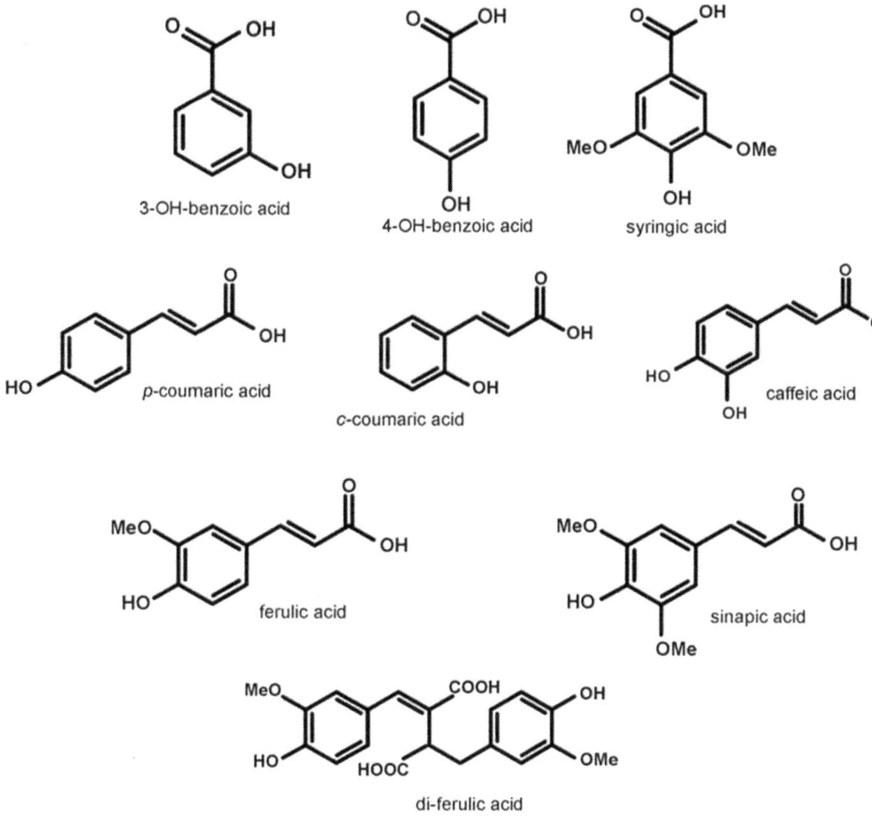

Fig. 8.3 Phenolic acids found in different varieties of Triticale. (Source: Kaszuba et al. 2021)

flavonoid content was found to be 0.019 mg EC/g. Triticale *Benni Haroun* variety showed 0.045 mg EAG/g total phenol content, whereas a 0.12 mg EC/g of total flavonoid content was reported (Benguella et al. 2020).

The total phenol content of nine different triticale seeds varies between 1.3 and 1.6 mg gallic acid equivalents (GAE)/g triticale, which appeared to be comparable to other cereals like wheat (1.5 GAE/g). During the milling process, the total phenol content in the flour obtained ranges from 0.7 to 1.1 mg GAE/g, which was reported to be similar to that of wheat (1.1 mg GAE/g) (Fraś et al. 2016). Besides the total phenol content in various triticale varieties was reported as1135.5 μg/g (db) (Kandil et al. 2012). The phenolic acids of whole triticale seeds are mainly conjugated (>90%) and include ferulic, protocatechuic, coumaric, as well as gallic acids (Kandil et al. 2012).

In addition, germinated grains are well-acknowledged to have higher nutritional as well as physiological value than non-germinated grains, pseudocereals and food products prepared from them. The effect of germination on polyphenols in triticale was well-researched. In a study done on different cereals, it was observed that wheat and rye showed the least up surge into phenolic content for the period of germination; however, the largest increase of about 2.5-fold was detected in triticale (Kruma et al. 2016). Phenolic compounds found in cereal grains are thought to influence the functional as well as nutritional characteristics of cereals (Tian et al. 2020), hence expanding their nutritive value. Germinated grains having enhanced bioactive components, may have the potential to fight serious health problems in obese and diabetic patients and has the ability to lower the chances of colon cancer (Donkor et al. 2012). Besides, the DPPH-free radical scavenging property of phenolic extracts of ungerminated triticale was 435 mMTE/100 g DW while after 24 h of germination, the DPPH scavenging activity of germinated grain showed a significant increase (Kruma et al. 2016). All through germination, hydrolases curb the endosperm for constituents with antiradical properties (Sharma and Gujral 2010).

However, in addition to increased total polyphenols, a substantial improvement in vitamin E content was also detected in different sprouted grains (Žilić et al. 2013). The enhanced vitamin E levels could be due to the formation of various biologically active components, for example, tocopherols as well as tocotrienols (Moongngarm and Saetung 2010) and/or release of bound vitamin E homologs from cellular constituents of grains all through the germination process (Ng et al. 2013).

Furthermore, it is also reported that the by-products of triticale processing contain good amounts of polyphenolic compounds. Triticale bran, straw, flakes, and leaves contain 9.9, 3.1, 1.7, and 3.7 mg/100 g of free phenolics while the bound phenolic content was 270.7, 356.1, 85.8, and 219.9 mg/100 g in triticale bran, straw, flakes, and leaves, respectively (Hosseinian and Mazza 2009). Triticale straw, bran, and leaves have been identified as important repositories of dietary phenols such as phenolic acids, proanthocyanidins as well as lignans. Ferulic acid is the most important phenolic acid and was predominantly present in bound form in the triticale cultivars. Low molecular weight proanthocyanidins accounted for more than 80% of proanthocyanidins in triticale bran; nonetheless, only 74% of proanthocyanidins were reported in the triticale straw.

Besides the polyphenolic bioactive compounds found in triticale, it also contains phenol lipids or resorcinolic fats, also known as alkylresorcinols or chemically

5-alkylresorcinols. Alkylresorcinol possesses many biological properties, such as antibacterial, antiparasitic, antioxidant, and anti-mutation (Ross et al. 2004). Alkylresorcinol obtained from dietary sources like triticale also tends to regulate γ-tocopherol, triglycerides as well as cholesterol concentrations in rat liver. It also helps in modulating the enzyme activity. Alkyl resorcinol can also control the activity of digestive enzymes, lower down oxidative stress, inhibit bacterial or fungal infections, and cut down the absorption of cholesterol (Bartłomiej et al. 2012). As a phenolic lipid, alkylresorcinol contains amphiphilic 1,3-dihydroxybenzene-derived products having an odd number of alkyl chains at the fifth position of the benzene ring and varying saturated alkyl tail lengths between 15 and 27 carbon atoms (Ross and Kochhar 2009). The different molecular weight alkylresorcinol found in varieties of triticale are 5-n-Heptadecylresorcinol (348 g/mole), 5-n-Nonadecylresorcinol (376 g/mole), 5-n-Heneicosylresorcinol (404 g/mole), 5-n-Tricosylresorcinol (432 g/mole), and 5-n-Pentacosylresorcinol (460 g/mole) (Tian et al. 2020).

Extraction parameters such as temperature and solid/solvent ratio of alkylresorcinol recovered from the bran of triticale were standardized using the optimization procedure. At a temperature of 24 °C and taking a solid-to-solvent ratio of 1:40 (w/v), the efficiency of alkylresorcinol reached 308 mg/100 g, while the performance of saturated and unsaturated alkylresorcinol reached 225 and 29 mg/100 g, correspondingly. High-performance liquid chromatography (HPLC) estimation revealed that the saturated alkylresorcinol comprises C17:0, C15:0, C19:0, C23:0, C21:0, and C25:0; however, the unsaturated alkylresorcinol was observed to be C23:1, C21:1, C19:1, and C17:1 (Agil and Hosseinian 2012).

Furthermore, in a study, the non-polar fats were recovered from the bran of triticale varieties by supercritical carbon dioxide extraction and comparisons were made with wheat. The study reported that the highest yield of non-polar fat (3.81 g/100 g) was observed in the bran of triticale cultivar Pronghorn while wheat showed a value of 3.26 g/100 g. Furthermore, a large difference in the supercritical carbon dioxide extraction of polar lipids from triticale as well as wheat was observed with 10% ethanol solvent and different time periods. Brans of triticale varieties showed the greatest polar lipid yield (approximately 0.22 g/100 g) at 4 h of treatment; however, wheat bran showed the greatest yield (0.18 g/100 g) at 6 h of extraction. Besides, the total alkylresorcinols content of the Ultima and Pronghorn cultivars of triticale was found to be 70 mg/100 g and 36 mg/100 g, respectively (Athukorala et al. 2010).

8.7 Health Implications

Triticale offers several health benefits owing to the biologically active compounds found in it. Studies have shown that triticale contains lunasin, a 43-aminoacid peptide. Lunasin is well recognized for its anticancer, antioxidant, and anti-inflammatory action (Galvez et al. 2001). Additionally, animal studies have shown that lunasin can penetrate target tissues when taken through oral means. The discovery that lunasin has high bioavailability, bioactivity, and thermostability after administration formed the foundation for recommendations regarding the addition of lunasin containing food products in the normal human diet (Nakurte et al. 2012). Furthermore, triticale has several health benefits, including its property to expand the efficiency of the human digestive tract, promote heart health, improve healing as well as metabolic rates, increase energy levels, guard the growing fetus, prevent and treat diabetes mellitus, promote the circulation of blood, and prevent asthma. Triticale provides relief from different skin ailments and contributes to stronger bones (Chopade et al. 2017).

The main cell wall polysaccharides found in the triticale is arabinoxylans, a type of pentosan, which have promising effects on human health and play a crucial role in many physiological processes occurring in the human body as well as in the disease prevention (Agil and Hosseinian 2014). 6–8% of arabinoxylans content is found in triticale grain (Bona et al. 2014). Besides, the molecular weight and the degree of substitution (arabinose/xylose ratio) of arabinoxylans have a significant influence on their physiological function (Chen et al. 2019). Arabinoxylan absorbs free radicals; in this manner, it helps in reducing oxidative stress as well as reducing the chances of cardiovascular illnesses as well as non-insulin-dependent diabetes mellitus. In addition, triticale seeds also contain a type of fiber called β-glucan. The β-glucan content of the triticale is 0.3–1.2 g. It has the ability to form highly viscous solutions in the human intestine, which are beneficial to the human body (El Khoury et al. 2012). In addition, no side effects were reported in a trial where participants adhere to a diet rich in β-glucans including oats or barley and their derivatives (Hallfrisch and Behall 2003). Plant-derived β-glucans are viscous fibers helpful in the treatment of various metabolic disorders.

Fructans are the other class of polysaccharides found in triticale grains. They support the immune functions in human body and have chemo-preventive effects against the development of colon cancer (de Moura et al. 2012). Besides, fructans encourage the growth of good microorganisms found in human gastrointestinal tract and improve relaxation (Szajewska et al. 2006). It is a prebiotic that helps in absorption of calcium, maintenance of the intestinal mucosal functional barrier, reduction of jeopardy of gastrointestinal tract-based infection. Other benefits include lowering blood sugar levels, lowering serum cholesterol, triglycerides as well as phospholipids (Kamanova et al. 2023). Triticale grain contains fructan content ranging from 1.6 g/100 g to 2.9 g/100 g (Rakha et al. 2011).

Triticale is an excellent reservoir of polyphenolic lipid compounds. One such polyphenol found in triticale is alkylresorcinols which is mostly found in concentrated form in the husk portion of the grain. Triticale bran is found to have the alkylresorcinols content ranging from 277.5 to 308.3 mg/100 g (Agil 2016). Alkylresorcinols possess anti-microbial, antioxidative, and anti-mutagenic effects (Kamanova et al. 2023). Researchers investigated the impact of triticale bran-derived alkylresorcinols incorporated into a fat-rich diet in the development of obesity and oxidative stress. In this study, triticale-derived alkylresorcinols was found to be a functional compound that helps prevent oxidative processes in obesity-related diets (Agil 2016).

Currently, attempts to increase food availability have led to novel approaches to broadening the application of triticale in the human dietary regime (Martinek et al. 2008). For instance, several cultivars of triticale have been utilized for the production of baked products like bread with appropriate quality characteristics (Tohver et al. 2005). The nutritive value of triticale seeds offers promise for the production of different food stuffs thereby improving its wide human consumption (Langó et al. 2017). Food products were developed using the bioactive components found in triticale. Yogurt enriched with insoluble dietary fiber (1.5% and 3.0% w/w) obtained from triticale showed excellent sensory attributes with high antioxidant properties. The developed yogurt showed better syneresis level, apparent viscosity, yield stress as well as thixotropic behavior (Miocinovic et al. 2018).

8.8 Processing

Processing of triticale has already been practiced in food industries in beverage sectors like malt, spirit, etc. It is also being extensively used for the development of bread, pasta, cookies, and biodegradable films and flour, thus having wide applicability in this sector and vast scope (Zhu 2018). Native protein present in triticale has good water absorption properties, require extent of gelation and viscosity. All these properties are directly influenced by agronomical factors, condition of storage, and the way bread is being processed (Tohver et al. 2005). Protein content is good in it but not suitable for the development of baked product. As particularly the content of gluten is low. This could be enhanced by allelic variation of genes for glutenin which is a high molecular weight subunit. The presence of mycotoxins in this makes it difficult to store for longer time period. Rakha et al. (2011) reported the presence of dietary fibers, Arabinoxylan, fructan and Klason lignin in triticale flour. The presence of very high activity of alpha amylase enzyme and low gluten content in triticale flour make it difficult to be used in baking industry as the rheological properties need to be improved (Leonova et al. 2022). Thus its mixing with other cereal grains is done to improve its baking properties. But the swelling properties of rye due to the presence of pentosans give it good texture and develop a viscous dough (Torbica et al. 2021). In spite of this', some varieties of triticale are very good functionally and can be blend up to 70% with different flour. However, some studies show only 30% substitution of its flour for the development of bakery products. This could be utilized in food industry thus has good scope in future. Fraś et al. (2021) compared the different forms (grain, flour, and bread) of wheat and triticale. They have used various varieties of triticale and compared it with standard wheat variety. They have concluded that specifically for baking only few varieties are of importance with good protein content (11.8% to 15.2%), aroma, and flavor. They have also observed that modern triticale varieties resemble wheat varieties or even better than wheat nutritionally except gluten content. The varieties suitable for breadmaking with high gluten content of triticale are SW 92280, Presto, Glu-1, and Heta (Tohver et al. 2000). Functionally, rye contains phenolic compounds, dietary fibers, alkylresorcinols, lignans, phytosterols, and protein. Processing of triticale flour by Torbica et al. (2021) shows that triticale flour has high protein: dietary fiber ratio and low starch which led to formation of dough with low torque. They have concluded that this property is very useful to enhance the shelf life of finished product. Malting and brewing is also practiced in triticale due to high alpha amylase activity and low gelatinization temperature (Özer 2021). High ash content of triticale flour sometime

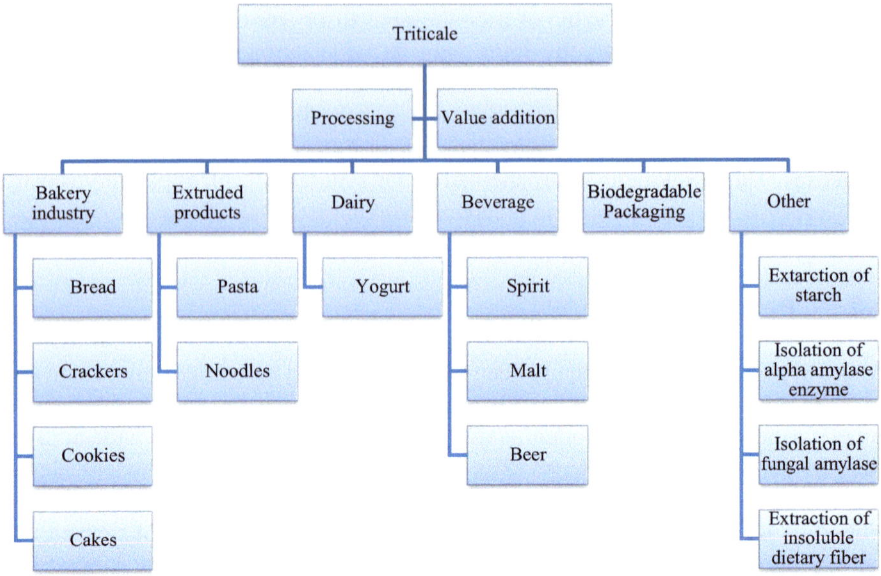

Fig. 8.4 Different ways of utilizing triticale in food industry

develops grayish color in noodles which is also considered as negative attribute. Figure 8.4 shows various aspects of its food applications.

8.9 Value Addition

Value addition of food products revolves around the use of multiple sources to enhance nutritional profile and marketability of that product. It is a customer-oriented approach where food manufacturer expects from the consumer that they will accept the product as it has more quality and is different from the rest of available food products. Generally, triticale flour is used in combination with other cereal flour. This type of incorporation increases the rheological properties as well as nutritional profile of that product. Research published by Doxastakis et al. (2002) concluded that addition of triticale flour up to 10% in wheat flour for the development of bread impart positive effect on qualitative traits of bread. This is due to the formation of strong gluten structure as the presence of triticale and wheat which gives bread a good volume, better tenderness, and crumb. Triticale is a good source of dietary fiber which includes 15% of total in it. According to Staffalo and his co-workers in 2004, yogurt with 1.3% enrichment of fiber can generate promising avenue and fetch higher consumer acceptance. Whereas Agil and Hosseinian (2012) have used 0.1 to 3.0 g triticale bran in 50 mL pasteurized milk for the development of yogurt.

There are different products shown in Table 8.4 prepared singly by triticale or have been used in combination with other cereal flour. The available literature revealed that this particular species is used mostly in bakery industry. Very less data is available on the applicability in other food products.

Table 8.4 Methods of processing and value addition of food products by utilizing triticale

Value addition/processing methods	Methodology	Findings/improvement	Source
Conversion of grains into high quality baker's grade flour	• Comparison between two varieties, i.e., Ramzes and Saur	• Method B was better that A	Kandrokov et al. (2019)
	• Processing technique	• Increase in yield of flour from 0.6% to 3.4% due to sieve purification	
	• Breaking and reduction • B) Breaking, sieving, sizing, and reduction	• Increase in baking absorption and decreased dough kneading time	
Development of breadsticks/rusk	• Replaced wheat flour from 10–60%	• In comparison to wheat rusk, the product is cracky and easily breakable	Leonova et al. (2022)
		• Mixed properly and develop pleasant taste and flavor	
		• 40% replacement was found to be suitable among all	
Triticale oat bread	• Triticale flour supplemented oat flour (2.5%, 5%, and 10%).	• Increased amino acid content	Fraś et al. (2018)
		• Enhanced crude fiber and beta glucan content	
		• Better baking property	
Comparative studies of multicereal, whole wheat, rye, and oat bread	• Bread development method	• Multicereal bread has highest quality in terms of protein (9%), fats, and fiber	Carocho et al. (2020)
	• All bread were prepared by mixing water (600 mL) + sourdough (200 g) and baked for 30 min at a constant temperature of 250 °C	• Rye bread contained 5.58% protein which was lowest among all. But have good mineral profile	
Bread	• Supplementation of wheat flour with rye flour from 10%, 20%, 30%, and 40%	• 30% substitution forms bread with good quality traits	
		• Maintained the taste and flavors and does not produce bad flavors	

(continued)

Table 8.4 (continued)

Value addition/processing methods	Methodology	Findings/improvement	Source
Triticale crackers	• Compared different varieties of triticale for its baking properties • Used blend of hard and soft flour for better results • 45 g of triticale flour and rest are other ingredients for the preparation of 100 g product • Crackers were baked for 5 min at 265 °C	• Successfully prepared standard quality crackers by using five different varieties of triticale	Pérez et al. (2003)
Rheological changes in wheat flour dough due to incorporation of soya, lupin, and triticale and development of bread	• Substitution of wheat flour at a level of 5% and 10% followed by stirring in farinograph bowl, fermentation, and baking for 45 min at 220 °C	• Increased water absorption of dough from 52.6% to 63.5% • Addition of triticale flour increases loaf volume from 33 to 41 BU and decrease with the addition of soya and lupin flour • Developed acceptable bread with good crumb structure, texture, and volume	Doxastakis et al. (2002)
Development fiber enriched yogurt with the addition of triticale flour	• Fortified yogurt with insoluble fiber from oat, triticale, and wheat at a level of 15 and 30 g/kg • Extracted insoluble by autohydrolysis of triticale flour	• Successfully fortified yogurt with insoluble fiber from oat, triticale, and wheat at a level of 15 and 30 g/kg • Best results obtained in triticale insoluble fiber at 30 g/kg level • Developed novel functional food with good consumer acceptability	Tomic et al. (2017)
Development of prebiotics (yogurt) and enhance antioxidant activity by using triticale bran	• Triticale bran (0.1, 0.3, 0.6, 0.9, 1.2, 1.5, 1.8, 2.0, and 3 g) in pasteurized milk (50 mL) • Microbial strains: *Lactobacillus delbrueckii subsp. Bulgaricus, Streptococcus salivarius subsp. thermophilus, Lactobacillus acidophilus, and Bifidobacterium lactis* • Incubation at 42 °C	• Promising source of prebiotics and antioxidative compounds • Developed product with 33.86 μmol trolox equivalent/g triticale bran activity • Stable product shelf life till 28 days	Agil and Hosseinian (2012)

8.10 Safety Considerations

Triticale has been approved as safe for human consumption across the world. However, as with any food product, there are certain safety considerations to keep in mind. Triticale is produced by cross-reactivity with wheat and rye which may raise the possibility of allergens. Triticale, like wheat and rye, also contain gluten which may trigger the adverse reaction in individual suffering from celiac disease. Furthermore, if triticale has been genetically modified (GMO) to introduce specific traits, the GMO variety would need to undergo safety assessments and regulatory approval before being considered safe for human consumption. The assessments might ensure that genetic modifications do not introduce any harmful substances or allergens. In many cases, the evaluation of GMOs for safety involves multiple aspects, including toxicological assessments, allergenicity testing, nutritional equivalence comparisons, and environmental impact assessments. Transgenic triticale genotypes with improved agricultural traits (e.g., better herbicide tolerance) have been produced. A systematic review on transgenic triticale genotype with improved agricultural traits, such as herbicide (glufosinate ammonium) tolerance has reported its safety to be used as food (Snell et al. 2012).

8.11 Future Challenges and Opportunities

A good supply of raw ingredients and direct connect with industry from farm level is the prerequisite which was necessary to maintain the market for the produced food grains for its maximum utilization. In this account, triticale fails to set the connect and expansion of market on global level due the existing surplus production in major cereal grains. It is also due to little efforts that has been put forward related to crop breeding for nutritional value improvement and higher industrial processing for new product development in comparison to the commercial cereal grains. In many countries of the world, triticale is still grown as feed crop due to more availability of rice and wheat beside lack of promotional activities related to triticale. Also triticale demands the exploration of its grains for the wider food development, but less expected results during milling, baking, and product development restricts its wider applications. Triticale food utilization is also impacted by non-grain variables that restrict grain availability, such as regional consumer preferences, competitiveness with other grains, and economic and marketing concerns. Triticale flour results in weaker dough because of the low gluten content, weak or deficient viscoelasticity of gluten, and strong alpha-amylase activity which is usually not considered good for most of the bakery product preparation. By blending the triticale flour with the other cereal grains flour can lead to obtain product with good functional characteristics for leavened and non leavened types of product. However, it is still an unexplored alternative for human food even though it is having higher protein content compared to many wheat varieties, thus can easily proliferate where climate and soil conditions limit the growth of commercial cereal grains. Its breeding is progressively increasing in European regions where wholemeal flour is utilized for the production of bread (Leon et al. 2008). It can be a very suitable alternative for the production of various unleavened type of products where strong elastic gluten network may not be required to hold the gases. Beside bran of triticale grain also hold potential to be explored for food enrichment as very less research on this part has been carried out yet. With the increasing global production, it may open new ways for its exploration as films preparation, for pasta preparation, cookie, yogurt, malt, spirits, etc. based upon nutritive values of grains. Further another challenge in processing of triticale grains to flour is the low milling yield as compared to commercial grain crops, regardless of triticale genotypes. Triticale grain and flour can be processed using a variety of processing methods in addition to milling, including germination, non-thermal plasma, irradiation, and high pressure to create a wide range of functionalities. Triticale production must offer enticing consumer qualities like a pleasing flavor and possible health benefits that are different from wheat in order to justify market development efforts in response to consumers' growing interest in the nutritional value and sustainability of food products.

8.12 Conclusion

Triticale is one of the under-explored grain crops which recently got attention due to its higher nutritional value and multiple uses with increasing production on global level in recent years. Due to the presence of wide number of components like arabinoxylans, starch, lignins, β-glucans, alkylresorcinols, vitamins, proteins, and polyphenols of triticale leads to potential of this grain crops for functional food production. But due to its limited availability and lower explored processing potential beside low gluten content slightly restrict its use in various product developments. Although higher protein and essential amino acid content expand its applications to be used in various functional food developments beside its application in edible film production and in generation of spirits and malt, etc. However as compared to other major cereal grains, it can grow in adverse soil and environmental conditions which will lead the higher prospect of its utilization in food industry in different ways for product commercialization and adoption by different group of individuals. But still more concentrated breeding programs with focused research are required to improve the nutritional value of various cultivars in industrial point for its utilization on commercial scale for new and innovative product development which may be highly accepted by the population.

References

Agil R (2016) Functional properties of triticale bran bioactives using in vitro and in vivo model systems. Doctoral dissertation, Carleton University

Agil R, Hosseinian F (2012) Dual functionality of triticale as a novel dietary source of prebiotics with antioxidant activity in fermented dairy products. Plant Foods Hum Nutr 67:88–93

Agil R, Hosseinian F (2014) Determination of water-extractable polysaccharides in triticale bran. J Food Compos Anal 34(1):12–17

Annison G, Choct M (1991) Anti-nutritive activities of cereal non-starch polysaccharides in broiler diets and strategies minimizing their effects. Worlds Poult Sci J 47:232–242

Athukorala Y, Hosseinian FS, Mazza G (2010) Extraction and fractionation of alkylresorcinols from triticale bran by two-step supercritical carbon dioxide. LWT Food Sci Technol 43(4):660–665

Bartłomiej S, Justyna RK, Ewa N (2012) Bioactive compounds in cereal grains–occurrence, structure, technological significance and nutritional benefits—a review. Food Sci Technol Int 18(6):559–568

Benguella R, Meziani S, Ghueffari I, Menadi N, Chenni FZ, Labga L, Saidani S, Barak S, Aissaouia M, Rahmoun N, Demmouche A (2020) Bioactive compounds in the peripheral layers of barley and Triticale species in the mature grain cultivated in Algeria. J Drug Del Ther 10(4):37–43

Berry CP, D'Appolonia BL, Gilles KA (1971) The characterization of triticale starch and its comparison with starches of rye, durum and HRS wheat. Cereal Chem 48:415–427

Bona L, Acs E, Lantos C, Tomoskozi S, Lango B (2014) Human utilization of triticale: technological and nutritional aspects. Commun Agric Appl Biol Sci 79(4):139–152

Buchholz M, Drotleff AM, Ternes W (2012) Thiamin (vitamin B1) and thiamin phosphate esters in five cereal grains during maturation. J Cereal Sci 56:109–114

Carocho M, Morales P, Ciudad-Mulero M, Fernandez-Ruiz V, Ferreira E, Heleno S, Rodrigues P, Barros L, Ferreira IC (2020) Comparison of different bread types: chemical and physical parameters. Food Chem 310:125954

Chen Z, Li S, Fu Y, Li C, Chen D, Chen H (2019) Arabinoxylan structural characteristics, interaction with gut microbiota and potential health functions. J Funct Foods 54:536–551

Chopade MU, Patil HS, More RB (2017) Cereals and their nutritional value for health benefits. Adv Innov Res 58:58–61

de Moura NA, Caetano BF, Sivieri K, Urbano LH, Cabello C, Rodrigues MA, Barbisan LF (2012) Protective effects of yacon (Smallanthus sonchifolius) intake on experimental colon carcinogenesis. Food Chem Toxicol 50(8):2902–2910

Dennett AL, Schofield PR, Roake JE, Howes NK, Chin J (2009) Starch swelling power and amylose content of triticale and Triticum timopheevii germplasm. J Cereal Sci 49:393–397

Dennett AL, Cooper KV, Trethowan RM (2013) The genotypic and phenotypic interaction of wheat and rye storage proteins in primary triticale. Euphytica 194:235–242

Dobreva S (2016) Triticale—past and future. Agric Sci Technol 8(4):271–275. https://doi.org/10.15547/ast.2016.04.051

Donkor ON, Stojanovska L, Ginn P, Ashton J, Vasiljevic T (2012) Germinated grains–sources of bioactive compounds. Food Chem 135(3):950–959

Doxastakis G, Zafiriadis I, Irakli M, Marlani H, Tananaki C (2002) Lupin, soya and triticale addition to wheat flour doughs and their effect on rheological properties. Food Chem 77(2):219–227

Dziki D, Hassoon WH, Kramek A, Krajewska A (2023) Grinding characteristics of new varieties of winter triticale grain. PRO 11:1477

El Khoury D, Cuda C, Luhovyy BL, Anderson GH (2012) Beta glucan: health benefits in obesity and metabolic syndrome. J Nutr Metab 2012:851362. https://doi.org/10.1155/2012/851362

FAO (2021) Food and Agriculture Organization of United States. FAOSTAT 2021. https://www.fao.org/faostat/en/#data/QCL

FAO (2022) World Food and Agriculture—statistical yearbook 2022. FAO, Rome. https://www.fao.org/faostat/en/#data/QV/visualize

Finch HJS, Samuel AM, Lane GPF (2002) Cereals. In: Lockhart and Wiseman's crop husbandry including grassland. Woodhead Publishing, Cambridge, pp 259–302. https://doi.org/10.1533/9781855736504.3.257

Fraś A, Gołębiewska K, Gołębiewski D, Mańkowski DR, Boros D, Szecówka P (2016) Variability in the chemical composition of triticale grain, flour and bread. J Cereal Sci 71:66–72

Fraś A, Gołębiewski D, Gołębiewska K, Mańkowsk D, Boros D, Gzowska M (2018) Triticale-oat bread as a new product rich in bioactive and nutrient components. J Cereal Sci 82(4). https://doi.org/10.1016/j.jcs.2018.05.001
Fraś A, Gołębiewska K, Gołębiewski D, Wiśniewska M, Gzowska M, Mańkowsk D (2021) Utilisation of triticale (X Triticosecale Wittmack) and residual oat flour in breadmaking. Czech J Food Sci 39(3):226–233
Galvez AF, Chen N, Macasieb J, de Lumen BO (2001) Chemopreventive property of a soybean peptide (lunasin) that binds to deacetylated histones and inhibits acetylation. Cancer Res 61(20):7473–7478
Gatel FO, Lavorel J, Fekete F, Grosjean J (1985) Feeding value of triticale for monogastrics: weaned pig-lets, growing-finishing pigs and broilers. In: Bernard M, Bernard S (eds) Genetics and breeding of triticale. Institut National de la Recherche Agronomique, Versailles, pp 659–670
Glamočlija N, Starčević M, Ćirić J, Šefer D, Glišić M, Baltić MZ, Marković R, Spasić M, Glamočlija D (2018) The importance of triticale in animal nutrition. ВетеринарскижурналРепубликеСрпске 18:73–94. https://doi.org/10.7251/VETJEN1801073G
Hallfrisch J, Behall KM (2003) Physiological responses of men and women to barley and oat extracts (NutrimX). I. breath hydrogen, methane, and gastrointestinal symptoms. Cereal Chem 80(1):76–79
Havensone G, Meija L, Balode L, Sturite I, Lejnieks A (2017) Glycaemic profile and insulin response after consuming triticale flakes. Proc Latvian Acad Sci B Nat Exact Appl Sci 71:434–439
Heger J, Eggum BO (1991) The nutritional values of some high-yielding cultivars of triticale. J Cereal Sci 14:63–71
Hosseinian FS, Mazza G (2009) Triticale bran and straw: potential new sources of phenolic acids, proanthocyanidins, and lignans. J Funct Foods 1:57–64. https://doi.org/10.1016/j.jff.2008.09.009
Ivanova A, Tsenov N (2014a) Production potencial of new triticale varieties grown in the region of Dobrudzha. Agric Sci Technol 6:243–246
Ivanova A, Tsenov N (2014b) Comparative evaluation of triticale cultivars grown in the region of Dobrudzha. Agric Sci Technol 6:387–391
Jonnala RS, Irmak S, MacRitchie F, Bean SR (2010) Phenolics in the bran of waxy wheat and triticale lines. J Cereal Sci 52(3):509–515. https://doi.org/10.1016/j.jcs.2010.07.013
Kamanova S, Yermekov Y, Shah K, Mulati A, Liu X, Bulashev B, Toimbayeva D, Ospankulova G (2023) Review on nutritional benefits of triticale. Czech J Food Sci 41(4):248–262. https://doi.org/10.17221/67/2023-CJFS
Kandil A, Li J, Vasanthan T, Bressler DC (2012) Phenolic acids in some cereal grains and their inhibitory effect on starch liquefaction and saccharification. J Agric Food Chem 60(34):8444–8449
Kandrokov RH, Pankratov GN, Meleshkina EP, Vitol IS, Tulyakov DG (2019) Effective technological scheme for processing triticale (Triticosecale L.) grain into graded flour. Foods Raw Mater 7(1):107–117
Kaszuba J, Kapusta I, Posadzka Z (2021) Content of phenolic acids in the grain of selected polish triticale cultivars and its products. Molecules 26(3):562. https://doi.org/10.3390/molecules26030562
Kirchev H, Popov N (2010) Influence of nitrogen fertilization on the content of crude protein and amino acids in the grain triticale (X. Triticosecale Wittmack). Plant Sci 47:441–445
Knudsen KEB (2014) Fiber and nonstarch polysaccharide content and variation in common crops used in broiler diets. Poult Sci 93:2380–2393
Kolev T (2010) Testing of the new Bulgarian varieties of triticale, grain. Plant Sci 47:14–16
Kolev T (2014) Effects of liquid fertilizer Amalgerol premium on productivity triticale. Sci Pap Agric Univ 58:219–223
Kruma Z, Tomsone L, Ķince T, Galoburda R, Senhofa S, Sabovics M, Straumite E, Sturite I (2016) Effects of germination on total phenolic compounds and radical scavenging activity in hull-less spring cereals and triticale. Agron Res 14(S2):1372–1383

Langó B, Bóna L, Ács E, Tömösközi S (2017) Nutritional features of triticale as affected by genotype, crop year, and location. Acta Aliment 46(2):238–245

Leon AE, Perez GT, Ribotta PD (2008) Triticale flours: composition, properties and utilization. Food 2(1):17–24

Leonova S, Badamshina E, Koshchina E, Kalugina O, Gareeva I, Leshchenko N (2022) Triticale flour in bakery and rusk products. Food Sci Technol Int 28(6):524–534

Lorenz K (1982) Triticale processing and utilization: comparison with other cereal grains. In: Wolff IA (ed) Handbook of processing and utilization in agriculture. CRC Press, Boca Raton, pp 277–327

Lorenz K, Reuter FW, Sizer C (1974) The mineral composition of triticales and triticale milling fractions by X-ray fluorescence and atomic absorption. Cereal Grains 5:534–542

Makowska A, Szwengiel A, Kubiak P, Tomaszewska-Gras J (2014) Characteristics and structure of starch isolated from triticale. Starch-Stärke 66:895–902. https://doi.org/10.1002/star.201300264

Martinek P, Vinterová M, Burešová I, Vyhnánek T (2008) Agronomic and quality characteristics of triticale (X TriticosecaleWittmack) with HMW glutenin subunits 5+ 10. J Cereal Sci 47(1):68–78

McGoverin CM, Snyders F, Muller N, Botes W, Fox G, Manley M (2011) A review of triticale uses and the effect of growth environment on grain quality. J Sci Food Agric 91(7):1155–1165

Mergoum M, Pfeiffer WH, Pena RJ, Ammar K, Rajaram S (2004) Triticale crop improvement: the CIMMYT programme. In: Mergoum M, Gomez-Macpherson H (eds) Triticale improvement and production, FAO plant production and protection paper no. 179. Food and Agriculture Organization of United Nations, Rome, pp 11–26

Mergoum M, Singh PK, Peña RJ, Lozano-del Río AJ, Cooper KV, Salmon DF, Gómez-Macpherson H (2009) Triticale: a new crop with old challenges. In: Carena M (ed) Cereals. Handbook of plant breeding, vol 3. Springer, New York. https://doi.org/10.1007/978-0-387-72297-9_9

Mikulski D, Kłosowski G (2015) Phytic acid concentration in selected raw materials and analysis of its hydrolysis rate with the use of microbial phytases during the mashing process. J Inst Brew 121:213–218

Miocinovic J, Tomic N, Dojnov B, Tomasevic I, Stojanovic S, Djekic I, Vujcic Z (2018) Application of new insoluble dietary fibres from triticale as supplement in yoghurt–effects on physico-chemical, rheological and quality properties. J Sci Food Agric 98(4):1291–1299

Moongngarm A, Saetung N (2010) Comparison of chemical compositions and bioactive compounds of germinated rough rice and brown rice. Food Chem 122(3):782–788

Naguleswaran S, Vasanthan T, Hoover R, Bressler D (2014) Amylolysis of amylopectin and amylose isolated from wheat, triticale, corn and barley starches. Food Hydrocoll 35:686–693

Nakurte I, Klavins K, Kirhnere I, Namniece J, Adlere L, Matvejevs J, Muceniece R (2012) Discovery of lunasin peptide in triticale (X Triticosecale Wittmack). J Cereal Sci 56(2):510–514

Navarro-Contreras AL, Chaires-González CF, Rosas-Burgos EC, Borboa-Flores J, Wong-Corral FJ, Cortez-Rocha MO, Cinco-Moroyoqui FJ (2014) Comparison ofprotein and starch content of substituted and complete triticales (× *Triticosecale*Wittmack): contribution to functional properties. Int J Food Prop 17:421–432

Ng LT, Huang SH, Chen YT, Su CH (2013) Changes of tocopherols, tocotrienols, γ-oryzanol, and γ-aminobutyric acid levels in the germinated brown rice of pigmented and nonpigmented cultivars. J Agric Food Chem 61(51):12604–12611

Özer E (2021) Uses and consumption problems of triticale. Theor Pract New Approach Cereal Sci Technol 347. https://doi.org/10.1016/j.foodchem.2004.01.079

Pattison AL, Appelbee M, Trethowan RM (2014) Characteristics of 641 modern triticale quality: glutenin and secalin subunit composition and mixograph properties. J Agric Food Chem 62:4924–4931

Pena RJ (2004) Food uses of triticale. In: Triticale improvement and production. FAO, Rome, pp 37–48. https://www.fao.org/3/y5553e/y5553e01.pdf

Perkowski J, Stuper K, Buśko M, Góral T, Jeleń H, Wiwart M, Suchowilska E (2012) A comparison of contents of group A and B trichothecenes and microbial counts in different cereal species. Food Addit Contam Part B Surveill 5:151–159
Pérez GT, León AE, Ribotta PD, Aguire A, Rubiolo OJ, Añón MC (2003) Use of triticale flours in cracker-making. Eur Food Res Technol 217(2):134–137. https://doi.org/10.1007/s00217-003-0729-9
Pettersson D, Aman P (1987) The variation in chemical composition of triticales grown in Sweden. Acta Agric Scand 37:20–26
Rakha A, Åman P, Andersson R (2011) Dietary fiber in triticale grain: variation in content, composition, and molecular weight distribution of extractable components. J Cereal Sci 54:324–331
Rakha A, Saulnier L, Aman P, Andersson R (2012) Enzymatic fingerprinting of arabinoxylan and β-glucan in triticale, barley and tritordeum grains. Carbohydr Polym 90:1226–1234
Rimpau W (1891) Kreuzungsprodukte landwirthschaftlicher Kulturpflanzen. Landwirthschaftlicher Jahrbucher 20:335–371
Rodríguez-Perez G, Cervantes-Ortiz JF, Gámez-Vázquez AJ, Reynaga-Franco FJ, Torres-Velázquez JR, Ávila-Perches MA (2023) Nutritional value in grains of triticale as an alternative in the food industry. Rev Mex Cienc Agríc 14(3):351–362
Ross AB, Kochhar S (2009) Rapid and sensitive analysis of alkylresorcinols from cereal grains and products using HPLC—coularray-based electrochemical detection. J Agric Food Chem 57(12):5187–5193
Ross AB, Åman P, Andersson R, Kamal-Eldin A (2004) Chromatographic analysis of alkylresorcinols and their metabolites. J Chromatogr A 1054(1–2):157–164
Salmon RE (1984) True metabolizable energy and amino acid composition of wheat and triticale and their comparative performance in Turkey starter diets. Poult Sci 63:1664–1666
Salmon DF, Mergoum M, Gomez Macpherson H (2004) Triticale production and management. In: Mergoum M, Gomez-Macpherson H (eds) Triticale improvement and production, FAO plant production and protection paper no. 179. FAO, Rome, pp 27–36
Salvucci E, Rossi M, Colombo A, Pérez G, Borneo R, Aguirre A (2019) Triticale flour films added with bacteriocin-like substance (BLIS) for active food packaging applications. Food Packag Shelf Life 19:193–199
Senol FS, Kan A, Coksari G, Orhan IE (2012) Antioxidant and anticholinesterase effects of frequently consumed cereal grains using in vitro test models. Int J Food Sci Nutr 63:553–559
Serna Saldivar SO (2010) Cereal grains: properties, processing and nutritional attributes. CRC Press, Taylor & Francis Group, Boca Raton
Sharma P, Gujral HS (2010) Antioxidant and polyphenol oxidase activity of germinated barley and its milling fractions. Food Chem 120(3):673–678
Sibian MS, Saxena DC, Riar CS (2017) Effect of germination on chemical, functional and nutritional characteristics of wheat, brown rice and triticale: a comparative study. J Sci Food Agric 97:4643–4651
Snell C, Bernheim A, Bergé JB, Kuntz M, Pascal G, Paris A, Ricroch AE (2012) Assessment of the health impact of GM plant diets in long-term and multigenerational animal feeding trials: a literature review. Food Chem Toxicol 50:1134–1148
Sobrova P, Adam V, Vasatkova A, Beklova M, Zeman L, Kizek R (2010) Deoxynivalenol and its toxicity. Interdiscip Toxicol 3:94–99
Stankowski S, Sobolewska M, Jaroszewska A, Michalska B (2017) Impact of form and dose of nitrogen fertilizers on the technological value of spring triticale. Folia Pomer Univ Technol Stetin Agric Aliment Pisc Zootech 336:167–178
Straumīte E, Galoburda R, Tomsone L, Kruma Z, Grāmatiņa I, Kronberga A, Stūrīte I (2017) Nutritional quality of triticale (× Triticosecale Wittm.) grown under different cropping systems. Proc Latvian Acad Sci B Nat Exact Appl Sci 71:481–485
Szajewska H, Ruszczyński M, Radzikowski A (2006) Probiotics in the prevention of antibiotic-associated diarrhea in children: a meta-analysis of randomized controlled trials. J Pediatr 149(3):367–372

Tian S, Sun Y, Chen Z, Zhao R (2020) Bioavailability and bioactivity of alkylresorcinols from different cereal products. J Food Qual 2020:5781356
Tohver M, Täht R, Rahnu I, Kann A (2000) Investigation of seed storage protein and bread-making quality of triticale. Acta Agronomica Hungarica 48(1):41–50
Tohver M, Kann A, Taht R, Mihhalevski A, Hakman J (2005) Quality of triticale cultivars suitable for growing and bread-making in northern conditions. Food Chem 89:125–132
Tomic N, Dojnov B, Miocinovic J, Tomasevic I, Smigic N, Djekic I, Vujcic Z (2017) Enrichment of yoghurt with insoluble dietary fiber from triticale—a sensory perspective. LWT 80:59–66
Torbica A, Belović M, Popović L, Čakarević J, Jovičić M, Pavličević J (2021) Comparative study of nutritional and technological quality aspects of minor cereals. J Food Sci Technol 58:311–322
Trofimova TA, Korzhov SI, Gulevsky VA, Obraztsov VN (2018) Assessing the degree of physical degradation and suitability of chernozems for the minimization of basic tillage. Eurasian Soil Sci 51(9):1080–1085
Waggle DH, Lambert MA, Miller GD, Farrell EP, Deyoe CW (1967) Extensive analyses of flours and millfeeds made from nine different wheat mixes. II. Amino acids, minerals, vitamins and gross energy. Cereal Chem 44:48
Wilson AS (1875) Wheat and rye hybrids. Trans Proc Bot Soc Edinburgh 12:286–288
Woś H, Brzeziński W (2015) Triticale for food—the quality driver. In: Eudes F (ed) Triticale. Springer, Cham, pp 213–232. https://doi.org/10.1007/978-3-319-22551-7_11
Yakushev VP, Mikhailenko IM, Dragavtsev VA (2015) Reserves of agro-technologies and breeding for cereal yield increase in The Russian Federation. Agric Biol 50(5):550–560
Zhu F (2018) Triticale: nutritional composition and food uses. Food Chem 241:468–479. https://doi.org/10.1016/j.foodchem.2017.09.009
Zhukov AM, Anosova MV, Popov IA, Churikova SY, Schedrin DS (2020) Triticale. The techniques of cultivation and processing. IOP Conf Ser Earth Environ Sci 422:012006
Žilić S, Mogol BA, Akıllıoğlu G, Serpen A, Babić M, Gökmen V (2013) Effects of infrared heating on phenolic compounds and Maillard reaction products in maize flour. J Cereal Sci 58(1):1–7

Buckwheat

9

Mukul Kumar, Deepak Kaushik, and Sakshi Kasodhan

Abstract

The pseudocereal, Buckwheat in scientific *Fagopyrum species*, is a dicotyledon, currently whopping great possibilities in the health food sector relevantly due to superior nutrition quality. It has valuable inception of oxybenzene compounds, phytosterols, fluorescent phototoxic fagopyrins, and roughage as fiber, lignans, vitamins, minerals, antioxidants, and unsaturated fatty acids in different plant parts. The biological value and productive properties of buckwheat lead to the development of healthy food products. The value-addition in the product of *Fagopyrum* will be enhanced by the incorporation of other compounds to it; in the case of buckwheat, there are numerous extruded products such as bread, biscuits, muffins, polenta, beer, sprouts, noodles, groats, pancake mix, confectioneries, cookies, soups, spirits, porridge, and other confectionery products. It contains an eclectic extent of bioactive compounds, such as polyphenolic flavones, triterpenoids, phenylpropanic acid, glycosides, phenols, polysaccharides, and bioactive peptides, as well as diastereomers of inositol and its derivatives. Some naturally occurring bioactive peptides found in seeds of buckwheat are multiple functional compounds that, in some cases, act as anti-nutritional factors, such as peptides having anti-microbial, hypotensive, and antioxidant activity, as well as trypsin inhibitors and anti-tumor proteins. The biological value of buckwheat proteins is considerably 100, possessing a significant amount of amino acids, and several tests revealed that it is more affluent in lysine, histidine, and tryptophan. The gluten-free proteins of buckwheat help to relieve people with celiac dis-

M. Kumar (✉) · S. Kasodhan
Department of Food Technology and Nutrition, Lovely Professional University, Phagwara, Punjab, India

D. Kaushik
Department of Biotechnology, Shoolini University, Solan, Himachal Pradesh, India

J. Singh et al. (eds.), *Cereals and Nutraceuticals*,
https://doi.org/10.1007/978-981-97-2542-7_9

eases. However, the RDA value or recommended intake of this grain is minimum (40 g/day) due to protein content present in buckwheat which also responsible for the allergic reaction. The occurrence rate of allergy with buckwheat is significantly ($p < 0.05$) lower than that related to the consumption of gluten-containing foods. It has been accepted as a functional or healthy food; therefore, it is suggested to improve human health and be used in treating diseases.

Keywords

Buckwheat · Bioactive compounds · Biological value · *Fagopyrum species* · RDA value

9.1 Origin and Background

Buckwheat is a dicotyledon fruit belonging to the *Fagopyrum* genus and *Polygonaceae* family, also categorized as "pseudocereal" as distant from the Gramineae family. In Hindi, Japanese, and Chinese, buckwheat is known as "kuttu," "soba," and "chaomai," respectively (Huda et al. 2021). In nine varieties of buckwheat generally, *Fagopyrum esculentum* (Common) and *Fagopyrum tataricum* (Tartary) are mainly cultivated in the temperate zones of the Northern hemisphere (Hunt et al. 2018; Kasar et al. 2021). Another variety that is less cultivated with major local value in Asia is *Fagopyrum dibotrys* (golden) or *acutatum* (tall), which is not a well-known buckwheat variety worldwide (Zhu 2016). The origin of this pseudocereal was seen in South China in the mid-sixth millennium B.C. (Graziano et al. 2022). Originally in the Eastern Himalayas, Tibet to some extent and the combined area of Yunnan and Sichuan, which has been spread out from southwest to the north of China, then from Tibet to Bhutan, Nepal, and India (Zhou et al. 2018; Zhang and Xu 2017). It was cultivated in Europe, South Russia, Ukraine, Germany, and Slovenia. Later, the Dutch scientist brought buckwheat to America; now, it is common in many countries (Woo et al. 2016). The Muli district of Sichuan Province, the Deqin district of Yunnan Province, and the Mangkang districts of Eastern Tibet found buckwheat in their provinces (Ohnishi 2004). The varieties' nomenclature has been performed according to the harvesting season, i.e., Kita-Wase-Soba, Shinano, and Miyazaki-Ootsubu, which belong to the summer, middle, and autumn types, respectively (Ikeda 2002). These factors dictate the quality of buckwheat, consisting of yield, the shape of grain and its uniform size, processing efficiency, grain color, flavor, safety standards, agricultural residues in the form of chemical compounds, and allergic reaction properties (Suzuki et al. 2020a, b). Moreover, these grains' nutrition and sensory attributes are also significant eminence characteristics (Farzana et al. 2022). The starch content of buckwheat has great potential for different Ready to Eat (RTE) and Ready to Cook (RTC) food preparations due to their distinctive structural and functional features (Zhu 2016).

9.2 Cultivation and Production

Buckwheat is a vital, highly nutritional, and non-poaceous crop that is one of the best-suited crops for higher altitudes, where crop growing season is very short (Babu et al. 2018), have sowing periods in the first half of May–June, requires only 10 to 12 weeks to mature and could grow on a wide range of soil in a cool and moist condition (Tömösközi and Langó 2017). It is suitable to cultivate these crops in altitude regions that are higher with low precipitation and degree of weather due to its numerous ecological adaptabilities (Huda et al. 2021). The cultivation of buckwheat, regarding soil requirements, is slightly acidic with a tolerance level of soil pH 4.8, light requirements, i.e., facultative short day, optimum temperature 15–20 °C, and requires high precipitation in which yield varies from 10–12 quintals per hectare (Radics and Mikóházi 2010). Due to global climatic conditions, crop productivity is conceited by the shortage of soil nutrition and degradation, compaction, and pollution, as nowadays, soil formation is 20–40 times more sluggish than soil degradation and erosion (Germ and Gaberščik 2016). During drying, the crop left out in the field has the most serious problem of being damaged by birds; even at maturity, rats are involved in the destruction of buckwheat crops (Unal et al. 2017).

The crop has also been cultivated as a protection crop, fertilizer crop, feedstuff crop, productiveness-inducing crop, honey crop, and pharmaceutical plant (Rehman et al. 2022). The conversion of protofagopyrins into fagopyrins as precursors is found at their peak during seed germination; and light also plays the crucial role in conversion (Kim and Hwang 2020; Huda et al. 2021). Buckwheat is grown in Japan, Korea, China, Lithuania, Estonia, Belarus, Moldova, Poland, Yugoslavia, Croatia, Slovenia, Austria, Italy, France, Nepal, Bhutan, India, Russia, Ukraine, Tajikistan, Kazakhstan, South Africa, Canada, the USA, Brazil, and other countries. The statistics of the United Nations Food and Agriculture Organization (FAO) showed that, during the past two decades, the world production of buckwheat decreased gradually towards 2010 before a slow increase from 2011 onwards (Zhu 2016). The largest producer of Buckwheat in 2021–2022 was the Russian Federation, with a production of 919,147 tonnes, and the second largest producer was China, with a production of 502,369 tonnes (FAOSTAT, April 2023; Zargar et al. 2023). The cultivation covered 16 lakh ha worldwide, with the production of approximately two million tons in 2019, which includes 47% Russian Federation, 30% China, 5% accumulated by the USA, Ukraine 4%, Kazakhstan 4%, and Japan 4% of global production (Graziano et al. 2022).

9.3 Consumption Patterns

Buckwheat is a kind of marvelous food to the hill agriculturalists of India; it is a good contender crop for sustainable diversification and intensification programs (Hurt 2022). It is consumed in most northern and western Indian states on fasting days in the form of various dishes. Although buckwheat is supposed to be a poor man's grain, the enormous health preferment potentials appeals to the rich towards the crops (Hema and Jairath 2017). Buckwheat is consumed as a substitute for rice after it is milled and boiled like rice. These grains are principally used for human consumption but also for cattle, pig farms, and poultry farm feeds. The rural population of the higher altitude region of India consumes buckwheat in the form of sprouts and pancakes, mainly in breakfasts (Babu et al. 2018). The prevalence of celiac disease and gluten intolerance cases has increased, thus necessitating a shift towards gluten-free products for susceptible consumers.

9.4 Physical Attributes of Grain

The length (l) of the buckwheat grains lies between 3 and 8 mm, but the most variety of grains are medium-sized, i.e., 3.5 to 6.5 mm. The average width (w) of grains is 2.75 to 4.5 mm for buckwheat commercially (Unal et al. 2017). The mean thickness (t) of grain is 2.5 to 4.0 mm, and the sphericities of commercial buckwheat were found to be 90%, whereas the general expected ratios of the buckwheat grains were 86%, respectively (Gao et al. 2023). The volume of a solo grain (V) ranged from 21 to 43 mm^3, and the density with void spaces (bulk density) of seeds ranged from 599 to 834 kg/m^3 (Neacsu et al. 2022). The parting force for the commercial variety has a mean value of 27.6 N, and the angle of repose (θ) of the buckwheat slant is approximately 22.0° (Aguiar et al. 2022). The peak coefficient of static friction is 0.46 for commercial buckwheat obtained on the elastic surface, followed by aluminum 0.43 for steel surfaces 0.40. The L*, a*, and b* values determined by Jing et al. (2019) ranged from 40 to 60, 10.5 to 11.5, and 12 to 24, respectively, for buckwheat seeds. The buckwheat starch has polymorphism and degree of crystallinity revealed by wide-angle X-ray scattering technique (WAXS) that are A-type and 26.5% to 35%, respectively (Gao et al. 2023).

9.5 Chemical Composition

Buckwheat kernel composition varies to an extent due to different variety and environmental factors in which a variety of compounds have a moisture content of 14%, and on the other hand water activity (a_w) ranged between 0.77 and 0.83%, then proteins 8.5% to16%, starch 62.3% to 63.8% which comprise of amylose and amylopectin, 20–30% and 70–85%, respectively, which depend on the different varieties and cellulose 10.2%, lipids 2–4%, minerals 2.2%, and phenols 1.5% on the dry basis (Gavrić et al. 2018; Garkina et al. 2021). Based on starch digestibility, it is divided into three types they are rapid digestible starch (RDS), slow digestible starch (SDS), and resistant starch (RS) (Christa and Soral-Śmietana 2008). It also constitutes fiber approximately 5–11% of which fractions of soluble, i.e., pectin and gums, and insoluble fiber, i.e., lignin and cellulose, are 3–7% and 2–4%, respectively, but in some cases, total α-galactosides is considered as soluble dietary fiber which increases by approximately 30% (Ruan et al. 2022). Common buckwheat

(*Fagopyrum esculentum*) has more dietary fiber than Tartary buckwheat (*Fagopyrum tataricum*) (Kreft et al. 2020; Sinkovič et al. 2022). The moisture content of buckwheat grain after primary treatment was approximately 8.2–8.6%, making it available for longer use and providing energy 332 kcal (Mishra and Jain 2019). The predominant protein section consists of albumin (40–50%), followed by glutelin (15–25%), and then some fractions are present in minor, like prolamin (2–5%) and globulin (3–8%) (Joshi et al. 2020). Figure 9.1 represents the protein fraction of buckwheat, as shown below.

There are eight fatty acids found in buckwheat, which is approximately 79.3% of the total acids, i.e., palmitic, stearic, oleic, linoleic, linolenic, arachidic, behenic, and lignoceric acid (Luthar et al. 2021). Lipids isolated from buckwheat grains have a greater contribution of germ, which comprises >22%, in which 2.5% is free of dry matter and 1.3% bound, while the hull contains the least amount, 0.8%. The total lipid content in the whole grain of *Fagopyrum esculentum* is 1.5–4.0% (Common), and *Fagopyrum tataricum* is 1.2–4.3% (Tartary), which is due to various subspecies of them (Subedi 2018). Furthermore, oleic acid (42%), linoleic acid (32%), and palmitic acid (16%) are the prevailing contributing components of fatty acids (Gavrić et al. 2018). They are well-endowed in K (potassium), Mg (magnesium), Ca (calcium), and Na (sodium) and are well thought-out in the vital nutritional source

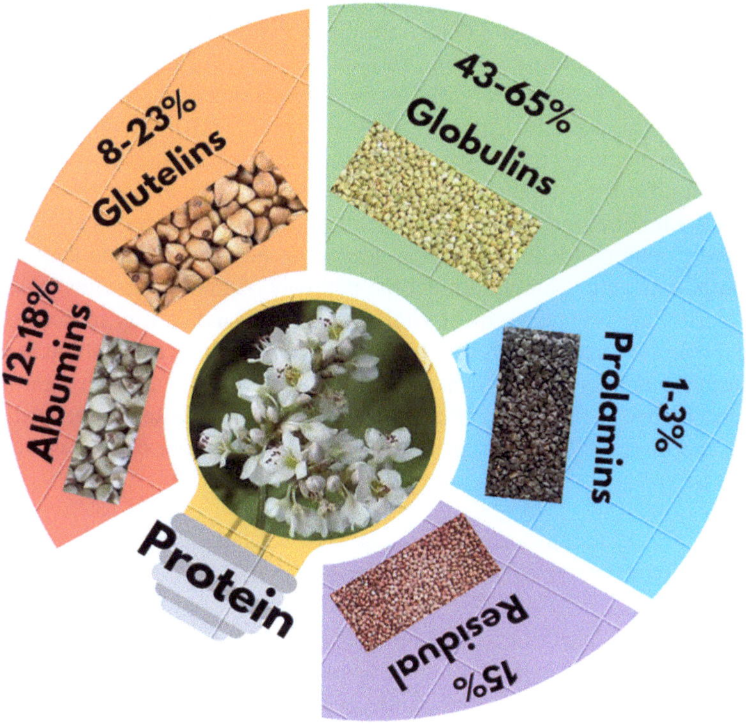

Fig. 9.1 Protein fractions of buckwheat (*Fagopyrum spp.*)

Table 9.1 Mineral and vitamin composition of buckwheat

Minerals	mg/100 g	References
Calcium	110	Podolska et al. (2021), Luthar et al. (2021), Gavrić et al. (2018), Garkina et al. (2021)
Iron	4	
Magnesium	390	
Phosphorous	330	
Potassium	450	
Copper	0.95	
Manganese	3.37	
Zinc	0.87	
Vitamins	*mg/1000 g*	*References*
Thiamine	3.5	Zhang et al. (2017), Gavrić et al. (2018), Garkina et al. (2021)
Riboflavin	1.5	
Pantothenic acid	11	
Choline	440	
Niacin	18	
Pyridoxine	1.5	
Tocopherols	40	

of microelements such as Fe (iron), Mn (manganese), and Zn (zinc) (Podolska et al. 2021). Rubidium and silver are common in buckwheat than in Tartary buckwheat although Se (Selenium), Zn (Zinc), Fe (Iron), Co (Cobalt), Ni (Nickel), Sb (Antimony), Cr (Chromium), and Sn (Tin) are the mineral elements which are more developed in Tartary buckwheat (Golob et al. 2016; Luthar et al. 2021). Furthermore, these seeds contain treasured vitamins, such as thiamine (B_1), riboflavin (B_2), niacin (B_3), pantothenic acid (B_5), and pyridoxine (B_6), as well as a few antioxidant vitamins like Vitamin E (tocopherol) and β-carotene (Garkina et al. 2021). Table 9.1 represents the mineral and vitamin composition of buckwheat, as shown below.

The flavor compounds in buckwheat, known as hexane aldehyde and other carbonyl compounds, are released amidst the pathway of lipoxidase enzyme. Sucrose and fagopyritol are two of the most vital suppliers of saccharinity in flour of buckwheat and the bitter taste-generating complexes such as quercetin, which is induced by rutinosidase activity (Luthar et al. 2020; Chiang et al. 2023). Iminosugar in buckwheat seed named d-fagomine, and they also found its diastereomers called epi-fagomine and di-epifagomine in groats, bran, leaves, and flour of buckwheat (Huda et al. 2021).

9.6 Antinutrients and Toxic Compounds

Buckwheat has few anti-nutritional components, like trypsin inhibitors of three varieties found in seeds, which inhibit trypsin and chymotrypsin (Tömösközi and Langó 2017). Phytic acid is extant in the peptide section of the embryo, and the aleurone layer of buckwheat seeds and its bran is a rich source of tannins, measured in significant content (Zhang et al. 2012). The dietary fiber in the hull acts as an anti-nutritional factor; however, it is mainly removed during the dehulling process. Polyphenols have a repercussion on protein digestibility after searing water treatment (Luthar et al. 2021). Phenolic substances and other anti-nutritional factors reduce the assimilation of proteins with in the small and large intestines. The quercetin compound in buckwheat grain may be responsible for α-glucosidase inhibitory activity (Kreft et al. 2020). Fagopyritols present in aleurone layer whereas, low concentrations was present in embryo cells of the seed. Therefore, it is difficult to extract from the embryo cell whereas negligible presence in the hull and endosperm of the buckwheat. However, its peripherals contain the light allergic compounds that may cause fagopyrism, the skin irritation upon sunlight exposure (Benković et al. 2014). Tartary buckwheat aroma eloquently diverges from the aroma of other buckwheat varieties due to the presence of salicylaldehyde, an explosive compound, which could be used as an indication for recognition of adulteration of common buckwheat with Tartary, which contains naphthalene (Janeš et al. 2010).

9.7 Bioactive and Functional Compounds

Figure 9.2 represents the presence of natural bioactive compounds in buckwheat, which show several therapeutic values. Buckwheat grains and hulls have a number of components, such as 178 bioactive compounds like bioflavonoids and flavones, polyphenols, D-fagomine, reduced tannins, phytosterols, and fagopyrins (Tömösközi and Langó 2017; Zou et al. 2023). The essential secondary metabolites produced by the biosynthetic pathway, called flavonoids, comprise six compounds such as

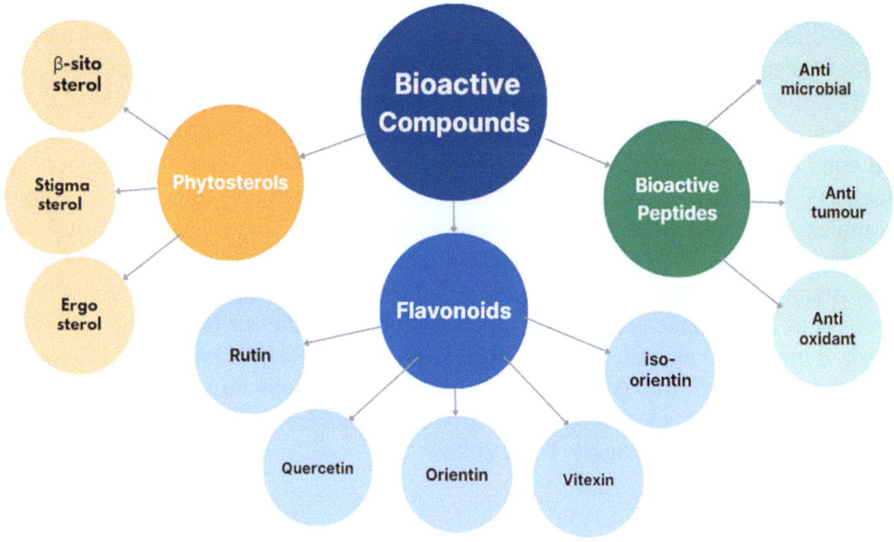

Fig. 9.2 Major bioactive compounds of buckwheat

rutinoside, quercetin, orientin, vitexin, iso-vitexin, and iso-orientin. Rutinoside is the only occurring flavonoid in pseudocereal buckwheat; henceforth, it is a substantial foundation of this flavonoid, ranging from 4% to 6%, preventing flour deterioration (Kwon et al. 2018; Yu et al. 2019; Klykov et al. 2020). The flavonoid biosynthesis pathway encoding genes for enzyme production in buckwheat, i.e., FePAL, FeC_4H, Fe_4CL, FeCHS, FeCHI, FeF_3H, FeFLS, and FeANS are sequenced (Li et al. 2010; Bobkov 2016). The anthocyanin of four types cyanidin 3-O-glucoside, 3-O-rutinoside, 3-O-galactoside, and 3-O-galactopyranosyl-rhamnoside secluded from the new growth of buckwheat sprouts by biosynthetic pathway derived from anthocyanidins by adding sugars is deliberated by extrinsic factors like radiation and temperature and intrinsic factors like hormones in the plant, 2° bio-transforming metabolites, and nutrients. The tissues of higher plants, including leaves, stems, roots, flowers, and fruits, have anthocyanins and clear, white-to-yellow counterparts occurring in plants called anthoxanthins (Kwon et al. 2018).

The carbolic compounds, D-chiro-inositol, plant sterols, stilbenes, resistant polysaccharides, and resistant amino acid, the thiamine-binding proteins are present in grains which are concentrated in seed coat and hull, i.e., the outer layer of seeds (Bobkov 2016; Wang et al. 2015; Noda et al. 2023). Tartary buckwheat leaves and grains have p-anisic acid and emodin, respectively, which lead to the anchorage of all three active sites of the ribonucleic acid-binding domain of the nucleus capsid phosphoprotein of covid epidemic (Raguindin et al. 2021). The class of phenolic compounds found in this pseudocereal are hydroxybenzoic, syringic, vanillic, gallic, caffeic acid derivatives of protocatechuic acid, ferulic, para-coumaric and coumaric acids (Kreft et al. 2020; Germ and Gaberščik 2016). *Fagopyrum tataricum* seeds have five plant sterols (esters), including stigmast-endione, palmitate sterol, ergosterol peroxide, stigmasterol, and β-sitosterol (Ruan et al. 2022). Some bioactive peptides naturally occurring in seeds of buckwheat are manifold functional compounds, such as anti-microbial peptides, trypsin inhibitors, anti-neoplastic proteins, hypotensive peptides, and antioxidant peptides (Zhou et al. 2015).

The functional compounds of these grains are γ-aminobutyric acid (GABA) and 2-OH-nicotinamine (Gimenez-Bastida and Zielinski 2015). The thiamine-binding proteins of buckwheat seeds can abide by low absorptions of denaturing agents, such as 2 M urea, which helps steady vitamin B_1 by forming an amino acid–vitamin complex. The food can be digested by a protein-digesting enzyme (protease), releasing the vitamin B_1. Therefore, it could be used to treat people who are deficient in this vitamin or not able to store it. Although Tartary buckwheat differs from those of rice grains and sesame seeds in subtype structure and immune features, the mechanism of binding thiamine is parallel (Ruan et al. 2022). Despite knowing the synthesis of flavonoids in this pseudocereal expansively, various biologically or natural synthetic inheritable factors and transcription features responsible for its synthesis trail have yet to be discovered (Joshi 2023).

9.8 Health Implications

Fagopyritols present in seed play a significant role in the treatment of type 2 diabetes mellitus (non-insulin-dependent) (NIDDM) and polycystic ovary syndrome (PCOS), both insulin response disorders; it also shows anti-tissue growth activity and may possess other beneficial bioactivities (Wang et al. 2015). The pervasiveness of diabetes mellitus and blood glucose level was found to be lower in people who intake buckwheat than in those who do not usually consume it as groats of buckwheat have shown a decreased glycemic index of approximately 34.7 specifies that it helps to prevent unnatural increment in blood sugar levels. The resistant polysaccharide can reduce the peril of colorectal cancer and depress blood plasma total cholesterol and triacylglycerol levels (Zhou et al. 2019). Lysin helps in lowering down the level of cholesterol, upsurges the fecal excretion of steroids, removal of bile acid, hinder gallstone formation, may sluggish mammary carcinogenesis, and overthrow colon carcinogenesis (Ge and Wang 2020; Ji et al. 2019; Li et al. 2019).

Buckwheat protein content is responsible for the increases in bile juice and gallstone decreases. Whereas the fiber content helps in curing constipation, overweightness, and malignance and used in treating type 2 diabetes (Wang et al. 2015). Buckwheat shows various health aids such as antioxidant, anti-carcinogenic, anti-hypertension, anti-diabetic, anti-hypolipidemic, and cognition-improving (Tömösközi and Langó 2017). Common and Tartary buckwheat tend to contain developed amounts of convinced bioactive components such as rutinoside, therefore having greater efficiency in avoiding various ailments (Zhu 2016). The peptides in kernels (buckwheat) and their hydrolysates/concentrates possess biotic activities, such as anti-bacterial, anti-neoplastic, hindering ACE (angiotensin-converting enzyme) and antioxidant activity (Zhang et al. 2017). It has a well-adjusted amino-acid composition, and functional properties include high water absorption capacity (WAC), enhanced emulsification, foaming, and other textural properties like chewiness, as well as inhibition of body fat accumulation (Yang et al. 2019). The quercetin in Tartary buckwheat can decrease physique heft, serum triacylglycerol, and LDL (low-density lipoprotein) and lower blood serum triglycerides and total cholesterol, as well as reduce blood glucose levels (Luthar et al. 2021).

Plant sterols in buckwheat seeds with lesser amounts lower blood cholesterol levels and play various nutritional and pharmaceutical effects in many chronic diseases (Wang et al. 2015). *Fagopyrum* spp. may have some valuable effects on digestion and bowel movement as the soluble and insoluble fiber in its flour has a binding capacity to materials such as peptides, which certainly protects humans from type 2 diabetes, hyperglycemia, hypercholesterolemia, cardiovascular diseases, obesity, and other chronic diseases especially colon cancer. The enhanced food produced from the structural parts of buckwheat, rich in dietary fiber and antioxidant substances, can be effectively used against atherosclerosis (Dziedzic et al. 2018). The hydrogen-bestowing assets of bioactive compounds significantly hydroxyl radical groups help reduce reactive oxygen species, providing effective health benefits like anti-aging, anti-cancerous, and protecting heart and cerebrovascular illnesses (Zhang et al. 2017).

9.9 Processing

Grain processing, responsible for the milling process to convert the grains into flour, essentially involves a physical process, i.e., cleaning of the kernel, with the appropriate value of free water content and then mechanical size reduction to the required constituent part size to produce flour and additional by-products, which require fractionation for the flour production, it not only to separate bran, germ, and endosperm from each other but also to pledge the accurate particle size of the endosperm obtained after milling. It has been categorized based on equipment and methods used for processing as the traditional and modern methods (Thielecke et al. 2021).

The outmoded or conventional method involves cleaning and kiln drying, after which the buckwheat is separated into numerous fractions before it is shelled. Further parting of the grain kernels, usually in several fractions, the detached husks and broken grains, takes place (Serna Saldívar and Hernández 2020). The wet husking process is performed without classifying into several fractions, which lessens the machinery requirements. The first flour section, which is abundant in starch, has a mild flavor and indistinct sweetness in taste; the second flour segment has a characteristic flavor and gives appropriate chewiness to the ensuing products; and the third flour fraction is abundant in characteristic flavor, but the consequential products do not have good masticating properties. Furthermore, a suggestively higher kernel yield can be reached with the modern developed process in contrast with the traditional process (Sasthri et al. 2020).

The modern method of milling is categorized as dry milling and wet milling, as the intent of dry milling is to attain a large quantity of flour. However, in wet milling, the unadulterated possible fraction of each component could be obtained (Haros and Wronkowska 2017). Thus, after harvesting, buckwheat grain leads to moisture reduction with the help of drying and then kept to about 15% moisture content (M.C.). Further, dried grains are stored at about 7 °C under 0.6–0.7 relative humidity (RH) surroundings (Ikeda 2002). The primary processing includes cleaning, dehulling, and milling, in that cleaning is the most essential or productive milling process as grain contains impurities and foreign materials. The cleaning section provides the removal of contaminants and impurities from the buckwheat, which must be performed before storage to increase the efficiency and functional value of the machinery and to create hygienic storage conditions (Skřivan et al. 2023). Next, dehulling is to detach the groats from the hulls by impact or abrasion of seed against emery stones (millstone), followed by circular screen separation of hulls with groats (Patel et al. 2019).

The milling separations comprised 23 fractions based on sizes like seven fractions of refined flour having less than 129 μm mesh size, three fractions of coarse flour having less than 183 μm mesh size, four fractions of semolina in different mesh sizes having less than 219–656 μm, and two fractions of bran having mesh size less than 505–656 μm, and remaining husk. A well-developed machinery, i.e., hullers, is required, especially for buckwheat. These are high-capacity and energy-efficient machines designed to keep broken to a minimum, giving a low broken rate value (Kaushal and Kumar 2022).

Sorting and grading of grains are performed with optical sorters, which remove subtle light and dark color defects, spots, damaged grains, and foreign materials but can even sort the grains based on shape (Haros and Wronkowska 2017). The specific featured machine is designed for grading buckwheat to provide a uniform dispersal of milled products and other by-products. The product of primary processing is not palatable directly; thus, preparation of the basic elements for secondary processing involves several operations where they are turned into a usable form, ultimately reaching the final step of processing, the technique for development designs and fabrication of different product requirement and structures involve these operations are such as administrating, transportation, dealing out and storage then, to analyze product quality entails acquaintance of its physicochemical, techno-functional, and nutritional properties (Unal et al. 2017). Secondary processing techniques for making it more digestible and palatable include bioprocessing like germination and fermentation, physical processing, and enzymatic hydrolysis or enzyme breakdown or other methods like baking, cooking, flaking, frying, puffing, popping, extrusion, fermentation, roasting, malting, and so on used for buckwheat-based product formulation attract consumers with highly bio-accessible bioactive components (Jin et al. 2021; Sofi et al. 2023).

9.10 Value-Addition

The pseudocereals have been significantly utilized in various value-added products and enhanced their utilization (Schoenlechner 2017). In processed forms, it has an inimitable flavor well known, not only in flour but also in seeds. Researchers observed a huge diversity of fagopyrum flour dishes in the northern region, especially in China (Ikeda 2002). The value-added food products are meant to be any secondary processing applied to any milled product that has been imperiled to further efforts or incorporated with additional products to increase the total nutritional value and bioavailability of the product (Sharma and Minhas 2023). The value-addition in the product of Fagopyrum will be enhanced by the incorporation of other compounds; in case of buckwheat, there are numerous extruded products such as bread, biscuits, muffins, polenta, beer, sprouts, noodles, groats, pancake mix, confectioneries, cookies, soups, spirits, porridge, and other confectionery products (Zhu 2016; Huda et al. 2021). Soba shochu (beverage) is a famous alcoholic drink made by the anaerobic fermentation of buckwheat seed grain with *Aspergillus oryzae* (Suzuki et al. 2020a, b). The products are gluten-free, rich sources of antioxidants, flavonoids, proteins, and insoluble dietary fiber as well as its starch has been used as a fat replacer, ingredient for extruded products, nanocomposite material, and fermentation substrate for alcoholic beverages. It may be concluded that buckwheat starch can be a unique source of specialty starch for innovative food and non-food applications (Zhu et al. 2015).

9.11 Safety Considerations

Allergic reactions may be caused by ingesting buckwheat and its products or by exposure to its dust, as it contains proteins that cause an allergy known as a hypersensitive reaction, which is not common (Lemière and Vandenplas 2017). Buckwheat proteins recognized as a normal allergen for binding to IgE antibodies such as 6–9 and 24 kDa. Especially, 16-kDa buckwheat protein is resistant to pepsin digestion, leading to be responsible for immediate hypersensitivity reaction causing anaphylactic shock, the serious symptom of immunoglobulin (IgE)-mediated hypersensitive response (Satoh et al. 2020). However, it is more potentially allergic to children. Fagopyrum grain grinding and further processing lead to allergic reactions like itching nose, sneezing, and even more serious symptoms like asthma to the industry workers (Md Zamri et al. 2023). A buckwheat hull presence in the cooked bread leads to severe anaphylaxis suffered by many allergic persons (Kowalski et al. 2022). However, the RDA value or recommended intake of buckwheat sprouts is less than 40 g/day as the consumption of buckwheat food, in their worst form, causes hemorrhagic disease, connected with a rapid decrease of blood pressure, i.e., hypertension shock through allergic reactions (Verma et al. 2020).

9.12 Future Challenges and Opportunities

Every food industry must have various challenges, especially during product development in the introduction period. However, using buckwheat could be more efficient for its nutritional quality in humans; thus, we must find a place for buckwheat-based food products in both traditional and modern food markets. Nowadays, the custom and its status are not enough to increase the utilization of this pseudocereal. To date, there are several unrequited interrogations concerning the nutritional and purposeful characteristics of buckwheat; therefore, we must obtain added information about its chemical composition, particularly about the health-implicating compounds, such as phytochemicals, hereditary, and environmental inconsistency. Due to the limited knowledge about the industrial conduct of different buckwheat varieties and the associations among chemical composition, such as protein subunit form, starch, and other polysaccharide present with their characteristics. It is accepted that the actual quality value of buckwheat is still unknown worldwide. Therefore, with the unceasing progress of knowledge, there is a chance of developing more erudite milling and processing machinery, which helps in health-promoting nutritive qualities and phytochemicals that could produce more broadly recognized and palatable buckwheat products for consumers. Thus, additional research is needed to inspect the venomousness and detrimental effects of different fragments of buckwheat, and further study for experimental clinic trials to examine the therapeutical benefits and pharmacological effects of bioactive compounds. Therefore, isolating the unexplored genes and transcription factors involved in the biosynthesis pathway and illustrating the regulation systems of flavonoid biosynthesis in buckwheat needed more focus and research.

9.13 Conclusion

Buckwheat is a valued source of carbohydrates, protein, unsaturated fatty acid, soluble and insoluble fiber, vitamins, ash, inositol, fagopyritols, flavonoids, β-sitosterols, and other bioactive compounds. Milling fractions of processed pseudocereal like buckwheat gives industries an opportunity for developing a new variety of functional or nutraceutical food with a particularly significant component that could be used to procure desired products of high nutritional value. The bioactive or functional compounds found as phenolic compounds, D–chiro–inositol, phytosterol, stilbenes, resistant starch, resistant protein, and the thiamine-binding proteins are present in buckwheat. Phytate is present in the protein section of the embryo and aleurone layer of buckwheat seeds, and its bran is a rich source of tannins, measured in a significant content which added anti-nutritional value. Flavonoid biosynthesis pathway with several genes encoded used in producing improved strains of *Fagopyrum* varieties with several health implications. Sucrose and fagopyritol are two of its most essential components to provide sweetness in buckwheat flour and the bitter taste-generating complexes such as quercetin, produced by rutinosidase activity. The biological value of buckwheat proteins is more due to holding a great profile of the amino acid, and several studies revealed that it is richer in histidine, lysine, and tryptophan than wheat, corn, barley, and egg. The value-addition in the product of Fagopyrum will be enhanced by the incorporation of another compound; in the case of buckwheat, there are numerous extruded products such as bread, biscuits, muffins, polenta, beer, sprouts, noodles, groats, pancake mix, confectioneries, cookies, soups, spirits, porridge, and other confectionery products. However, the major consideration is to protect the buckwheat crop from the deterioration is conservation, cultivation, and enhancement.

References

Aguiar EV, Santos FG, Centeno ACLS, Capriles VD (2022) Defining amaranth, buckwheat and quinoa flour levels in gluten-free bread: a simultaneous improvement on physical properties, acceptability and nutrient composition through mixture design. Foods 11(6):848

Babu S, Yadav GS, Singh R, Avasthe RK, Das A, Mohapatra KP, Tahashildar M, Kumar K, Prabha M, Devi MT, Rana DS (2018) Production technology and multifarious uses of buckwheat (Fagopyrum spp.): a review. Indian J Agron 63(4):415–427

Benković ET, Žigon D, Friedrich M, Plavec J, Kreft S (2014) Isolation, analysis and structures of phototoxic fagopyrins from buckwheat. Food Chem 43:432–439

Bobkov S (2016) Biochemical and technological properties of buckwheat grains. In: Molecular breeding and nutritional aspects of buckwheat. Academic Press, London, pp 423–440

Chiang JH, Hua XY, Yu AHM, Peh EWY, See EE, Jeyakumar Henry C (2023) A review on buckwheat and its hypoglycemic bioactive components in food systems. Food Rev Int 39(9):6362–6386

Christa K, Soral-Śmietana M (2008) Buckwheat grains and buckwheat products–nutritional and prophylactic value of their components–a review. Czech J Food Sci 26(3):153–162

Dziedzic K, Górecka D, Szwengiel A, Sulewska H, Kreft I, Gujska E, Walkowiak J (2018) The content of dietary fibre and polyphenols in morphological parts of buckwheat (Fagopyrum tataricum). Plant Foods Hum Nutr 73:82–88

FAOSTAT (2023) World Food and Agriculture – Statistical Yearbook. Rome 231–236. https://doi.org/10.4060/cc8166en
Farzana T, Hossain FB, Abedin MJ, Afrin S, Rahman SS (2022) Nutritional and sensory attributes of biscuits enriched with buckwheat. J Agric Food Res 10:100394
Gao L, Wang H, Wan C, Wang P, Eeckhout M, Gao J (2023) Suitable nitrogen fertilizer application drives the endosperm development and starch synthesis to improve the physicochemical properties of common buckwheat grain. Int J Biol Macromol 235:123837
Garkina PK, Kurochkin AA, Frolov DI, Shaburova GV (2021) Chemical composition and physicochemical properties of extruded buckwheat. IOP Conf Ser: Earth Environ Sci 640(2):022–037
Gavrić T, Čadro S, Gadžo D, Đikić M, Bezdrob M, Jovović Z et al (2018) Influence of meteorological parameters on the yield and chemical composition of common buckwheat (Fagopyrum esculentum Moench). Agric Forestry/Poljoprivreda i šumarstv 64(4)
Ge RH, Wang H (2020) Nutrient components and bioactive compounds in tartary buckwheat bran and flour as affected by thermal processing. Int J Food Prop 23(1):127–137
Germ M, Gaberščik A (2016) The effect of environmental factors on buckwheat. In: Molecular breeding and nutritional aspects of buckwheat. Academic Press, London, pp 273–281
Gimenez-Bastida JA, Zielinski H (2015) Buckwheat as a functional food and its effects on health. J Agric Food Chem 63(36):7896–7913
Golob A, Gadžo D, Stibilj V, Djikić M, Gavrić T, Kreft I, Germ M (2016) Sulphur interferes with selenium accumulation in Tartary buckwheat plants. Plant Physiol Biochem 108:32–36
Graziano S, Agrimonti C, Marmiroli N, Gullì M (2022) Utilisation and limitations of pseudocereals (quinoa, amaranth, and buckwheat) in food production: a review. Trends Food Sci Technol 125:154–165
Haros CM, Wronkowska M (2017) Pseudocereal dry and wet milling: processes, products and applications. In: Pseudocereals: chemistry and technology. Wiley, Oxford, pp 140–162
Hema Y, Jairath M (2017) Growth and potential of buckwheat in mountain regions. CABI 8(2):113–117
Huda MN, Lu S, Jahan T, Ding M, Jha R, Zhang K, Zhang W, Georgiev MI, Park SU, Zhou M (2021) Treasure from garden: bioactive compounds of buckwheat. Food Chem 335:127653. https://doi.org/10.1016/j.foodchem.2020.127653
Hunt HV, Shang X, Jones MK (2018) Buckwheat: a crop from outside the major Chinese domestication centres? A review of the archaeobotanical, palynological and genetic evidence. Veg Hist Archaeobot 27:493–506
Hurt RD (ed) (2022) A companion to American agricultural history. Wiley, Hoboken
Ikeda K (2002) Buckwheat composition, chemistry, and processing. Adv Food Nutr Res 44:395–434
Janeš D, Prosen H, Kreft I, Kreft S (2010) Aroma compounds in buckwheat (Fagopyrum esculentum Moench) groats, flour, bran, and husk. Cereal Chem 87(2):141–143
Ji X, Han L, Liu F, Yin S, Peng Q, Wang M (2019) A mini-review of isolation, chemical properties and bioactivities of polysaccharides from buckwheat (Fagopyrum Mill). Int J Biol Macromol 127:204–209
Jin J, Okagu OD, Yagoub AE, Udenigwe CC (2021) Effects of sonication on the in vitro digestibility and structural properties of buckwheat protein isolates. Ultrason Sonochem 70:105348
Jing Y, Li X, Hu X, Ma Z, Liu L, Ma X (2019) Effect of buckwheat extracts on acrylamide formation and the quality of bread. J Sci Food Agric 99(14):6482–6489
Joshi BK (2023) Buckwheat (Fagopyrum esculentum Moench and F. tataricum Gaertn.). In: Neglected and underutilized crops. Academic Press, London, pp 151–200
Joshi DC, Zhang K, Wang C, Chandora R, Khurshid M, Li J, He M, Georgiev MI, Zhou M (2020) Strategic enhancement of genetic gain for nutraceutical development in buckwheat: a genomics-driven perspective. Biotechnol Adv 39:107479
Kasar C, Thanushree MP, Gupta S, Inamdar AA (2021) Milled fractions of common buckwheat (Fagopyrum esculentum) from the Himalayan regions: grain characteristics, functional properties and nutrient composition. J Food Sci Technol 58:3871–3881

Kaushal P, Kumar N (2022) Processing of cereals. In: Agro-processing and food engineering: operational and application aspects. Springer, Singapore, pp 415–454
Kim J, Hwang KT (2020) Fagopyrins in different parts of common buckwheat (Fagopyrum esculentum) and Tartary buckwheat (F. tataricum) during growth. J Food Compos Anal 86:103354
Klykov A, Chaikina E, Anisimov M, Borovaya S, Barsukova E (2020) Rutin content in buckwheat (Fagopyrum esculentum Moench, F. tataricum (L.) Gaertn. and F. cymosum Meissn.) growth in the Far East of Russia/Vsebnost rutina v ajdi (Fagopyrum esculentum Moench, F. tataricum (L.) Gaertn. in F. cymosum Meissn.). Folia Biol Geol 61(1):61–68
Kowalski S, Mikulec A, Mickowska B, Buksa K (2022) Nutritional properties and amino acid profile of buckwheat bread. J Food Sci Technol 59(8):3020–3030
Kreft I, Zhou M, Golob A, Germ M, Likar M, Dziedzic K, Luthar Z (2020) Breeding buckwheat for nutritional quality. Breed Sci 1:67–73
Kwon SJ, Roy SK, Choi J, Park J, Cho S, Sarker K, Woo SH (2018) Recent research updates on functional components in buckwheat. J Agric Sci 34:1–8
Lemière C, Vandenplas O (2017) Occupational allergy. In: Middleton's allergy essentials. Elsevier, London, pp 361–375
Li X, Park NI, Xu H, Woo SH, Park CH, Park SU (2010) Differential expression of flavonoid biosynthesis genes and accumulation of phenolic compounds in common buckwheat (Fagopyrum esculentum). J Agric Food Chem 58(23):12176–12181
Li J, Yang P, Yang Q, Gong X, Ma H, Dang K et al (2019) Analysis of flavonoid metabolites in buckwheat leaves using UPLC-ESI-MS/MS. Molecules 24(7):1310
Luthar Z, Germ M, Likar M, Golob A, Vogel-Mikuš K, Pongrac P et al (2020) Breeding buckwheat for increased levels of rutin, quercetin and other bioactive compounds with potential antiviral effects. Plan Theory 9(12):1638
Luthar Z, Golob A, Germ M, Vombergar B, Kreft I (2021) Tartary buckwheat in human nutrition. Plan Theory 10(4):700
Md Zamri ASS, Saruddin MZ, Harun A, Abd Aziz SF, Aizad Za'bah AK, Dapari R et al (2023) Factors associated with occupational asthma among food industry workers: a systematic review. PLoS One 18(6):e0287040
Mishra M, Jain S (2019) A Comparative study on nutritional profile and antinutrients of buckwheat fractions (Fagopyrum esculentum). Int J Curr Microbiol App Sci 8(3):3384–3393
Neacsu M, De Lima Sampaio S, Hayes HE, Duncan GJ, Vaughan NJ, Russell WR, Raikos V (2022) Nutritional content, phytochemical profiling, and physical properties of buckwheat (Fagopyrum esculentum) seeds for promotion of dietary and food ingredient biodiversity. Crops 2(3):287–305
Noda T, Ishiguro K, Suzuki T, Morishita T (2023) Tartary buckwheat bran: a review of its chemical composition, processing methods and food uses. Plan Theory 12(10):1965
Ohnishi O (2004) On the origin of cultivated buckwheat. Adv Buckwheat Res 18:16–21
Patel A, Devraja HC, Sharma P, Singh RRB (2019) Food technology II 1:5–65
Podolska G, Gujska E, Klepacka J, Aleksandrowicz E (2021) Bioactive compounds in different buckwheat species. Plants 10(5):961
Radics L, Mikóházi D (2010) Principles of common buckwheat production. Eur J Plant Sci Biotechnol 4(1):57–63
Raguindin PF, Itodo OA, Stoyanov J, Dejanovic GM, Gamba M, Asllanaj E, Minder B, Bussler W, Metzger B, Muka T, Glisic M (2021) A systematic review of phytochemicals in oat and buckwheat. Food Chem 338:127982
Rehman U, Bahar F, Saad A, Bhat M, Singh L, Mahdi S et al (2022) Trends, challenges, and prospects of underutilized food crops under temperate ecologies: a review. Pharma Innov J 11(2):503–512
Ruan J, Zhou Y, Yan J, Zhou M, Woo SH, Weng W et al (2022) Tartary buckwheat: an under-utilized edible and medicinal herb for food and nutritional security. Food Rev Intl 38(4):440–454
Sasthri VM, Krishnakumar N, Prabhasankar P (2020) Advances in conventional cereal and pseudo-cereal processing. In: Innovative processing technologies for healthy grains. Wiley, Hoboken, pp 61–81

Satoh R, Jensen-Jarolim E, Teshima R (2020) Understanding buckwheat allergies for the management of allergic reactions in humans and animals. Breed Sci 70(1):85–92
Schoenlechner R (2017) Pseudocereals in gluten-free products. In: Haros CM, Schonlechner R (eds) Pseudocereals: chemistry and technology. Wiley, Hoboken, pp 193–216
Serna Saldívar SO, Hernández DS (2020) Dietary fiber in cereals, legumes, pseudocereals and other seeds. In: Welti-Chanes J, Serna-Saldívar S, Campanella O, Tejada-Ortigoza V (eds) Science and technology of fibers in food systems. Springer, Cham, pp 87–122
Sharma N, Minhas R (2023) Value-added food products from buckwheat (Fagopyrum) species: a functional food. Pharma Innov 12(5):2109–2113
Sinkovič L, Deželak M, Meglič V (2022) Macro/microelements, nutrients, and bioactive components in common and Tartary buckwheat (Fagopyrum spp.) grain and stone-milling fractions. LWT 161:113422
Skřivan P, Chrpová D, Klitschová B, Švec I, Sluková M (2023) Buckwheat flour (Fagopyrum esculentum Moench)—a contemporary view on the problems of its production for human nutrition. Foods 12(16):3055
Sofi SA, Ahmed N, Farooq A, Rafiq S, Zargar SM, Kamran F et al (2023) Nutritional and bioactive characteristics of buckwheat, and its potential for developing gluten-free products: an updated overview. Food Sci Nutr 11(5):2256–2276
Subedi N (2018) Changes in phytochemical properties of buckwheat varieties on malting (Doctoral dissertation), 11–27. http://202.45.146.37:8080/jspui/handle/123456789/111
Suzuki T, Hara T, Hara T, Katsu K (2020a) Effect of storage temperature on occurrence of secondary dormancy in buckwheat seeds. Seed Sci Technol 48(2):257–267
Suzuki T, Noda T, Morishita T, Ishiguro K, Otsuka S, Brunori A (2020b) Present status and future perspectives of breeding for buckwheat quality. Breed Sci 70(1):48–66
Tömösközi S, Langó B (2017) Buckwheat: its unique nutritional and health-promoting attributes. In: Gluten-free ancient grains. Woodhead Publishing, pp 161–177
Thielecke F, Lecerf JM, Nugent AP (2021) Processing in the food chain: do cereals have to be processed to add value to the human diet? Nutr Res Rev 34(2):159–173
Unal H, Izli G, Izli N, Asik BB (2017) Comparison of some physical and chemical characteristics of buckwheat (Fagopyrum esculentum Moench) grains. CyTA-J Food 15(2):257–265
Verma KC, Rana AS, Joshi N, Bhatt D (2020) Review on common buckwheat (Fagopyrum esculentum Moench): a potent Himalayan crop. Ann Phytomed 9:125–133
Wang C, Hu D, Liu X, She H, Ruan R, Yang H, Yi Z, Wu D (2015) Effects of uniconazole on the lignin metabolism and lodging resistance of culm in common buckwheat (Fagopyrum esculentum M.). Field Crop Res 180:46–53
Woo SH, Roy SK, Kwon SJ, Cho SW, Sarker K, Lee MS, Chung KY, Kim HH (2016) Concepts, prospects, and potentiality in buckwheat (Fagopyrum esculentum Moench): a research perspective. In: Molecular breeding and nutritional aspects of buckwheat, vol 1, pp 21–49
Yang J, Gu Z, Zhu L, Cheng L, Li Z, Li C, Hong Y (2019) Buckwheat digestibility affected by the chemical and structural features of its main components. Food Hydrocoll 96:596–603
Yu JH, Kwon SJ, Choi JY, Ju YH, Roy SK, Lee DG et al (2019) Variation of rutin and quercetin contents in Tartary buckwheat germplasm. Fagopyrum 36(2):51–65
Zargar SM, Hami A, Manzoor M, Mir RA, Mahajan R, Bhat KA et al (2023) Buckwheat OMICS: present status and future prospects. Crit Rev Biotechnol:1–18. https://doi.org/10.1080/07388551.2023.2229511
Zhang Q, Xu JG (2017) Determining the geographical origin of common buckwheat from China by multivariate analysis based on mineral elements, amino acids and vitamins. Sci Rep 7(1):9696
Zhang ZL, Zhou ML, Tang Y, Li FL, Tang YX, Shao JR, Xue WT, Wu YM (2012) Bioactive compounds in functional buckwheat food. Food Res Int 49(1):389–395
Zhang L, Li X, Ma B, Gao Q, Du H, Han Y, Li Y, Cao Y, Qi M, Zhu Y, Lu H (2017) The tartary buckwheat genome provides insights into rutin biosynthesis and abiotic stress tolerance. Mol Plant 10(9):1224–1237
Zhou X, Wen L, Li Z, Zhou Y, Chen Y, Lu Y (2015) Advance on the benefits of bioactive peptides from buckwheat. Phytochem Rev 14:381–388

Zhou M, Kreft I, Suvorova G, Tang Y, Woo SH (eds) (2018) Buckwheat germplasm in the world. Academic Press, London

Zhou Y, Zhao S, Jiang Y, Wei Y, Zhou X (2019) Regulatory function of buckwheat-resistant starch supplementation on lipid profile and gut microbiota in mice fed with a high-fat diet. J Food Sci 84(9):2674–2681

Zhu H, Wang H, Zhu Y, Zou J, Zhao FJ, Huang CF (2015) Genome-wide transcriptomic and phylogenetic analyses reveal distinct aluminum-tolerance mechanisms in the aluminum-accumulating species buckwheat (Fagopyrum tataricum). BMC Plant Biol 15:1–3

Zhu F (2016) Buckwheat starch: structures, properties, and applications. Trends Food Sci Technol 49:121–135

Zou L, Wu D, Ren G, Hu Y, Peng L, Zhao J et al (2023) Bioactive compounds, health benefits, and industrial applications of Tartary buckwheat (Fagopyrum tataricum). Crit Rev Food Sci Nutr 63(5):657–673

Quinoa

10

Amrita Poonia, Ananya Bhardwaj, and Nadimintri Chandana Sri Sai

Abstract

Quinoa, also known as Chenopodium quinoa, is a pseudocereal or pseudograin that has its origins about 7000 years ago in the Andes. It can withstand extremely harsh climatic conditions rather well. It does well in conditions of salinized soil and dryness. Because of its exceptional biological and nutritional qualities, quinoa has been dubbed "the grain of the 21st century". Quinoa is known as "The Golden Grain" because of its outstanding balance of essential amino acids and plenty of macro- and micronutrients such as lipids, fibre, vitamins, and minerals. The high quality of its protein has led to its recognition as a complete food. In addition, it has been regarded as an oil crop due to its noteworthy vitamin E content and intriguing omega-6 ratio. Numerous phytochemicals are also present in it, such as phenolics, phytosterols, phytoecdysteroids, saponins, and bioactive peptides. These substances might have positive impacts on digestive, cardiovascular, and metabolic health. Apart from its health benefits, it has several functional properties like solubility, water-holding capacity (WHC), gelation, emulsifying, and foaming that allow it a range of uses in food. Quinoa starch has several physicochemical properties (such viscosity and freeze stability) that give starch new uses and functional properties. This cultivar is a promising substitute for cereals.

Keywords

Quinoa · Pseudocereals · Phytochemicals · Saponins · Gluten-free · Betalains · Mother grain

A. Poonia (✉) · A. Bhardwaj · N. C. S. Sai
Department of Dairy Science and Food Technology, Institute of Agricultural Sciences, Banaras Hindu University, Varanasi, Uttar Pradesh, India
e-mail: amrita12@bhu.ac.in

J. Singh et al. (eds.), *Cereals and Nutraceuticals*,
https://doi.org/10.1007/978-981-97-2542-7_10

10.1 Origin and Background

The scientific name for quinoa is Chenopodium quinoa Willd. It is an annual dicotyledonous broadleaf plant in the genus Chenopodium and family Chenopodiaceae. German botanist Carl Ludwig Willdenow originally described this tetraploid crop plant in 1797. Situated in the genus Chenopodium, it is a distinct pseudocereal and functionally diploid grain with a genetic composition of $2n = 36$, making it 1 of about 200 species.

Quinoa is one of the oldest crops on the American continent, and the Quechua and Aymara peoples of the rural areas still depend heavily on it as a food crop. Quinoa has a rich cultural and historical past. It was a staple diet for the Incas. According to archaeological findings, quinoa has been grown for 5000 years in Chile and roughly 7000 years in Peru. Throughout the whole Andrean region, including Bolivia, Chile, Peru, Ecuador, and Columbia, quinoa was commonly grown. Quinoa was a staple crop of pre-Columbian tribes in the Andes. It was referred to as "chisiya mama", which means "mother grain" in Quechua, and was a staple of the Inca civilization's diet. Its well-balanced amino acid profile, higher protein content than most cereals, and superior nutritional value all contribute to this appellation.

However, because of saponins-plant glycosides that can give quinoa-based meals a harsh taste-quinoa has not gained much international attention.

Quinoa has gained popularity and attention in recent years, particularly in Bolivia and Peru, because of its rich nutritional profile and capacity to adapt to a variety of settings. Quinoa seeds are small, spherical, flat, and come in a variety of colours, including white, grey, black, yellow, and red. They are sometimes referred to as the "golden grain". Like maize and potatoes, it is a staple meal eaten for breakfast. Presently, Bolivia and Peru represent 88% of the global quinoa output, making them the leading exporters.

10.2 Production and Cultivation Practices

Quinoa is a tetraploid crop plant first described by the German botanist Carl Ludwig Willdenow in 1797. It has been widely cultivated for over 8000 years in the regions of South American Andes, Columbia, Ecuador, Peru, Bolivia, and Chile. However, it faced challenges after the arrival of the Spanish, leading to the decline of quinoa cultivation and consumption, as foreign crops like wheat and barley replaced it.

10.2.1 Geographical Distribution

Mostly quinoa is cultivated in South America and in the United States (Colorado and California), China, Europe, Canada, and India. It is also cultivated experimentally in Finland and the United Kingdom.

Quinoa cultivation has a wide geographic reach, primarily concentrated in South America, especially within and around the Andes Mountain range. Key quinoa-producing countries in South America include Ecuador, Peru, northern Argentina, and Bolivia, with the Atacama Desert acting as a natural barrier to its distribution in Chile (Wilson 1990).

In recent years, quinoa has gained popularity outside of South America, with successful cultivation in Europe, North America, Asia, and Africa. Several European countries have initiated projects to explore quinoa's potential as a versatile crop.

10.2.2 Environmental Conditions

Because of its genetic variability, quinoa can be grown in both alkaline and acidic soils with pH levels up to 4.5, anywhere from sea level to over 4000 m above sea level (Mujica et al. 2001; Jacobsen 2003). Quinoa can withstand temperatures as high as −8 °C to 38 °C at specific phenological stages; nevertheless, 15–20 °C is the ideal range for growth (Mujica et al. 2001). But quinoa can tolerate temperatures as high as −8 °C, particularly during the branching stage; temperatures higher than 38 °C result in pollen infertility and blossom abortion (Mujica et al. 2001). According to Tapia (2001), quinoa thrives normally in relative humidity ranges of 40% to 100%. Quinoa has remarkable flexibility and resilience.

It thrives in diverse environments; it can grow in marginal conditions and tolerate stresses such as cold, salt, and drought. This adaptability makes quinoa a valuable crop for areas with challenging climates, including the high-altitude regions of Andean highlands, Himalayas, and North Indian Plains.

10.2.3 Classifications and Varieties

Quinoa was first classified mainly by the colour of the plant and its fruits; later, the morphological traits of the plant were taken into consideration. Quinoa samples that were gathered from Bolivia, Peru, and Ecuador have been divided into 17 races; it is possible that there are still unidentified races. A few examples of the variables that affect the diversity of quinoa varieties are height, location, growing methods, and patterns of precipitation. Amarilla de Marangani and Blanca de Juninan, the two main commercial quinoa cultivars, are grown mostly in Peru's southern Andes (Angeli et al. 2020; Vega-Galvez et al. 2010). These cultivars produce tiny, flat, oval-shaped seeds that are usually pale yellow in hue but can also be pink or black.

The unique adaptations of quinoa to specific geographical regions within the Andes have resulted in the development of five major ecotypes, each associated with subcentres of diversity. These ecotypes differ in their branching morphology and their ability to thrive under varying rainfall conditions, ranging from areas with abundant precipitation of 2000 mm/year to regions experiencing severe drought stress with as little as 150 mm/year (Valencia-Chamorro 2003). These ecotypes are as follows:

Table 10.1 Ecotypes of quinoa (Valencia-Chamorro 2003)

Ecotypes	Location	Growth altitude	Varieties	Characteristics
Sea level	Central and Southern Chile	<500	Chilean varieties	Unbranched, long day plants, yellow, bitter seeds
Valley	Andean and Inter-Andean Valley in Colombia, Ecuador, and Peru	2000–4000	Blanca de Junín, Rosada de Juni, Amarilla de Matangani, Dulce de Quitopamba, Dulce de Lazo	Big plants, branched, short growth period
Subtropical	Bolivian subtropical forest	2500–3000		Plants with intense green colour that tums orange as they mature, small seeds, white or orange
Salar	Salares in Bolivia, northern Chile, and Argentina	3700–3800	Real	Plants adapted to salty and alkalinic soils, bitter seeds, high saponin content
Antiplanic	Area around Lake Titicaca	3500–4000	Cheweca, Kancolla, Blanca de	Short plants with straight stems, short growth period, resistant to frost

(i) Sea level: Quinoa from coastal locations that are either at sea level or low lying (found in central and southern Chile)
(ii) Valley: Quinoa from the Inter-Andean Valleys (found in Peru, Ecuador, and Colombia)
(iii) Subtropical: Yungas quinoa, which grows in the subtropical forest of Bolivia
(iv) Salar: "Salares" quinoa in Argentine, Bolivian, and northern Chilean salt flats
(v) Antiplanic: Quinoa from the highlands (found in Bolivia and Peru) (Table 10.1)

10.2.4 Botanical Characteristics of Quinoa

Plant: Quinoa exhibits a wide range of growth forms, from herbaceous annuals to suffrutescent and arborescent perennials. The most well-known and cultivated species are Chenopodium Quinoa and Chenopodium album. Quinoa is a gynomonoecious annual plant with an erect stem, growing heights in the range of 1–3 m tall.

Stem: The cylindrical stem has a diameter of 3.5 cm, and its colour varies. It can be straight or branching. It becomes red from white, yellow, or light brown according on the cultivar.

Quinoa inflorescence: Quinoa inflorescence is produced in the axils of lower leaves and rises from the top of the plant in the shape of a panicle that can range in

Fig. 10.1 Quinoa plant (pixtastock.com-73620739)

length from 15 to 70 cm. It is divided into two types with a primary axis that gives rise to secondary axes:

(i) Glomerulate: Tiny clusters of flowers that come from tertiary axis, known as glomeruli.
(ii) Amaranthiform: These cells have glomeruli that mostly originate from secondary axes.

Leaves: Because betacyanins are present, quinoa leaves are polymorphic, with lower leaves being rhomboidal and higher leaves lanceolate (Hunziker 1943) (Fig. 10.1).

Roots: Quinoa resists dryness in part because of its strong tap-root system, which reaches up to 1.5 m below the surface (Gandarillas 1979).

Fruit: Quinoa fruit is an achene that can be conical, cylindrical, or ellipsoidal in shape. It has multiple layers, with the perigonium, pericarp, and episperm arranged from outside to inside. The pericarp is where the saponins are concentrated (Risi and Galwey 1984).

Seeds: When exposed to moisture, seeds can germinate quickly-in just a few hours. Black seed colour predominates over red and yellow, which predominate over white seed colour (Risi and Galwey 1984).

10.2.5 The Planting Season and Cultivation Practices

The planting season for quinoa exhibits regional variations, typical occurring between August and December in the Andean highlands and, in some instances, spanning from January to march. An attribute that underscores quinoa's versatility is its ability to thrive in diverse soil types, including marginal soils characterized by a wide pH range (Jacobsen and Stolen 1993; Tapia 1979). Planting methods encompass row sowing, grouping, mixed cultivation, broadcasting, and transplanting. To facilitate effective hoeing, a recommended row spacing of 25–50 cm is preferred. Ideal quinoa cultivation demands a level and well-drained seedbed, while seeds should be sown at depths of 1–2 cm within a finely structured, moist seedbed (Jacobsen 2003). The adoption of mechanized agriculture practices in row planting further streamlines weeding and overall crop management.

10.2.6 Harvesting

Depending on the cultivar, the leaves of the quinoa plant change colour to either yellow or red when it is time for harvest (Garcia et al. 2015). It is essential when quinoa plants reach maturity, the grains begin to fall from the panicle, thus it is important to determine the best time to harvest the crop (Garcia et al. 2015). When the perigonium opens to reveal the seeds inside the panicle or the grains become hard and dry, breaking even under the pressure of a fingernail, quinoa is said to be physiologically ripe (Aroni 2005). After flowering, this developmental stage is usually attained in 70–90 days. Depending on the quinoa variety, the time required for quinoa plants to reach full maturity can vary between plants may require between 5 and 8 months. Harvesting is usually done manually using tools like sickle and cutting the plant 10–15 cm above the soil (Garcia et al. 2015).

The yield of quinoa can vary significantly, spanning from 45 to 500 g/m^2. This yield variation is influenced by factors such as quinoa variety and the prevailing cultivation conditions (Valencia-Chamorro 2003). Following the harvest, quinoa plants can naturally dry in the field, and the process of threshing can be carried out either manually or mechanically (Salas 2003). For commercial purposes, additional ventilation may be employed to reduce the grain moisture content to below 10% (Garcia et al. 2015).

10.2.7 Quinoa's Remarkable Adaptability

Varieties of quinoa are remarkably adaptable, allowing them to be successfully grown in conditions that are frequently regarded as unsuitable for the production of high-protein and mineral-rich seed (Karyotis et al. 2003). Agriculture is more versatile due to its adaptability, which is particularly helpful in high-altitude regions like the Himalayas and the North Indian Plains (Bhargava et al. 2005). One of the few crop plants that can thrive in salinity is quinoa, particularly in the salinity-influenced regions of northern Chile and southern Bolivia (Jacobsen et al. 2000; Tagle and Planella 2002).

10.2.8 Response to Salinity and Fertilization

Salinity has a major effect on quinoa's quality, yield, and plant growth, among other factors. This effect extends to halophytic crops like quinoa. When salinity is present, there is a decrease in plant growth, total seed output, number of seeds, fresh weight, and dry weight of seeds. Interestingly, quinoa seeds' protein content rises and their overall carbohydrate content falls at greater saline levels (Koyro and Eisa 2007). Although too much nitrogen is available can result in a lower yield because of delayed maturation and higher lodging, quinoa responds well to nitrogenous fertilizers (Oelke et al. 1992).

However, according to recent research, quinoa responds positively to nitrogen fertilization, and as nitrogen rates rise, grain output does not seem to diminish (Berti et al. 2000; Schulte-auf'm-Erley et al. 2005). Applying nitrogen has also been shown to increase quinoa seed output and protein content (Johnson and Ward 1993). On the other hand, high phosphorus and potash treatments mainly promote vegetative development without raising seed output (Etchevers and Avila 1979). While it is noteworthy that this result may be ascribed to elevated soil potassium levels in tropical Andean soils, Gandaillas (1982) argues that quinoa does not demonstrate a significant reaction to either phosphorus or potassium.

10.2.9 Resilience to Frost and Drought

Quinoa is resistant to frost before it flowers, but if there are any frost events after that, it could suffer a great deal of harm, especially because of how vulnerable its flowers are to the cold. The crop's capacity to thrive with minimal water requirements, even in areas with yearly rainfall levels between 200 and 400 mm, is indicative of its tolerance to drought (Valencia-Chamorro 2003). It is important to remember that overwatering seedlings can have negative consequences such damping off and stunting (Oelke et al. 1992). Quinoa's low water requirement highlights its drought tolerance, making it an excellent crop to grow throughout large parts of India where farmers mostly depend on seasonal rainfall patterns, and there is little chance of guaranteed irrigation.

10.2.10 Weed Control and Pest Management

To maximize the grain output of quinoa, weed control is essential. It is important to seed as soon as possible since quinoa grows slowly in the first 2 weeks following emergence and can be overtaken by weeds that develop quickly during this time. The competitive advantage that early sowing gives quinoa over weeds is crucial because hand weeding is usually done instead of the use of herbicides. Although downy mildew (*Peronospora farinosa*) is less common in dry, cold climates, it can be an issue in humid, warm climates where quinoa is grown. However, pests and diseases are generally not a serious worry.

10.2.11 Global Quinoa Production

Grain yields of quinoa vary greatly throughout geographical locations. As low as 108 kg/ha can be achieved with some cultivars (like the KU-2 cultivar in Washington State, North America) and as high as 9667 kg/ha with others (like the Longli cultivar in China, Asia). Quinoa is widely grown throughout South America, with Ecuador, Bolivia, and Peru producing the majority of the crop. Beyond its typical growing regions, the United States, Europe, and Africa (especially Morocco and Egypt) have seen an increase in interest in quinoa farming in recent decades.

Production of Quinoa in Europe: Europe has had success with quinoa farming, especially in Denmark and Turkey. Certain cultivars in Europe, such as Pasto, Puno, and Titicaca, have adapted well to a variety of environmental circumstances. Quinoa yields in Europe have generally exceeded those in its area of origin, with some cultivars producing up to 3200 kg per hectare.

North American Quinoa Production: Quinoa cultivation in North America has been limited to certain regions, including Washington and Utah states in the United States and Canada. Yields in North America have been relatively lower compared to other regions, ranging from 108 kg/ha (KU-2 cultivar) to 563 kg/ha (Cherry Vanilla cultivar).

South American Quinoa Production: In South America, local quinoa varieties have been cultivated. Argentina recorded high quinoa yields, with the 2-Want cultivar performing well. Colombia also showed promising results with the Narino cultivar achieving a grain yield of 4317 kg/ha. However, Bolivia had lower yields for the Real Blanca cultivar (380 kg/ha).

Asian Quinoa Production: Quinoa cultivation has expanded to various Asian countries. China saw high yields with the Longli cultivar (9667 kg/ha). India and the United Arab Emirates (UAE) also reported high yields with specific cultivars (e.g., IHBT/Ac-1, IHBT/Ac-2, Q5). Yields were lower in Syria, Uzbekistan, Pakistan, possibly due to drought, heat stress, and salt-affected soils.

10.3 Consumption Pattern

Quinoa was traditionally grown on little rural plantations, mostly as a border crop or for local use. It was frequently a staple diet in underprivileged areas. Quinoa has become more and more popular around the world, with nations like China, Australia, India, Canada, and the United States promoting its production. For example, in the 1970s, Bolivia passed a rule requiring commercially produced wheat-based items like pasta and bread to contain a minimum of 5% quinoa flour. Nowadays, a variety of quinoa products, such as flour, flakes, tortillas, pancakes, and puffed grains, are sold commercially in Bolivia and Peru.

Quinoa can be found in health food stores and utilized as a crop for game covers in the United Kingdom. Denmark has explored quinoa as an alternative for individuals with coeliac disease and for general consumers, incorporating it into products like bread, cakes, and biscuits.

Quinoa has historically been used in many different recipes, either as rice substitute or as whole grains in soups, stews, and broths (Ruales 1998). Quinoa was ground and then combined with other flours, such as rice flour, to create flour with a desired texture and flavour. This flour was then used to construct a variety of breads, including krispina, gritty bread (Weber 1978). Long-used puffed quinoa can be eaten on its own, mashed into a quick cereal, or combined with other foods to make energy bars. However, because to exposure to high temperatures and prolonged processing times, the puffing process may result in some nutritional loss, including protein, oleic, and linoleic acids.

Quinoa is a versatile food that can be eaten in a variety of ways, such as a whole grain like rice, ground into flour for a variety of uses, or added to soups, cereals, and other foods. Because of its abundance of nutrients, quinoa flour can be used to replace wheat flour, which lacks protein, in items like processed dishes, bread, noodles, spaghetti, and sweet biscuits. It has practical qualities like stability, emulsion capacity, gelation, and water absorption. Even in the beverage business, quinoa has found a home, especially in the creation of malted drink recipes. Another way to ferment quinoa is to make chicha beer (Simmonds 1965). Quinoa seeds can be sprouted for salads or popped like popcorn. Additionally, quinoa has been employed for medicinal purposes, including treating inflammation, pain relief, urinary tract disinfection, fractures, internal haemorrhaging, and as an insect repellent. Interestingly, quinoa is being investigated as a possible crop for NASA's Controlled Ecological Life Support System (CELSS) to facilitate long-duration space flights (Schlick and Bubenheim 1996). In addition, it is an excellent fodder crop that helps developing nations in Asia and Africa, especially in harsh conditions.

10.4 Chemical Composition

Quinoa grain is a pseudocereal that mostly consists of starch and a minor amount of sugars, and it has high carbohydrate content. Its high proportion of high-quality protein makes it a whole grain with great nutritional value. Together with fibre, vitamins B complex, E, and C, minerals including calcium, magnesium, iron, potassium, phosphorus, manganese, zinc, copper, and sodium are also included in its composition. It has higher quantities of methionine, lysine, and total protein than typical cereals including rice, maize, barley, and wheat. With fatty acids resembling those in that grain, its lipid makeup is comparable to that of soybean oil (Chauhan et al. 1992; Ogungbenle 2003; Spehar 2007; James 2009; Vega-Galvez et al. 2010).

As a "functional food", quinoa can help lower the chance of developing a number of illnesses. Its functional qualities stem from the fibre, vitamins, minerals, fatty acids, antioxidants, and plant hormones that are present. These nutrients help sustain human nutrition, especially when it comes to protecting cell membranes, which has been demonstrated to improve brain function. Grain has the characteristics that set it apart from other plant diets for human nutrition and health maintenance (Vega-Galvez et al. 2010; Repo-Carrasco-Valencia and Serna 2011) (Table 10.2).

Table 10.2 Approximate composition in genotypes of quinoa grain (g 100 g^{-1} dm)

Component	References				
	Kozioł (1992)	Dini et al. (1992)	Wright et al. (2002)	Ogungbenle (2003)	US Department of Agriculture (USDA)—Agricultural Research Service (2011)
Protein	16.5	13.7	16.7	15.2	16.3
Lipids	6.3	14.5	5.5	7.1	7.0
Ashes	3.8	3.5	3.2	1.4	2.7
Carbohydrates	69.0	65.7	74.7	65.6	74.0
Fibre	3.8	2.6	10.5	10.7	7.0

10.4.1 Proteins

Quinoa seeds have a protein level that ranges from 8% to 22%, which is often higher than that of typical grains like rice, wheat, and barley. However, its protein content is less than half that of other legumes. On a dry basis, the protein concentration ranges from 14% to 20% (g/100 g). The embryo contains the majority of its protein (Valencia-Chamorro 2003). Albumins and globulins account for 44–77% of total protein in pseudocereals like quinoa, which is a larger percentage than prolamins (0.5–7.0%). Because quinoa has little to no prolamin, it is a gluten-free grain. For those with gluten intolerance, including those with celiac disease, quinoa offers a flavourful, affordable, and nutritious food source that is simple to prepare (Valencia-Chamorro 2003).

The primary proteins found include prolamins, which are present in small amounts, albumin (2S), globulin (11S), also known as chenopodin, which primarily makes up the majority of storage proteins. Varying species have varying ratios of proteins. It was discovered that the digestion of quinoa protein was on par with other premium food proteins (Comai et al. 2007).

The oligomeric protein globulin 11S, also known as chenopodin, has a 320 kDa quaternary structure. It is made up of monomers joined by disulphide bonds that are composed of an acid and a basic polypeptide with molecular masses of 30–40 kDa and 20–25 kDa, respectively. When compared to the entire protein content of quinoa seeds, chenopodin has a comparatively low sulphur amino acid concentration. However, it is high in glutamine-glutamic acid, asparagines-aspartic acid, arginine, serine, leucine, and glycine (Brinegar and Goudan 1993; James 2009). In decreasing conditions, the type 2S protein fraction displays a diverse array of polypeptides ranging in molecular mass from 8 to 9 kDa. It has been demonstrated that the amino acid content of this protein is high in histidine, cysteine, and arginine—all of which are essential for proper nourishment, especially for growing bodies. The 2S fraction thus appears to contribute to the higher level of sulphur amino acids in quinoa, demonstrating great potential as dietary supplementation, especially for children (Brinegar et al. 1996; James 2009). Quinoa grains or derivatives serve as both an enrichment and a food supplement. Quinoa contains a good balance of all the essential amino acids, with levels that are equivalent to those defined by the Food and

Agriculture Organization (FAO). It is high in sulphur amino acids and lysine. Quinoa is considered a complete protein source because it is a high-quality protein. It contains an unusually high amount of lysine, a limiting amino acid that is uncommon in plants. Moreover, its inclusion is very beneficial to legumes, which are often lacking in cysteine and methionine (Valencia-Chamorro 2003).

According to studies, the protein efficiency ratio (PER) of both raw and debittered quinoa is 78–93% that of casein, according to reports on the nutritional evaluation of quinoa protein. (Valencia-Chamorro 2003).

Proteins are involved in many bodily processes, including the synthesis and upkeep of tissues, the production of hormones, enzymes, and antibodies, the production of energy, and the control of metabolic processes. Amino acids give the body sulphur compounds in addition to nitrogen. These proteins are involved in the transport of phospholipids, fat-soluble vitamins, cholesterol, and triglycerides in the form of lipoprotein (Alves et al. 2008).

10.4.2 Carbohydrates

Quinoa's primary carbohydrate content ranges from 52% to 69% (dry basis) starch. The overall dietary fibre content is nearly same to that of cereals (7–9.7% dm). Compared to perisperm, the levels in the embryo are greater. The soluble fibre content is reported to range from 1.3 to 6.1% (dm). About 3% of simple sugars, namely maltose, D-galactose, and D-ribose, are present in quinoa together with minor amounts of fructose and glucose (James 2009). Quinoa starch granules are found either as individual particles or in clusters of elliptical or spherical composite structures. Their diameter, which ranges from 0.6 to 2.2 mm and is smaller than that of the starch present in most cereal grains, gives them a polygonal form.

The minuscule size of the starch granules can be usefully employed when blending with synthetic polymers for biodegradable packaging applications. It is the perfect thickener for sauces, condiments, and soups because of its low gelation temperature, storage durability at low temperatures, and other uses where resistance to retrogradation is desired. Additionally, it can be utilized to create a texture that is creamy and silky, akin to fats (Tari et al. 1990; Lorenz 1990; James 2009; Vega-Galvez et al. 2010). Its structure is extremely branched, with a minimum degree of polymerization of 4600 glucose units, a maximum of 161,000 units, and a weighted average of 70,000 units (Praznik et al. 1999). Depending on the botanical source, the chain might have a length of 500–6000 glucose units.

According to Kozioł (1992) and James (2009), quinoa starch has an amylose percentage that ranges from 3% to 22%. This is greater than certain varieties of barley, less than that of wheat or maize, and similar to certain varieties of rice. This fraction's average degree of polymerization, according to Tang et al. (2002), was 920, a significantly lower number than barley's degree of polymerization, which was 1660. Reports state that quinoa starch gelatinizes at a relatively low temperature, between 62.6 and 67 °C. This is greater than the temperatures of wheat, rice, and barley starch, but lower than those of amaranth and waxy barley. It is more water-absorbing, has a greater maximum viscosity, and has a stronger swelling power than starch derived from wheat and barley. Reduced pasting temperature and slower rate of quinoa starch degradation result in higher hot paste stability, or increased shear resistance during heating.

Because dietary fibre is indigestible in the small intestine, it provides several health benefits. As a result, the high quinoa fibre content has been shown to enhance digestibility by aiding in the large intestine's ability to absorb other nutrients found in quinoa (Ogungbenle 2003). Because the amount of dietary fibre that is free of gluten is thought to be insufficient, experts advise patients with celiac disease to consume more whole grains that are high in fibre. This will help to compensate for the patients' partial lack of dietary fibre intake (Alvarez-Jubete et al. 2010). The extrusion procedure is the primary method used to reduce the total and insoluble fibre content of quinoa grains during processing; this effect is noteworthy for specific quinoa cultivars only. Nonetheless, during this process, the amount of soluble dietary fibre rose. When grains are exposed to high temperatures and shear stress, chemical bonds are broken, resulting in particles that are more soluble and smaller.

Because there is a higher concentration of soluble dietary fibre during extrusion, some insoluble fibre components change into soluble fibre (Repo-Carrasco-Valencia and Serna 2011).

Because quinoa carbs reduce free fatty acids and have positive hypoglycaemic effects, they can be considered nutraceuticals. Research on celiac disease patients revealed that quinoa had a somewhat lower glycaemic index than gluten-free breads and pastas. In addition, quinoa produced considerably lower triglyceride levels when compared to gluten-free bread and lower levels of free fatty acids when compared to gluten-free pasta. There have been reports of quinoa's additional nutraceutical benefits.

10.4.3 Lipids

Quinoa's high quality and high quantity of lipid fraction have led to it being regarded as an alternative oilseed crop. Between 2.0% and 9.5% of quinoa's weight is made up of fat. It is abundant in antioxidants like alpha- and gamma-tocopherol and rich in important fatty acids like linoleic and a-linolenic acid. The primary saturated fatty acid in quinoa, according to Ando et al. (2002), was palmitic acid, which accounted for around 10% of all the fatty acids present. The unsaturated fatty acid composition of quinoa oil was found to be comparable to that of soy oil, consisting of 87.2% to 87.8% total fatty acids, of which 19.7% to 29.5% were oleic, 49.0% to 56.4% were linoleic, and 8.7% to 11.7% were linolenic.

The health benefits associated with polyunsaturated fatty acids, including improved immunity, cellular membrane function, prostaglandin metabolism, heart disease, and increased insulin sensitivity have led to a rise in their prominence (Ogungbenle 2003; Repo-Carrasco et al. 2003; James 2009). Since gamma-tocopherol has antioxidant properties, the observed level was marginally greater than that of maize oil, giving quinoa oil a long shelf life. A-tocopherol, which includes vitamin E, is also highly significant since it functions as a natural antioxidant at the level of the cell membrane, shielding fatty acids from damage caused by free radicals (Ruales and Nair 1993a; Ng et al. 2007; Ryan et al. 2007; James 2009).

The naturally occurring vitamin E in quinoa flour allowed its lipids to remain constant for 30 days, as reported by Ng et al. (2007). The potential of quinoa as a new oilseed has been reinforced due to the presence of this micronutrient, which acts as a natural antioxidant in the grain. This is especially true when developing new food applications that highlight the antioxidant properties found in this raw material. Along with tocopherols, the components found in the food's unsaponifiable lipid fraction are squalene and phytosterols. All higher species produce squalene, which is an intermediary in the manufacture of cholesterol and is known to have health benefits for humans. Tocotrienols, which are a biological precursor to the entire family of steroids, have several significant roles in addition to their potent antioxidant action. These roles include supporting a healthy cardiovascular system and potentially preventing cancer (James 2009). Oils, seeds, and grains are rich sources of phytosterols, which are naturally occurring components of plant cell membranes. Their biological effects are diverse, encompassing anti-inflammatory, antioxidant, and anticarcinogenic properties.

10.4.4 Vitamins

Micronutrients like vitamins and minerals are also abundant in quinoa. The substances that are vital to both human and animal health are vitamins. They must come from outside food sources because the human body is unable to synthesize them. The phrase "provitamins" refers to chemicals that the human body can turn into vitamins, specifically sterols and carotenoids. Vitamins are separated into two groups based on how soluble they are: fat-soluble and water-soluble. These are some of the most often used compounds to increase food items' nutritional content (Mahan and Escott-Stump 2010; Vega-Galvez et al. 2010).

Although there is little information on the vitamin content of quinoa, some writers have found significant quantities of pyridoxine (B6) and folic acid in the grain, with values in 100 g meeting the needs of both adults and children. Quinoa contains more riboflavin (B2) and α-tocopherol than rice, barley, or wheat. Quinoa has higher quantities of riboflavin (B2), pyridoxine (B6), and folic acid than most cereals, including wheat, oats, barley, rice, and maize, but lower levels of thiamine (B1) than oats or barley. Quinoa seeds are a good source of vitamin E (Valencia-Chamorro 2003). Quinoa seeds are higher in calcium, phosphorus, iron, fibre, and B-vitamins than barley, oats, rice, corn, or wheat, according to observations made by the Food and Agriculture Organization (FAO) (Dini et al. 2005).

Ascorbic acid levels need to be managed carefully since this vitamin is very susceptible to oxidation. According to reports, the ascorbic acid content of quinoa ranges from 0 to 63.0 mg/100 g. Because the vitamin content figures are based on dry measurements and both the bitter and sweet kinds of quinoa are usually cleaned before cooking, they could be misleading. The bitter variety is polished (abrasively peeled) in this procedure to eliminate saponins, which changes the vitamin content of the raw grain (Kozioł 1992).

10.4.5 Minerals

Minerals are inorganic micronutrients that cannot be created by living things, in contrast to proteins, fats, and carbs. They play crucial roles in the body's metabolism and biology. Decreased bioavailability or low intake can lead to physiological imbalances and the impairment of essential activities. Calcium, phosphorus, iron, potassium, sulphur, sodium, magnesium, zinc, copper, selenium, and chromium are some of the most well-known minerals. To meet daily demands, these micronutrients must be obtained from an external, full, and balanced diet. To maximize their bioavailability, particular emphasis should be paid to the right food combinations, preparation methods, and components (Cozolino 2009; Vega-Galvez et al. 2010). Quinoa is an excellent way to get your minerals. Compared to typical cereals, it has higher levels of calcium, magnesium, iron, and zinc. Quinoa contains very high levels of iron. The quinoa seeds' mineral content is somewhat reduced by polishing and washing them; iron, zinc, and potassium concentrations range from 12% to 15%, while copper and magnesium loss is 27% and 3%, respectively. Compared to most other cereals, rice (0.5%) and wheat (1.8%), quinoa has greater ash content (3.4%). Because of this, quinoa grain has a high mineral content. Compared to the regularly used grain, there is a considerable increase in calcium and iron content.

10.5 Anti-nutrients and Toxic Compounds

The primary issue with using plants as a dietary supplement is the existence of some substances known as "anti-nutritional factors" that are produced from secondary plant metabolism. "Anti-nutritional factors" refers to a group of substances found in many different types of plant foods that, when eaten, lower their nutritional value, obstruct their ability to be digested, absorb, or use nutrients, and may even be harmful to health if consumed in high enough concentrations.

10.5.1 Saponins

Quinoa has a bitter flavour due to the presence of anti-nutritional substances called saponins (Mora-Ocación et al. 2022). The broad class of triterpenoids glycosides known as saponins, which are present in many plants, is crucial in defending seeds against fungus and insect damage (Simmonds 1965). They are members of the family Leguminosae. Hederagenin, 30-O-methyl-espergulagenate, phytolaccagenic acid, and oleanolic acid are the four major types of sapogenins found in quinoa. Glucose, arabinose, and galactose are the main types of carbohydrates found (Dini et al. 2001; Zhu et al. 2002; Souza et al. 2004; Farro 2008).

They are naturally harmful, and how dangerous they are varies depending on the kind of saponin and how it is absorbed. Oral intake of saponins can have a fatal dose that is three to a thousand times higher than intravenous injection (George 1965). Certain saponins are 3000 times more poisonous than others, with a fatal dose by oral administration ranging from 1.9 to 6000 mg saponin/kg body weight in mice. They are found in the outer layer of the seed and range from 0.1% to 5% (Chaudhary et al. 2023) in concentration. Because these substances are soluble in water, washing makes them easy to remove. Washing quinoa grains is not only a sanitary measure, but it also guarantees that the bitterness that is linked to saponins is eliminated and the amount of saponin is reduced. Additional pharmacological, fungal, and antibiotic qualities that saponins have are documented by Basu and Rastogi (1967), Agarwal and Rastogi (1974), Chandel and Rastogi (1980), and Nonaka (1986). Saponins are of pharmaceutical relevance because of their hypocholesterolaemic properties (Oakenfull and Sidhu 1990) and capacity to alter intestinal permeability, which may facilitate the absorption of specific medications (Basu and Rastogi 1967). They contain poisonous qualities that can cause red blood cell haemolysis and are soluble in both methanol and water. The kind of saponin and the recipient organism's susceptibility determine its toxicity. However, although they are extremely toxic to cold-blooded creatures, they are only mildly hazardous to mammals when eaten orally, and they have no negative effects on the protein's nutritional value. Though some saponins decrease iron and zinc absorption, they can still form complexes with those minerals even though there is no evidence that complexes with vitamins A, E, and D3 have formed (Ruales and Nair 1993b; Dini et al. 2001; Farro 2008; Jancurova et al. 2009). Their insecticidal, fungicidal, and antibiotic activities are known, even if some are regarded as hazardous. They have pharmacological potential because they alter the small intestine's permeability, which may help some medications be absorbed and lower plasma or serum cholesterol by releasing neutral steroids and bile acid in the faeces. These are precursors in the pharmaceutical business that are used to make hormones, steroids, anti-inflammatories, expectorants, diuretics, and contraceptives. These are utilized in soaps, detergents, shampoos, toothpastes, and hair lotions and are also referred to as foaming agents with water due to their surfactant and emulsifier properties (Dini et al. 2001; Farro 2008; Vega-Galvez et al. 2010).

Not only are saponins present in seeds, but they are also present in other areas of the plant, serving as a barrier against the harsh external environment. The primary sapogenin seen in the leaves was hederagenin, whereas oleanolic acid was detected in the seeds. Saponins make up 86% of the total saponins in the grain, and their higher concentration in the pericarp indicates that most of this material may be removed through peeling. In comparison to soybeans and other legumes, quinoa grains have a significantly lower concentration of saponins (Chauhan et al. 1992; Mastebroek et al. 2000; Ando et al. 2002; James 2009). Because of its distinct bitter flavour, this material is typically eliminated or decreased from the grain's pericarp in order to improve sensory quality and consumer acceptance. The removal process involves choosing sweet varieties or using one of the following methods: heat treatment, extrusion, roasting, mechanical abrasion, or a combination of the two (Koziol

1990; Dini et al. 2001; Brady et al. 2007; Comai et al. 2007; Spehar 2007; Farro 2008; Jancurova et al. 2009).

10.5.2 Phytic Acid

Phytic acid's strong negative charge has led to its classification as an anti-nutritional agent. It has the ability to chelate proteins, carbohydrates, enzymes, and bivalent minerals like calcium, iron, magnesium, zinc, and copper, reducing their bioavailability. It is the majority of these seeds' primary phosphate reserve and may be found in peels from most cereals and legumes at concentrations of 1% to 3% of dry matter. In addition to its outer layers, the endosperm of quinoa grains has a uniform distribution of phytic acid.

Bran had higher phytate content than flour or peel, accounting for 38–41% of the total phytate from whole seeds, according to Chauhan et al. (1992). Quinoa has more phytic acid than grains do, and neither the absorption of iron nor the incorporation of calcium into the bones were shown to be negatively impacted (Kozioł 1992). About 30% less phytic acid was present, thanks to the procedure that involves brushing and rinsing quinoa seeds to remove saponins. Steeping, germination, and fermentation can all considerably lower the amount of phytic acid present. Processes like fermentation and cooking that encourage phytase activation have higher degradation efficiencies (Ruales and Nair 1993b; Oliveira et al. 2003; Khattab and Arntfield 2009). To determine whether or not the phytic acid in quinoa affects iron availability, more research on humans is still needed.

10.5.3 Trypsin Inhibitor

Protease inhibitors are widely distributed proteins that combine with proteolytic enzymes to produce extraordinarily stable complexes. A trypsin inhibitor is present in the digestive tract and inhibits the action of trypsin, which breaks down proteins. This causes the pancreas to produce more enzymes, which causes the organ to become larger and grow less. Due to their thermolability and tendency to be destroyed under typical home or commercial food preparation circumstances, such anti-nutritional substances seldom affect people (Khattab and Arntfield 2009; Lopes et al. 2009). When different legumes, such as beans and peas, were subjected to heat treatments like boiling, roasting, autoclaving, and microwave for a duration of 18 to 22 h, Khattab and Arntfield (2009) observed a complete reduction of this component. This fact indicates that methods commonly used in home food preparation can be utilized to inactivate this anti-nutrient in quinoa (Borges et al. 2010). Protease inhibitors are detected in quinoa seeds at a concentration of less than 50 parts/million. Trypsin inhibitors are present in quinoa in trace concentrations, far less than in cereals that are typically ingested; therefore, they are not a significant cause for worry (Vega-Galvez et al. 2010). According to Jancurova et al. (2009) and Ruales and Nair (1993b), the anti-nutrient content of eight quinoa varieties ranged from

1.36 to 5.04 unit trypsin inhibitor, which is significantly lower than that of soybeans, beans, and lentils.

10.5.4 Nitrates

All plants contain nitrates (NO_3), which are a vital source of nitrogen for healthy plant growth. When absorption surpasses their metabolic requirements, certain plants store these materials in their roots and shoots. Petioles and stems are where the greatest accumulation occurs, but high quantities are observed throughout leaves, particularly in mesophyll cells. Because phloem supplies amino acids to the reproductive organs, fruits, and seeds, they have low nitrate levels. According to Beninni et al. (2002) and Santos (2006), the nitrate found in vegetables can originate from utilized fertilizers or can form on the substrate through nitrification or mineralization. It is well known that nitrates affect how the thyroid gland and vitamin A metabolism work in the human body. These can react with secondary and tertiary amines to generate potentially cancer-causing N-nitrous compounds, or they can be converted to nitrites, which after being absorbed cause cyanosis because metmyoglobin is formed (Beninni et al. 2002; Santos 2006; Lopes et al. 2009). They are hence strong carcinogens to people. Lopes et al. (2009) observed that the nitrate levels in quinoa wholemeal were twice as low as those in fresh produce, including spinach, lettuce, radish, and beets. This indicates that quinoa poses no nutritional disadvantages. Therefore, Lopes et al. (2009) state that a consumer's safety would not be endangered by the amount of flour they typically consume.

10.5.5 Oxalates

Oxalates are harmful compounds that pose a serious threat to human health. Common sources of it include vegetables such as spinach, beets, chard, rhubarb, tomatoes, almonds, and cocoa. It is not metabolized by humans and is excreted in urine. A high intake of oxalate in the diet affects the absorption of minerals and trace elements, which is important in the development of hyperoxaluria, a risk factor for the formation of calcium oxalate stones in the kidneys (Santos 2006; Jancurova et al. 2009; Lopes et al. 2009). This is because oxalates can form insoluble complexes with divalent cations in the gastrointestinal tract. Many negative effects have been linked to oxalic acid in both humans and animals. These include decreased mineral bioavailability, gastrointestinal distress, muscle contraction or tetany associated with other neurological symptoms, impaired blood clotting ability, and potential harm to excretory organs as a result of cell substance deposited with high concentrations of crystalline calcium oxalate (Lopes et al. 2009). Vegetables belonging to the Polygonaceae, Amaranthaceae, and Chenopodiaceae families are the ones that have the highest oxalate contents. Naturally, oxalate accumulates in all plant tissues of these families, including the hypocotyl seeds, leaves, stems, and roots.

Research, however, shows that little is known about the levels of oxalic acid in various species within these families—quinoa, for example—in the literature.

These anti-nutritional elements can, however, be neutralized or reduced to acceptable levels, making the grain fit for processing and usage in the home or industry.

10.6 Bioactive and Functional Components

The seeds of the quinoa plant are the main edible part of the plant; they contain a lot of phenolic compounds that have anti-inflammatory and anti-cancer properties. Quinoa is rich in macro- and micronutrients, but it also contains a significant amount of bioactive phytochemicals that are beneficial to human health. Quinoa's main constituents are phenolics, betalains, glycine betaine, and triterpenoids (which include saponins, phytosterols, and phytoecdysteroids). These molecules serve as chemical barriers against infections and insects and can be either lipophilic or hydrophilic. Among the principal lipophilic bioactive phytochemicals included in quinoa are saponins, phytosterols, phytoecdysteroids, and phytotates. The hydrophilic substances with special health-promoting properties are glycine betaine and betalains.

Most of the quinoa's secondary metabolites are hydrophilic phenolic compounds, which include tannins, flavonoids, and phenolic acids (Graf et al. 2015; Tang and Tsao 2017). Owing to their antioxidative potential, which inhibits oxidation by avoiding free radical chain reactions and prevents diseases associated with oxidative stress, plant polyphenols, and phenolic content play a significant role and are beneficial to human health. Quinoa has a higher phenolic content than other grains and beans, and this content is strongly associated with its ability to scavenge free radicals. According to Zhang and Tsao (2016), eating a diet high in polyphenols on a regular basis may help avoid diabetes, cancer, cardiovascular disease, and neurological illnesses. It has been proposed that these substances may help lower the risk of diabetes, neurological diseases, and cardiovascular illnesses. Compared to other grains and beans, quinoa has a higher phenolic content, which is strongly associated with its ability to scavenge free radicals. Eating a diet high in polyphenols on a regular basis may help prevent diabetes, heart disease, and cancer (Zhang and Tsao 2016). Numerous investigations have revealed that quinoa extracts included significant levels of rutin, isorhamnetin, kaempferol, and the acids ferulic, sinapic, and gallic. These identified substances were found to have an inhibitory effect on the motility and proliferation of prostate cancer cells.

10.6.1 Saponins

Consumption of quinoa should be limited since saponins, which are bitter compounds predominantly present in the outer layers of seeds, are the cause of this. Saponins have a variety of biological actions, including anti-inflammatory, antioxidant, antiviral, anti-cancer, and anti-fungal properties, while being anti-nutritional factors (Basu and Rastogi 1967; Chandel and Rastogi 1980; Nonaka 1986). Because of their ability to modify intestinal permeability, which may aid in the absorption of particular drugs, and their hypocholesterolaemic qualities (Oakenfull and Sidhu 1990), saponins are significant in the pharmaceutical industry (Basu and Rastogi 1967).

10.6.2 Phytosterols

The amount of phytosterol in quinoa has not been thoroughly studied; 100 g of quinoa seeds can have up to 118 mg of phytosterol. Villacr'es et al. (2013) state that b-sitosterol, campesterol, brassicasterol, and stigmasterol are the main phytosterol components of quinoa seeds. The quantities of campesterol (15.6 mg/100 g), stigmasterol (3.2 mg/100 g), and b-sitosterol (63.7 mg/100 g) in quinoa seeds were higher than those in barley, rye, millet, and corn, per a study by Ryan et al. Phytosterols are lipophilic compounds with a structure similar to that of cholesterol. Plants are their major habitat. Research, meta-analyses, and epidemiological studies all support the importance of phytosterols in lowering cholesterol levels in the human body (Graf et al. 2015; Marangoni and Poli 2010). Ho and Pal (2005) claim that phytosterols can reduce blood cholesterol levels by competing with cholesterol for intestinal absorption and possibly by inhibiting the hepatic and intestinal production of lipoproteins. Moreover, research has shown that phytosterols have anti-inflammatory, antioxidative, and anticarcinogenic qualities (Ryan et al. 2007).

10.6.3 Phytoecdysteroids

Among plants that humans can eat, quinoa seeds have the highest content of phytoecdysteroids. Ecdysteroids are polyhydroxylated steroid hormones that, in addition to their well-known functions, such as regulating insect moult, may also have pharmacological and metabolic effects in mammals (Foucault et al. 2011; Dinan et al. 2009). The research found that the total phytoecdysteroid concentration of quinoa was between 138 and 570 μg/g. The most prevalent ecdysteroids are those found in numerous plants, such as quinoa, such as 20-hydroxyecdysone. 20-hydroxyecdysone (20HE), which makes up between 62% and 90% of all phytoecdysteroids, is the most prevalent of at least 13 different phytoecdysteroid types that have been identified from quinoa seeds (Graf et al. 2016).

10.7 Health Implications

The health benefits of any nutritional component are the result of its digestion and subsequent absorption by the human body. Due to its low glycaemic index, high quantities of macro- and micronutrients that frequently exceed those found in typical cereals, and gluten-free status, quinoa has great nutritional, bioactive, and functional qualities that may support consumer wellbeing. Eating quinoa seems to be linked to an exceptionally low risk of obesity, diabetes, hypertension, and some gastrointestinal disorders. This summarizes the possible health advantages of quinoa grains and contributes to our understanding of the function of quinoa and its constituents in human health. Furthermore, the nutritional composition of quinoa is altered by processing, which further alters its health advantages.

10.7.1 Quinoa Against Celiac Disease

In people who are genetically vulnerable to the condition, gluten consumption causes immune-mediated enteropathy, which is the hallmark of celiac disease (Makharia et al. 2022). The gluten proteins in the diet are the cause of celiac disease. Consequently, there is an increasing need for cereal-free alternatives made from gluten-free grains as more people are diagnosed with celiac disease. One strategy to avoid celiac disease is to strictly avoid gluten-containing cereals. However, a long-term gluten-free diet therapy may not ensure enough nutrient intake and may lead to several nutritional deficiencies. As a result of quinoa's position as a pseudo-cereal grain, it has been recommended as a part of the gluten-free diet.

A 6-week gluten-free diet supplemented with 50 g of quinoa per day was well tolerated and did not worsen this disease, per an in vivo research. There was a slight hypocholesterolaemic effect as well as an improvement in the histology and serological parameters. In conclusion, there is ongoing debate regarding whether quinoa is safe for celiac disease sufferers to eat. Clinical research is still required to confirm whether a particular quinoa variety is safe for individuals with celiac disease, even though in vitro experiments offer an initial safety assessment of many quinoa variations. The long-term consequences of quinoa eating in in vivo research need to be further explored and studied.

But managing celiac disease is difficult because of obstacles and difficulties that keep patients from adhering to this diet, such as high cost, risk of cross-contamination, and scarcity in the market (Makharia et al. 2022). Future studies should focus on creating different quinoa-based food products to satisfy the everyday needs of celiac disease patients, from the standpoint of dietary intervention. A study suggests that quinoa can be given to celiac disease patients as a gluten-free substitute for cereal grains. Additionally, this study discovered that consuming 50 g of quinoa daily for 6 weeks improved gastrointestinal characteristics and slightly decreased levels of triglycerides, LDL, HDL, and total cholesterol.

10.7.2 Quinoa Against Diabetics

These days, the global prevalence of diabetes mellitus particularly type 2 diabetes (T2DM) has become a serious issue that affects how people metabolize fat, protein, and carbohydrates. Commercial anti-diabetic medications that target the current therapeutic targets of type 2 diabetes have demonstrated efficacy in lowering levels of glycosylated haemoglobin and blood glucose. But because these medications may also have negative consequences, researchers are looking more closely at alternate approaches to glycaemic management (Zamudio and Segura Campos 2022). Quinoa and other whole-grain diets are crucial for managing and preventing diabetes. Quinoa has anti-diabetic properties primarily through inhibition of α-amylase, α-glucosidase, and dipeptidyl peptidase IV (DPP-IV) activities. DPP-IV is thought to be responsible for the cleavage of incretins, which results in a decrease in postprandial insulin secretion. These include glucagon-like peptide-1 (GLP-1) and glucose-dependent insulinotropic polypeptide (GIP). α-amylase and α-glucosidase are the enzymes that hydrolyse starch to oligosaccharides and disaccharides, trisaccharides, and oligosaccharides to glucose. Additionally, it has been found that quinoa might be a good source of a number of bioactive peptides. Under papain hydrolysis and in vitro GI digestion simulation, the inhibitory effects of quinoa protein concentrates have been investigated against DPP-IV, α-amylase, and α-glucosidase. These enzymes, which are involved in the breakdown of incretin and the digestion of dietary carbs, may be inhibited by the peptides produced from quinoa proteins (Zamudio et al. 2022).

Studies indicate that patients with diabetes may benefit from including quinoa proteins as an active component in their diet. Functional foods and dietary supplements can contain bioactive peptides produced from quinoa proteins. Quinoa's phenolic compounds have the ability to impede the action of α-glucosidase, an enzyme involved in the breakdown of carbohydrates. Polyphenols in quinoa have been shown to have an impact on the decrease in blood sugar levels after meals (Zhang et al. 2022). Quinoa can therefore be used to assist manage type 2 diabetes and enhance the profile of metabolic risk factors. Moreover, in vivo experiments have demonstrated the quinoa grain's anti-diabetic properties. The impact of substituting foods high in complex carbs with quinoa on blood glucose oscillations in elderly individuals with prediabetes was assessed by Díaz-Rizzolo et al. (2022). Because quinoa is low in carbohydrates and high in unsaturated fats and protein, it has the potential to lower postprandial glycaemia and prevent the development of type 2 diabetes. Quinoa therapy also lessens the harm that elevated glucose levels do to the liver and kidneys, which helps with obesity and diabetes. Quinoa's ability to prevent diabetes may potentially be impacted by the methods used during processing. The hypoglycaemic effect of sprouted quinoa yoghurt was greater in vitro than that of non-sprouted yoghurt due to the considerable influence of germination on glucose and lipid homeostasis in diabetic mice (Obaroakpo et al. 2020). In order to increase the quantity of certain components linked to quinoa's anti-diabetic properties, it is crucial to use the right processing methods and environments. This will help produce quinoa-based products that are good for diabetic patients. These patients have the option to use them in place of whole or some cereal staples. Nevertheless, additional clinical trials are required to confirm the anti-diabetic properties of quinoa prepared in various ways.

10.7.3 Antioxidant Activity of Quinoa

Free radicals are produced by human metabolism, but too many of them can cause oxidative stress in the body. Radicals known as reactive oxygen species (ROS) are created when oxygen is present. Excessive production of ROS is associated with a number of detrimental effects, such as lipid peroxidation in cellular membranes, protein denaturation, carbohydrate oxidation, deoxyribonucleic acid degradation, and pigment disintegration (Hoang et al. 2019). The degree to which a substance reduces the effects of reactive oxygen species (ROS) is measured by its antioxidant activity. Antioxidant activity has been measured using DPPH radical scavenging ability, ferric reducing antioxidant power (FRAP), oxygen radical absorbance capacity (ORAC), and other in vitro experiments.

Conventional natural antioxidants, such as flavonoids, carotenoids, polyphenolic compounds, and tocopherols, may offer defence against chronic oxidation-related illnesses like cancer, heart disease, and cerebrovascular disease (Espitia-Hernández et al. 2022). Because phenolic compounds have a beneficial effect on perceived antioxidant activity, quinoa is well known for being a good source of natural antioxidants. Quinoa phenolics have the ability to block the actions of α-amylase and α-glucosidase, as well as prevent oxidative damage to the organs involved in the digestion of carbohydrates. This suggests that these compounds may be able to prevent hyperglycaemia and its associated consequences (Hemalatha et al. 2016). It is noteworthy that the protein isolates from quinoa were shown to have antioxidant properties, and their enzymatic hydrolysates showed remarkable ability to scavenge free radicals. According to Zheng et al. (2019), three polypeptides that were isolated from quinoa albumin RGQVIYVL, ASPKPSSA, and QFLLAGR showed strong antioxidant potential.

In conclusion, quinoa's high-protein content and abundance of phenolic compounds contribute to its exceptional antioxidant activity, suggesting that it may find application in the functional food market as a nutraceutical component. Due to the role of fibre and polyphenols, adding quinoa flour to wheat bread, for instance, can demonstrate reduced in vitro digestibility with a lower estimated glycaemic index value and higher antioxidant activity. The quinoa-fortified bread's superior nutritional qualities align with the industry's development trend (Xu et al. 2019). To confirm its actions and health advantages, more research on humans and animal models should be done. Processing techniques like milling, germination, fermentation, and cooking can also significantly affect quinoa's antioxidant activity. Antioxidant activity and levels can be increased in quinoa processed by germination, fermentation, and baking, but they are reduced in quinoa milled and sieved. Research suggests that items made from whole quinoa grains have more nutritional value than those made from milled grains.

10.7.4 Anti-inflammatory Activity of Quinoa

Inflammation serves as a buffer between the immune system and harmful chemicals and microbes. Conversely, inflammatory reactions can be responsible for serious tissue damage and the development of acute or chronic inflammatory disorders. Quinoa grain constituents have potent anti-inflammatory qualities. In one study, high saponin content quinoa seed extracts inhibited the release of inflammatory cytokines, such as interleukin-6 (IL-6) and tumour necrosis factor-α (TNF-α), in lipopolysaccharide-stimulated macrophage cells and reduced the production of inflammatory mediators, such as nitric oxide. It has been proposed that quinoa saponins have anti-inflammatory properties (Yao et al. 2014). Additionally, in the activated immune response, quinoa protein and its hydrolysates made with pancreatin, pepsin, and papain shown strong anti-inflammatory activity (Shi et al. 2019). Chenopodin, the main protein present in quinoa seeds, has anti-inflammatory qualities. The same research group found that the very low-charge globulins and quinoa seed albumins had a stronger anti-inflammatory impact than the low-charge and high-charge globulin fractions (Capraro et al. 2021). These findings could be the foundation for the quinoa seed's development as a chemical source with anti-inflammatory properties.

10.7.5 Quinoa Against Cardiovascular Diseases (CVDs)

Globally, CVDs are the primary cause of death. Whole grain-rich diets have been linked to decreased mortality and CVD risk, according to several epidemiological studies. Triglyceride levels, total cholesterol, and low-density lipoprotein-cholesterol (LDL) are risk factors for cardiovascular disease. A brief investigation on a dietary intervention showed that regular consumption of quinoa-enriched bread (20 g of quinoa flour) changed human glucose response but had minimal impact on CVD risk variables (Li et al. 2018). An in vivo study examined how eating biscuits high in quinoa affected older persons' risk factors for CVD. The findings demonstrated that beneficial changes in circulating cholesterol concentrations, body weight, and body mass index occurred, resulting in a decreased risk of CVD (Pourshahidi et al. 2020). These findings suggested that quinoa eating may help avoid CVDs. In addition to its high nutritional content and gluten-free status, quinoa has been demonstrated to help high-risk customers, such as children, the elderly, lactose intolerant people, and those with anaemia, diabetes, obesity, dyslipidaemia, and celiac disease. Quinoa has a noteworthy nutritional and physiological edge over other grains due to its abundance of phytochemicals. These benefits have been linked to the protein, fibre, vitamins, and minerals, as well as the fatty acid content.

10.8 Processing

The three main steps in the quinoa production chain are grain cultivation, processing, and value-added product development. Before quinoa grains are cooked and made into food, they usually go through a number of post-harvest procedures, such as stacking, drying, threshing, venting, and storing. Additionally, germination, fermentation, milling, and high-temperature treatments are utilized as major processing procedures to enhance the nutritional value, sensory attributes, and edibility of the product (Han et al. 2019; Nickel et al. 2016; Carrizo et al. 2020). But these processing methods frequently have noticeable negative effects and harmful nutritional changes. Thus, it is essential to apply ideas from food science and technology to comprehend how processing modifies the chemical composition of quinoa.

10.8.1 Post-harvest Processing

Post-harvest processing of quinoa involves several essential steps to prepare the grain for consumption and further utilization. These steps include:

Drying or Stacking: Following harvesting, quinoa plants are stacked as soon as they are cut. This can be accomplished in a number of ways, including Chucus (cone-shaped mounds), Arcos (cross stacking), and Taucas (panicles placed in a single direction) (Bojanic 2011; Quiroga et al. 2015).

Threshing: Threshing is the process of separating the quinoa grains from the panicle. This can be done manually, semi-mechanically (using tractors or trucks to pass over the panicles), or fully mechanically, with the latter two methods providing higher processing yields (Bojanic 2011; Quiroga et al. 2015).

Winnowing:

(a) Traditional method: Dependent on the presence of moderate winds with low yield of about 400 kg/day (Bojanic 2011; Kozioł 1992).
(b) Improved manual method: Employing a prototype machine that regulates air flow resulting in a higher yield compared to the traditional method, with yield of 600 kg/h (Bojanic 2011; Kozioł 1992).
(c) Mechanical method: It uses a designed machine for winnowing, providing yield of 500 to 800 kg/h that is higher than the yield of the two mentioned methods (Bojanic 2011; Kozioł 1992).

Storage of Grains: Grains should be kept in polyethylene bags or woven llama wool (Bojanic 2011; Quiroga et al. 2015).

1. In order to achieve quality standards, quinoa seeds must undergo additional processing. Involving a number of processes, such as initial sorting and impurity removal, drying, saponin removal, sorting by size, separating grains according to colour, and removing any remaining contaminants (Bojanic 2011; Quiroga et al. 2015).

(i) In order to achieve quality standards, quinoa seeds must undergo further processing. Involving a number of processes, such as initial sorting and impurity removal, drying, saponin removal, sorting by size, separating grains according to colour, and removing any remaining contaminants (Bojanic 2011; Quiroga et al. 2015).

(ii) **Removal of Saponins**

Saponin removal is the first step in processing quinoa, whether it is for flour or whole grains. This has historically been accomplished by washing the grain under flowing water, occasionally with an alkaline treatment (Simmonds 1965). However, this method utilizes a significant quantity of water and contaminates rivers and lakes (Ruales 1998). Turbulence may increase the washing process's efficiency (Ruales 1998). An alternate strategy to deal with these problems is to physically remove saponins by abrading the saponin-rich pericarp with a tangential abrasive dehulling device (TADD) (Reichert et al. 1986). The best approach seems to be to combine dry abrasive polishing and washing because it reduces nutrient losses and removes saponin-rich pericarp debris.

10.8.2 Consumption Processing Techniques

Various processing techniques are employed in the food manufacturing industry, and their application to quinoa can impact its nutritional value either positively or negatively. The methods to preserve the nutrients and raise the nutritional value of processed quinoa-based products have been the focus of recent research. Milling, germination, fermentation, and high-temperature procedures like baking, roasting, frying, puffing, and extrusion are among the most often used processing methods.

10.8.2.1 Milling and Sieving

In order to improve the sensory qualities of whole grains during processing, milling is a commonly used technique to separate unwanted or harmful chemicals, such as saponins in the pericarp that cause bitterness and anti-nutritional effects (Ahmed et al. 2019). On the other hand, it might cause phenolic chemicals' bioactivity and concentrations in ground goods to decrease. Wet and dry milling are two categories of milling techniques:

The Dry Milling: The goal of dry milling, which is frequently used on cereal grains like wheat or maize, is to separate the anatomical components of the grain. Endosperm material with a high carbohydrate and protein content is produced when the pericarp and germ are removed, which lowers the fibre and fat content. Since quinoa grains are tiny, after saponin is removed, they are usually ground intact. As a result, pericarp may be present in whole-grain quinoa flour if saponins were eliminated by abrasion or may vary in concentration if saponins were eliminated prior to milling. It may be possible to obtain usable grain fractions using roller milling.

Wet Milling: Although technique has demonstrated potential in removing beneficial elements from quinoa grains, wet milling invariably results in some nutrient loss. Effective wet milling techniques that make use of the whole quinoa kernel and reduce nutritional loss have been studied recently. In order to create new methods for producing high-yield quinoa flour with increased nutrient content and better nutritional quality, future research can clarify the connection between milling conditions and the nutritional profile of quinoa.

10.8.2.2 Germination

In seeds, germination is a biological process that creates new molecules through interconversion. Based on the rate at which water is absorbed, this can be split into three phases, each of which has a unique impact on the nutritional qualities of quinoa.

(i) In the first stage, seeds are soaked in water and they absorb water causing cellular materials and matrices to hydrate fully. This process leads to reduction in saponins, thereby reducing the bitter taste of quinoa grains.
(ii) The activation of endogenous enzymes, the release of reserve chemicals, and the encouragement of radicle growth comprise the second stage.
(iii) As seeds continue to absorb water, macromolecules like proteins and lipids are hydrolysed and metabolized in the third stage.

Since many minerals are transported to radicles and act as coenzymes for protein and carbohydrate catalysis, ash content, which is closely correlated with mineral content, tends to decrease during soaking or germination (Aguilar et al. 2019; Bhinder et al. 2021). Additionally, bioactive substances including anti-nutrients and total polyphenols are changed during germination. Additionally, the germination process lessens the bitterness and astringency of quinoa and influences the starch digestibility in vitro; this is mostly because of changes in the amylose chain length distributions that occur during the germination treatment (Ma et al. 2021). To sum up, germination could be a productive way to create quinoa products that are slow to digest starch. In summary, the activation of endogenous enzymes during germination leads to changes in nutrients, the breakdown of starch and non-starch polysaccharides, and an increase in soluble dietary fibres, amino acids, reducing sugars, and peptides.

This procedure yields quinoa malts that can be used to supplement high-quality products distributed to broader populations with additional nutrients, hence encouraging the use of quinoa. Malting, sometimes called sprouting, is the process of activating the natural enzymes that grains create during their germination in a moist environment, hence changing the chemical and structural makeup of grains. It has been noted that malted quinoa flour somewhat lowers the viscosity of wheat flour pastes. Malted grain flour, often known as "power flour" or "amylase-rich flour," has been shown to improve the palatability of porridges for infant weaning (Mosha and Svanberg 1983). It is essential to comprehend the changes that take place during germination in order to produce quinoa-based products that have targeted nutrient enrichments. The possibility of germination to produce minimally processed, highly nutritious goods that can reduce the risk of diseases like hyperglycaemia and hyperlipidaemia has sparked interest in the impacts of germination on cereals in recent years. To increase quinoa seed germination rates, more study should concentrate on integrating germination with cutting-edge technology like microwaves, ultrasonography, or magnetic fields.

10.8.2.3 Fermentation

Traditional fermentation techniques are widely employed to process cereal grains in regions like Asia, Africa, Latin America, and the Pacific Islands. Fermentation plays a crucial role in enhancing the nutritional content, prolonging shelf life, and improving the sensory attributes of food products.

Using the grain as a substrate—either in a liquid or solid state—lactic acid bacteria and other microorganisms change the structure and composition of the grain during the fermentation process. Fermentation offers a way to increase in vitro protein digestibility and micronutrient availability while lowering levels of anti-nutrients like tannins and phytic acid from a nutritional standpoint. Microbial enzyme activity that breaks down the grain's cell wall structure and releases bound phenolic compounds during fermentation also modifies the profile of phenolic compounds (Melini and Melini 2021).

Fermentation is typically used to increase the amounts of beneficial chemicals and biological activity, like antioxidant activity, in quinoa. The quality of fermented quinoa products, which have the potential to become indispensable parts of the human diet and functional foods targeted at averting chronic illnesses, is greatly influenced by the selection of starter strains. However, fermentation technology is still mostly used in homes and laboratories and has not yet reached a commercial level. Products made from fermented quinoa frequently go through a number of steps and are frequently made in conjunction with other methods like milling, germination, and baking. For example, combining fermentation with germination has been shown to improve the end product's nutritional value. Increased peptide, free amino acid, and soluble dietary fibre content can result from sourdough fermentation of sprouted flours, but anti-nutritional elements such phytic acid, condensed tannins, and trypsin inhibitors can be significantly reduced (Montemurro et al. 2019). Furthermore, the concentration of phenolic compounds can be significantly increased by enzymatic hydrolysis in combination with fermentation, namely components such as quercetin, procyanidin B2, procatechuic acid, and catechin (Zhang et al. 2021). The order in which these processes are applied also has some impact on quinoa's nutrient content.

Bread, biscuits, and beverages are just a few of the food items that may be made by fermentation and have significant health benefits. For instance, gluten-free bread made with fermented quinoa flour might enhance the nutritional value and acceptance of conventional maize or rice bread. The need for gluten-free bread that not only complies with dietary requirements but also has a good nutritional content and delicious flavour has increased due to the growing incidence of celiac illness (Jagelaviciute and Cizeikiene 2021).

10.8.2.4 Cooking

Cooking is a common heat-moisture process used to prepare different cereal grains, such as quinoa, for human consumption. This is accomplished by efficiently breaking covalent and hydrogen bonds between molecules under the combined effect of high temperature and moisture, which subsequently results in modifications to the nutritional makeup of quinoa grains. Significantly, boiling lowers the amount of trypsin inhibitors present and denaturates proteins, which lowers the amount of in vitro protein digestibility (Cisneros-Yupanqui et al. 2020). Quinoa's chemical composition varies according on the cooking method used, falling into two main categories:

(i) **Wet Cooking:** This category primarily aims to gelatinize the starch found in starchy grains like quinoa and their flours. This process yields favourable textural changes in the food product and enhances its digestibility. Both normal wet cooking and pressure cooking of whole quinoa grains yield flours with notably low pasting viscosity, whether they are simply dried or drum dried. The reduced pasting viscosities are likely attributed to polymer degradation. Examples of wet cooking methods include boiling and steaming.

(ii) **Dry and Semi-Dry Cooking:** Dry cooking processes are often preferred over wet methods because they eliminate the need for post-cooking drying to produce a shelf-stable product. Examples of dry and semi-dry cooking methods include puffing and extrusion.

Although cooking quinoa grains can cause certain nutrients to be lost, it also improves nutrient bioavailability and the nutritional value of dishes made with quinoa, especially when processing conditions are optimized. A comprehensive comprehension of how cooking affects the nutritional quality of quinoa can enable producers to create suitable cooking methods that optimize quinoa's value for the benefit of customers.

10.8.2.5 Baking or Roasting

According to De Pilli and Alessandrino (2020), baking is a sophisticated unit activity that involves a variety of physicochemical and biological events that affect how digestible and palatable baked items are. Changes in the phenolic composition and an increase in antioxidant capability occur during baking.

A prior investigation demonstrated that quinoa grains, when pan-roasted at 100 °C, exhibited elevated total phenolic and flavonoid contents as well as increased antioxidant activity. Furthermore, prolonged roasting times were associated with higher values in these parameters (Babiker et al. 2022). Research has also explored the influence of different roasting conditions on the antioxidant compounds in quinoa seeds. Optimal conditions for achieving the highest antioxidant activity were identified at 190 °C for 30 min. By encouraging the release of phenolic compounds and the creation of MRPs (Maillard Reaction Products), these circumstances lessened the degradation of heat-sensitive antioxidants. Similarly, during baking, the Maillard reaction increased the bound phenolic concentration. Phenolic content, antioxidant activity, and MRP levels improved with increasing roasting temperature; roasting time had a negligible effect (Carciochi et al. 2016; Gu et al. 2021).

10.8.2.6 Puffing

Puffing is a rapid, high-temperature processing technique applied to quinoa grains to create expanded products that are not only nutritionally improved but also offer extended shelf life and enhanced palatability. Various methods, such as fluidized bed air, sand, gun, extrusion, and microwave, have been employed for puffing, each exerting different effects on the physicochemical and nutritional attributes of quinoa.

Commercially available puffed quinoa products are the result of gun puffing (Hoseney 1994), a process involving the pressurized heating of grains with small water content, followed by instantaneous pressure release through a trip valve. This vaporizes the water within the grain, causing it to expand, particularly gelatinizing the starch and potentially disrupting bran layers.

Notably, gun and extrusion puffing tend to yield superior puffed quinoa compared to microwave puffing (Zapana et al. 2020). Sand puffing, known for its cost-effectiveness, has also been employed to produce puffed quinoa snacks. The temperature during sand puffing plays a crucial role, with low-moisture addition resulting in high yield, expansion, and improved sensory qualities. When opting for extrusion as the puffing medium, careful selection of quinoa varieties is necessary (Kowalski et al. 2016). After puffing, sifting can be used to separate the bitter, saponin-rich bran from the puffed grain. This method has been shown to promote the release of flavonoids and raise the concentration of fat, ash, and carbohydrates. But it tends to lower levels of important amino acids and protein content, especially when it comes to dry-heat puffing (Paucar-Menacho et al. 2018). Therefore, the utilization of simple, low-cost puffing methods or their combination with other pretreatments like extrusion or microwave shows promise for the large-scale manufacturing of nutritious, tasty, and ready-to-eat quinoa products.

10.8.2.7 Extrusion

Extrusion cooking, characterized by high temperatures and short durations, is a widely employed method for semi-dry grain processing (Hoseney 1994). Its advantages lie in its quick and continuous nature compared to other high-temperature food processing techniques.

In the context of quinoa, extrusion processes involve subjecting it to elevated pressure and temperatures, coupled with brief periods of shearing. This treatment alters the starch and protein content, leading to the creation of diverse textures. The physicochemical changes during quinoa extrusion encompass several key reactions:

(a) Starch undergoes gelatinization and dextrinization, proteins denature, complex formation transpires between amylose and lipids, and vitamins experience partial denaturation.
(b) The food material undergoes melting and plasticization.
(c) Expansion occurs through the rapid evaporation of moisture.

In the extrusion process, pearled quinoa is introduced into an extruder, where it undergoes a mechanical and thermal transformation into semi-solid dough. Subsequently, the dough is extruded through the machine and sheared at the outlet using a rotary cutter, resulting in the final product (Przybylski et al. 1994). Essentially, extrusion cooking employs a closely fitted screw within a barrel to facilitate this process. Pearled quinoa, along with minimal water, enters the extruder, undergoing mechanical and thermal transitions to form semi-solid dough. The dough is then pushed through a die, rapidly vaporizing water and expanding the product. A rotary cutter shears the product at the outlet to obtain the final form (Przybylski et al. 1994). This method is commonly utilized to manufacture expanded snack foods.

Quinoa's chemical makeup is significantly altered throughout the extrusion process. Because amylose-lipid complexes are formed during this process, the amount of protein and ash increases while the amount of starch and lipids decreases (Espinosa-Ramirez et al. 2021). But, aside from valine, it might result in the loss of vital amino acids, which could lower the protein's nutritional value (Kuktaite et al. 2022).

Additionally, insoluble dietary fibre content decreases, while soluble dietary fibre content increases due to shear-induced damage to fibre-rich cell walls (Espinosa-Ramirez et al. 2021). Moreover, the extreme heat generated during extrusion cleaves and dehydrates quinoa saponins, ultimately reducing their overall content (Brady et al. 2007).

Overall, the application of extrusion technology in quinoa processing holds significant potential for enhancing its nutritional and techno-functional characteristics.

10.9 Valorization of Quinoa

The surge in quinoa production and consumption has spurred the development of various industrialized quinoa products. Quinoa flour, a versatile product derived from quinoa seeds, finds application in a wide array of flour-based products, including bread, pasta, sponge cakes, and biscuits.

10.9.1 Quinoa Protein

Quinoa has high nutritional value and rich source of lipids, proteins, vitamins, fibre, minerals and excellent balance of amino acids and thus opens avenues for the extraction of valuable compounds to create value-added products (Rao and Poonia 2022). The 15% malted quinoa flour added to the quinoa-based ice cream by the authors improved its sensory qualities, physicochemical characteristics, and bioactive components. About 7.1 g of protein, 29.0 g of carbs, and 184 calories are found in 100 g of quinoa flour. These items include starches, bioactive substances, protein isolates and concentrates, and oil extraction. Quinoa's gluten-free nature makes it suitable for producing gluten-free spaghetti (Caperuto et al. 2000), offering a valuable option for individuals with celiac disease or gluten sensitivity. Quinoa flour is also used in the production of quinoa noodles, catering to those with dietary restrictions and offering unique sensory attributes. Similar studies have been reported by Rao and Poonia (2023). Quinoa can be a crucial component of many other value-added products. The authors looked at the quality attributes of functional Kulfi enhanced with malted quinoa flour. Quinoa's high fibre content and gluten-free nature make it a beneficial food for people with digestive issues and celiac disease.

10.9.2 Quinoa Starch

Very small granule size, low amylose-amylopectin ratio, low gelatinization temperature, high paste viscosity, and exceptional freeze-thaw resilience are just a few of the remarkable properties of quinoa starch. It has potential value because of these qualities. The low-calorie fat replacement potential of quinoa starch is among its most inventive uses. Food items can be made more lubricating by adding small granule starches, like those found in quinoa and amaranth, which simulate fat in the mouthfeel (Whistler and BeMiller 1997).

10.9.3 Quinoa Saponins

Quinoa saponins, often considered waste, have potential applications in various industries due to their surface-active properties. These include use in soaps, detergents, shampoos, cosmetics, beer production, and fire extinguishers (Johnson et al. 1986).

10.9.4 Quinoa Flakes

Quinoa flakes are made by crushing the grain between two converging rollers and then drying it. Research has indicated that the ultimate properties of the flakes are contingent upon the type of quinoa used, as distinct varieties exhibit varying levels of flexibility (Przybylski et al. 1994). Quinoa flakes find utility in diverse food products such as juices, soups, pies, or cakes, thanks to their shorter cooking time compared to other grains (Przybylski et al. 1994).

10.9.5 Tempeh

Quinoa tempeh production has been achieved by utilizing cooked whole grains of both sweet and bitter quinoa varieties, especially after saponin removal (Penaloza et al. 1992).

10.9.6 Instant Infant Porridge

Instant quinoa infant porridge, a convenient product, is created by Ruales from Ecuador and co-workers (Ruales 1998), with the following method, i.e., precooking quinoa flour with hot water and subsequent drum drying (Ruales and Nair 1992). This shelf-stable flour can be reconstituted by consumers into porridge with boiling water.

10.9.7 Food Packages

Quinoa's protein and starch content makes it suitable for producing edible films and serving as an emulsion stabilizer agent, particularly in Pickering emulsions (González et al. 1989).

10.9.8 Quinoa Incorporation

Incorporating quinoa into food products has been shown to extend shelf life and reduce microbial spoilage, likely due to a slower starch degradation rate and the presence of bioactive compounds, such as polyphenols, which can inhibit mould growth (Wilson 1990).

10.10 Safety Considerations

Quinoa is generally safe to consume for the vast majority of people and has gained popularity worldwide due to its nutritional and functional benefits. However, like any food product, there are some safety considerations to kept in mind before employing it in wider applications:

1. **Saponin Content:** Quinoa naturally contains saponins, bitter compounds that act as a natural pesticide. While saponins are generally safe, they can give quinoa a bitter taste and may cause digestive discomfort in some individuals. The saponins are toxic in nature if consumed above the safe levels. To reduce saponin content, it is advisable to wash quinoa seeds thoroughly before cooking. Some commercially available quinoa has already been pre-rinsed to remove saponins.
2. **Autochthonous Pathogens**: Nonetheless, it has been found that autochthonous pathogens with thermal stability are present in quinoa grains, which may have a significant impact on the safety of fermented quinoa products. Creating the right fermentation environment is crucial to limiting the growth of microorganisms and avoiding health risks.
3. **Allergies:** While quinoa is not a common allergen, there have been a few reported cases of quinoa allergies. Individuals with known allergies to related plants, such as beets or spinach, should be cautious when trying quinoa for the

first time. Allergic reactions to quinoa can include hives, itching, and difficulty breathing, so anyone with known allergies should consult with a healthcare provider before consuming quinoa.

4. **Oxalates:** Oxalates are naturally occurring substances present in a variety of foods, including quinoa. Kidney stones can develop in certain individuals due to high oxalate intake. Should an individual have a medical history of renal problems or be at risk for kidney stones, their intake of quinoa should be limited or should be done so only after consulting a healthcare provider.
5. **Digestive Issues:** Some individuals may experience digestive discomfort, such as bloating or gas, after consuming quinoa. This can be due to its high fibre content. Starting with smaller servings and gradually increasing intake can help some people adapt to quinoa's fibre content more easily.
6. **Heavy Metal Contamination:** Like many crops, quinoa can absorb heavy metals from the soil in which it is grown. It is essential to source quinoa from reputable suppliers and regions with lower heavy metal levels to minimize the risk of contamination.
7. **Cross-Contamination:** Cross-contamination can occur if quinoa is processed or stored alongside other grains or products that contain allergens. For individuals with allergies, it is crucial to check labels and choose quinoa products that are certified allergen-free.
8. **Storage:** Proper storage of quinoa is essential to prevent spoilage and the development of mould or pests.
9. **Variety and Source:** Some quinoa varieties may have different nutritional profiles and cooking requirements. One should follow the specific cooking instructions for the type of quinoa, and consider purchasing quinoa from trusted sources to ensure quality and safety.

In summary, while quinoa is generally considered safe and nutritious, it is essential to be aware of potential issues related to saponins, allergies, oxalates, digestive comfort, heavy metal contamination, cross-contamination, and proper storage. By taking these precautions and consuming quinoa in moderation, most people can enjoy its numerous health benefits without any safety concerns.

10.11 Future Challenges and Opportunities

Quinoa is a nutritious and versatile grain which has gained global popularity due to its remarkable health benefits and adaptability to various unfavourable conditions. As we look to the future, there are several challenges and opportunities associated with the use of quinoa.

10.11.1 Challenges

1. **Environmental Sustainability:** Quinoa cultivation has expanded rapidly, leading to concerns about the environmental impact. Issues such as soil degradation, water use, and pesticide use need to be addressed to ensure sustainable production.
2. **Diversity:** Maintaining genetic diversity is crucial to quinoa's resilience against pests, diseases, and changing environmental conditions. Efforts must be made to preserve and expand quinoa's genetic diversity.
3. **Cultivation Practices:** Scaling up quinoa production can lead to monoculture and intensive farming practices that may harm the ecosystem. Promoting sustainable farming methods, such as crop rotation and organic farming, is essential.
4. **Market Access:** Small-scale quinoa farmers often struggle to access international markets due to trade barriers, pricing fluctuations, and lack of infrastructure. Enhancing market access and fair trade practices can benefit these farmers.
5. **Climate Change:** Quinoa is relatively resilient to drought and cold, but climate change may alter the regions where it can be grown. Research and adaptation strategies are needed to address potential challenges related to changing weather patterns.

10.11.2 Opportunities

1. **Nutritional Benefits:** Quinoa is rich in protein, fibre, and essential nutrients. Its health benefits make it a valuable food source for addressing malnutrition and improving public health.
2. **Alternative Crops:** Quinoa can be grown in diverse climates and is often used as a rotational or intercrop with other staples like potatoes or maize. This can enhance soil fertility and reduce the need for chemical fertilizers.
3. **Crop Improvement:** Ongoing research into quinoa's genetics and breeding programmes can lead to improved varieties that are more resistant to pests, diseases, and environmental stressors.
4. **Global Demand:** The increasing global demand for healthy and plant-based foods offers a significant opportunity for quinoa producers to expand their markets and increase exports.
5. **Cultural Exchange:** Quinoa has cultural significance for indigenous communities in South America. Promoting sustainable quinoa production can help preserve and celebrate these cultural traditions.
6. **Value-Added Products:** Quinoa can be used in a variety of products beyond grains, such as flour, flakes, and even snacks. These value-added products can open up new market opportunities.
7. **Research and Education:** Continued research on quinoa's health benefits and agronomic practices, along with education and training for farmers, can enhance the crop's productivity and sustainability.

Quinoa presents both challenges and opportunities in the realm of agriculture, nutrition, and sustainable development. Addressing environmental concerns, improving cultivation practices, and expanding market access are essential to ensuring that quinoa continues to be a valuable and sustainable food source in the future.

10.12 Conclusion

Quinoa's low water requirements and saline surroundings make it a possible agricultural substitute for cereals in adverse settings, particularly in desert locations. In addition to withstanding difficult growing circumstances, it is resilient to pests, birds, insects, and a number of illnesses that can result in large output losses that are brought on by fungi, bacteria, nematodes, and viruses. These days, its great functional and nutritional value is drawing a lot of interest, particularly from health-conscious consumers. This crop contains a wide range of bioactive chemicals that have been identified, including as phytosterols, phytoecdysteroids, bioactive peptides, phenolic compounds, and saponins. These substances have favourable and helpful impacts on various bodily systems, which support human health and lower the risk of certain chronic illnesses and lifestyle choices. The need to develop technology aimed at boosting quinoa's production and productivity is driven by the crop's limited resources and growing demand. It may be cultivated in a wide range of conditions due to its high genetic variability and flexibility. It would be necessary to inform farmers and customers alike about quinoa in order to raise awareness about the crop. Quinoa is a fascinating cash crop with a lot of potential to improve marginalized small farmers' livelihoods.

References

Aguilar J, Miano AC, Obregon J, Soriano-Colchado J, Barraza-Jauregui G (2019) Malting process as an alternative to obtain high nutritional quality quinoa flour. J Cereal Sci 90:102858

Agarwal SK, Rastogi RP (1974) Triterpenoid saponins and their genins. Phytochemistry 13:2623–2645

Alvarez-Jubete L, Wijngaard H, Arendt EK, Gallagher E (2010) Polyphenol composition and in vitro antioxidant activity of amaranth, quinoa buckwheat and wheat as affected by sprouting and baking. Food Chem 119(2):770–778. ISSN 0308-8146. https://doi.org/10.1016/j.foodchem.2009.07.032

Alves LF, Rocha MS, Gomes CCF (2008) Avaliação da qualidade protéica da Quinua Real (Chenopodium quinoa Willd) através de métodos biológicos. e-Scientia 1(1)

Ando H, Chen Y, Tang H, Shimizu M, Watanabe K, Mitsunaga T (2002) Food components in fractions of quinoa seed. Food Sci Technol Res 8(1):80–84

Ahmed J, Thomas L, Arfat YA (2019) Functional, rheological, microstructural and antioxidant properties of quinoa flour in dispersions as influenced by particle size. Food Res Int 116:302–311

Angeli V, Miguel Silva P, Crispim Massuela D, Khan MW, Hamar A, Khajehei F, Graeff-Honninger S, Piatti C (2020) Quinoa (*Chenopodium quinoa* Willd.): an overview of the potentials of the "golden grain" and socio-economic and environmental aspects of its cultivation and marketization. Foods 9(2):216

Aroni JC (2005) Cosecha y poscosecha. In: Serie de módulos publicados ensistemas de producción sostenible en el cultivo de la quinoa. Fundación Proinpa, Fundación Autapo, Embajada Real de los Países Bajos, La Paz, pp 87–102
Babiker EE, Uslu N, Ghafoor K, Al-Juhaimi F, Ozcan MM, Ahmed IAM (2022) Variations in bioactive properties, fatty acid compositions, and phenolic compounds of quinoa grain and oils roasted in a pan. J Food Process Preserv 46(1):e16161
Basu N, Rastogi RP (1967) Triterpenoid saponins and sapogenins. Phytochemistry 6:1249–1270
Beninni ERY, Takahashi HW, Neves CSVJ, Fonseca ICB (2002) Teor de nitrato em alface cultivada em sistemas hidrop^onico e convencional. Hort Brasileira 20(2):183–186
Berti M, Wilckens R, Hevia F, Serri H, Vidal I, Mendez C (2000) Fertilizacion nitrogenada en quinoa (*Chenopodium quinoa* Willd.). Ciencia e Investigacion Agraria 27(2):81–90
Bhargava A, Rana T, Shukla S, Ohri D (2005) Seed protein electrophoresis of some cultivated and wild species of Chenopodium. Biol Plan 49(4):505–511
Bhinder S, Kumari S, Singh B, Kaur A, Singh N (2021) Impact of germination on phenolic composition, antioxidant properties, antinutritional factors, mineral content and Maillard reaction products of malted quinoa flour. Food Chem 346:128915
Bojanic A (2011) Quinoa: an ancient crop to contribute to world food security. FAO Fiat Panis, Rome
Borges JT, Bonomo RC, Paula CD, Oliveira LC, Cesário MC (2010) Physicochemical and nutritional characteristics and uses of quinoa (Chenopodium quinoa Willd.). Temas Agrarios 15(1):9–23
Brady K, Ho CT, Rosen RT, Sang SM, Karwe MV (2007) Effects of processing on the nutraceutical profile of quinoa. Food Chem 100(3):1209–1216
Brinegar C, Goundan S (1993) Isolation and characterization of chenopodin, the 11S seed storage protein of quinoa (Chenopodium quinoa). J Agric Food Chem 41(2):182–185
Brinegar C, Sine B, Nwokocha L (1996) High-cysteine 2S seed storage proteins from quinoa (Chenopodium quinoa). J Agric Food Chem 44(7):1621–1623
Caperuto LC, Amaya-Farfan J, Camargo CRO (2000) Performance of quinoa *(Chenopodium quinoa* Willd.) flour in the manufacture of gluten-free spaghetti. J Sei Food Agric 81:95–10
Capraro J, Benedetti SD, Heinzl GC, Scarafoni A, Magni C (2021) Bioactivities of pseudocereal fractionated seed proteins and derived peptides relevant for maintaining human well-being. Int J Mol Sci 22(7):3543
Carciochi RA, D'Alessandro LG, Manrique GD (2016) Effect of roasting conditions on the antioxidant compounds of quinoa seeds. Int J Food Sci Technol 51(4):1018–1025
Carrizo SL, de LeBlanc ADM, LeBlanc JG, Rollán GC (2020) Quinoa pasta fermented with lactic acid bacteria prevents nutritional deficiencies in mice. Food Res Int 127:108735
Chandel RS, Rastogi RP (1980) Triterpenoid saponins and sapogenins: 1973–1978. Phytochemistry 19:1889–1908
Chauhan GS, Eskin NAM, Tkachuk R (1992) Nutrients and antinutrients in quinoa seed. Cereal Chem 69(1):85–88
Chaudhary N, Walia S, Kumar R (2023) Functional composition, physiological effect and agronomy of future food quinoa (Chenopodium quinoa Willd.): a review. J Food Compos Anal 118:105192
Cisneros-Yupanqui M, Lante A, Mihaylova D, Krastanov AI, Vilchez-Perales C (2020) Impact of consumption of cooked red and black *Chenopodium quinoa* Willd. over blood lipids, oxidative stress, and blood glucose levels in hypertension-induced rats. Cereal Chem 97(6):1254–1262
Comai S, Bertazzo A, Bailoni L, Zancato M, Costa CVL, Allegri G (2007) The content of proteic and nonproteic (free and protein-bound) tryptophan in quinoa and cereal flours. Food Chem 100(4):1350–1355
Cozolino MF (2009) Biodisponibilidade de nutrientes, 3ª edn. Manole, Barueri, SP, 1172 p
Espitia-Hernández P, Chavez Gonzalez ML, Ascacio-Valdés JA, Dávila-Medina D, Flores-Naveda A, Silva T, Chácon XR, Sepúlveda L (2022) Sorghum (Sorghum bicolor L.) as a potential source of bioactive substances and their biological properties. Crit Food Sci Nutr 62(8):2269–2280. https://doi.org/10.1080/10408398.2020.1852389

De Pilli T, Alessandrino O (2020) Effects of different cooking technologies on biopolymers modifications of cereal-based foods: impact on nutritional and quality characteristics review. Crit Rev Food Sci Nutr 60(4):556–565

Díaz-Rizzolo DA, Acar-Denizli N, Kostov B, Roura E, Sisó-Almirall A, Delicado P, Gomis R (2022) Glycaemia fluctuations improvement in old-age prediabetic subjects consuming a Quinoa-based diet: a pilot study. Nutrients 14(11):2331. https://doi.org/10.3390/nu14112331

Dinan L, Whiting P, Scott AJ (2009) Taxonomic distribution of phytoecdysteroids in seeds of members of the chenopodiaceae. Biochem Syst Ecol 26(5):553–576. ISSN 0305-1978. https://doi.org/10.1016/S0305-1978(98)00005-2

Dini A, Rastrelli L, Saturnino P, Schettino O (1992) A compositional study of Chenopodium quinoa seeds. Mol Nutr Food Res Die Nahrung 36(4):400–404

Dini I, Schettino O, Simioli T, Dini A (2001) Studies on the constituents of Chenopodium quinoa seeds: isolation and characterization of new triterpene saponins. J Agric Food Chem 49(2):741–746

Dini I, Schettino O, Simioli T, Dini A (2005) Studies on the constituents of Chenopodium quinoa seeds: isolation and characterization of new triterpene saponins. J Agric Food Chem 49(2):741–746

Espinosa-Ramirez J, Rodriguez A, De la Rosa-Millan J, Heredia-Olea E, Perez-Carrillo E, Serna-Saldivar SO (2021) Shear-induced enhancement of technofunctional properties of whole grain flours through extrusion. Food Hydrocoll 111:106400

Etchevers BJ, Avila TP (1979) Factores que afectan el crecimiento de quinua (Chenopodium quinoa) en al centro-sur de Chile. Pap 10th Lat Am Meet Agric Sci 14:13–29

Farro PCA (2008) Desenvolvimento de filmes biodegradáveis a partir de derivados do grão de quinoa (Chenopodium quinoa Willdenow) da variedade "Real". [Tese]. Campinas: Faculdade de Engenharia de Alimentos, Universidade Estadual de Campinas, 303 p

Foucault AS, Mathe V, Lafont R, Even P, Dioh W, Veillet S, Tome D et al (2011) Quinoa extract enriched in 20-hydroxyecdysone protects mice from dietinduced obesity and modulates adipokines expression. Obesity (Silver Spring) 20(2):270–277

Gandaillas H (1982) Quinoa production. IBTA-CIID (Trans. by Sierra-Blanca Association, Denver, CO, 1985)

Gandarillas H (1979) Botanica. Quinua y Kaniwa. Cultivos Andinos. In: Tapia ME (ed) Serie Libros y Materiales Educativos, vol 49. Instituto Interamericano de Ciencias Agricolas, Bogota, pp 20–44

Garcia M, Condori B, Castillo CD (2015) Agroecological and agronomic cultural practices of quinoa in South America. In: Quinoa: improvement and sustainable production, vol 25. Wiley, Hoboken, p 46

George AJ (1965) Legal status and toxicity of saponins. Food Cosmet Toxicol 3:85–91

González JA, Roldán A, Gallardo M, Escudero T, Prado FE (1989) Quantitative determinations of chemical compounds with nutritional value from inca crops: Chenopodium quinoa ('quinoa'). Plant Foods Hum Nutr 39(4):331–337

Graf BL, Rojas-Silva P, Rojo LE, Delatorre-Herrera J, Baldeon ME, Raskin I (2015) Innovations in health value and functional food development of quinoa (Chenopodium quinoa Willd.). Compr Rev Food Sci Food Saf 14(4):431–445

Graf BL, Rojo LE, Delatorre-Herrera J, Poulev A, Calfio C, Raskin I (2016) Phytoecdysteroids and flavonoid glycosides among Chilean and commercial sources of Chenopodium quinoa: variation and correlation to physicochemical characteristics. J Sci Food Agric 96(2):633–643

Gu R, Chang X, Bai G, Li X, Di Y, Liu X, Sun L, Wang Y (2021) Effects of household cooking methods on changes of tissue structure, phenolic antioxidant capacity and active component bioaccessibility of quinoa. Food Chem 350:129138

Han YM, Chi JW, Zhang MW, Zhang RF, Fan SH, Dong LH, Huang F, Liu L (2019) Changes in saponins, phenolics and antioxidant activity of quinoa (Chenopodium quinoa willd) during milling process. LWT 114:108381. https://doi.org/10.1016/j.lwt.2019.108381

Hemalatha P, Bomzan DP, Rao BVS, Sreerama YN (2016) Distribution of phenolic antioxidants in whole and milled fractions of quinoa and their inhibitory effects on α-amylase and α-glucosidase activities. Food Chem 199:330–338. https://doi.org/10.1016/j.foodchem.2015.12.025

Ho SS, Pal S (2005) Margarine phytosterols decrease the secretion of atherogenic lipoproteins from $HepG_2$ liver and $CaCO_2$ intestinal cells. Atherosclerosis 182(1):29–36

Hoang TV, Vo KTX, Rahman MM, Choi SH, Jeon JS (2019) Heat stress transcription factor OsSPL7 plays a critical role in reactive oxygen species balance and stress responses in rice. Plant Sci: Int J Exp Plant Biol 289:110273. https://doi.org/10.1016/j.plantsci.2019.110273

Hoseney RC (1994) Principles of cereal science and technology, 2nd edn. American Association of Cereal Chemists, St Paul

Hunziker AT (1943) Los especies alimenticias de *Amaranthus* y *Chenopodium* cultivadas por los Indios de America. Rev Argen Agron 30:297–353

Jacobsen SE (2003) The worldwide potential of quinoa (*Chenopodium quinoa* Willd.). Food Rev Int 19(1–2):167–177

Jacobsen SE, Stolen O (1993) Quinoa—morphology, phenology and prospects for its production as a new crop in Europe. Euro J Agron 2(1):19–29

Jacobsen S, Quispe H, Mujica A (2000) Quinoa: an alternative crop for saline soils in the Andes. CIP Progr Rep 1999–2000:403–408

Jagelaviciute J, Cizeikiene D (2021) The influence of non-traditional sourdough made with quinoa, hemp and chia flour on the characteristics of gluten-free maize/rice bread. LWT 137:110457

James LEA (2009) Quinoa (Chenopodium quinoa Willd.): composition, chemistry, nutritional, and functional properties. Adv Food Nutr Res 58:1–31

Jancurova M, Minarovičova L, Dandar A (2009) Quinoa—a review. Czech J Food Sci 27:71–79

Johnson DL, Ward SM (1993) Quinoa. In: Janick J, Simon JE (eds) New crops. Wiley, New York, pp 222–227

Johnson IT, Gee JM, Price KR, Curl CL, Fenwick GR (1986) Influence of saponins on gut permeability and active nutrient permeability in vitro. J Nutr 116:2270–2277

Karyotis T, Iliadis C, Noulas C, Mitsibonas T (2003) Preliminary research on seed production and nutrient content for certain quinoa varieties in a saline-sodic. Soil J Agron Crop Sci 189:402–408

Khattab RY, Arntfield SD (2009) Nutritional quality of legume seeds as affected by some physical treatments. 2. Antinutritional factors. LWT Food Sci Technol 42(6):1113–1118

Kowalski RJ, Medina-Meza IG, Thapa BB, Murphy KM, Ganjyal GM (2016) Extrusion processing characteristics of quinoa (*Chenopodium quinoa* Willd.) var. Cherry Vanilla. J Cereal Sci 70:91–98

Koyro H, Eisa S (2007) Effect of salinity on composition, viability and germination of seeds of Chenopodium quinoa Willd. Plant Soil 302:79–90

Koziol MJ (1990) Chap VIII: Composición química. In: Wahli C (ed) Quinua hacia su cultivo comercial. Latinreco S.A., Quito, pp 137–159

Kozioł MJ (1992) Chemical composition and nutritional evaluation of quinoa (Chenopodium quinoa Willd.). J Food Compos Anal 5:35–68

Kuktaite R, Repo-Carrasco-Valencia R, de Mendoza CCH, Plivelic TS, Hall S, Johansson E (2022) Innovatively processed quinoa (*Chenopodium quinoa* Willd.) food: chemistry, structure and end-use characteristics. J Sci Food Agric 102(12):5065–5076

Li L, Lietz G, Bal W, Watson A, Morfey B, Seal C (2018) Effects of quinoa (Chenopodium quinoa Willd.) consumption on markers of CVD risk. Nutrients 10(6):777. https://doi.org/10.3390/nu10060777

Lopes CO, Dessimoni GV, Da Silva MC, Vieira G, Pinto NAVD (2009) Aproveitamento, composição nutricional e antinutricional da farinha de quinoa (Chenoipodium quinoa). Alimentos e Nutrição 20(4):669–675

Lorenz K (1990) Quinoa (Chenopodium quinoa) starch - physico-chemical properties and functional characteristics. Starch 42(3):81–86

Ma ZM, Guan X, Gong B, Li C (2021) Chemical components and chain-length distributions affecting quinoa starch digestibility and gel viscoelasticity after germination treatment. Food Funct 12(9):4060–4071

Mahan LK, Escott-Stump S (2010) Krause: Alimentos, nutrição e dietoterapia, 12ª edn. Elsevier, Rio de Janeiro, 1351 p

Makharia GK, Singh P, Catassi C, Sanders DS, Leffler D, Ali RAR, Bai JC (2022) The global burden of coeliac disease: opportunities and challenges. Nature reviews. Gastroenterol Hepatol 19(5):313–327. https://doi.org/10.1038/s41575-021-00552-z

Marangoni F, Poli A (2010) Phytosterols and cardiovascular health. Pharmacol Res 61:193–199

Mastebroek HD, Limburg H, Gilles T, Marvin HJP (2000) Occurrence of sapogenins in leaves and seeds of quinoa (Chenopodium quinoa Willd.). J Sci Food Agric 80(1):152–156

Melini F, Melini V (2021) Impact of fermentation on phenolic compounds and antioxidant capacity of quinoa. Fermentation 7(1):20

Montemurro M, Pontonio E, Gobbetti M, Rizzello CG (2019) Investigation of the nutritional, functional and technological effects of the sourdough fermentation of sprouted flours. Int J Food Microbiol 302:47–58

Mora-Ocación MS, Morillo-Coronado AC, Manjarres-Hernández EH (2022) Extraction and quantification of Saponins in Quinoa (Chenopodium quinoa Willd.) Genotypes from Colombia. Int J Food Sci 2022:7287487. https://doi.org/10.1155/2022/7287487. PMID: 35265708; PMCID: PMC8901361

Mosha AC, Svanberg U (1983) Preparation of weaning foods with high nutrient density using flour of germinated cereals. Food Nutrit Bull 5:10–14

Mujica A, Jacobsen SE, Ezquierdo J, Marathee JP (2001) Resultados de la Prueba Americana y Europes de la Quinua. FAO, UNA, CIP, Puno

Ng S, Anderson A, Coker J, Ondrus M (2007) Characterization of lipid oxidation products in quinoa (Chenopodium quinoa). Food Chem 101(1):185–192

Nickel J, Spanier LP, Botelho FT, Gularte MA, Helbig E (2016) Effect of different types of processing on the total phenolic compound content, antioxidant capacity, and saponin content of Chenopodium quinoa Willd grains. Food Chem 209:139–143. https://doi.org/10.1016/j.foodchem.2013.09.155

Nonaka M (1986) Variable sensitivity of Trichoderma viride to Medicago sativa saponins. Phytochemistry 25:73–75

Oelke EA, Putnam DH, Teynor TM, Oplinger ES (1992) Alternative field crops manual. University of Wisconsin Cooperative Extension Service, University of Minnesota Extension Service, Centre for Alternative Plant and Animal Products, Madison

Paucar-Menacho LM, Duenas M, Penas E, Frias J, Martinez-Villaluenga C (2018) Effect of dry heat puffing on nutritional composition, fatty acid, amino acid and phenolic profiles of pseudocereals grains. Polish J Food Nutr Sci 68(4):289–297

Penaloza W, Davey CL, Hedger JN, Kell DB (1992) Physiological studies on the solid-state quinoa tempe fermentation, using on-line measurements of fungal biomass production. J Sei FoodAgric 59:227–235

Pourshahidi LK, Caballero E, Osses A, Hyland BW, Ternan NG, Gill CIR (2020) Modest improvement in CVD risk markers in older adults following quinoa (Chenopodium quinoa Willd.) consumption: a randomized-controlled crossover study with a novel food product. Eur J Nutr 59(7):3313–3323. https://doi.org/10.1007/s00394-019-02169-0

Praznik W, Mundigler N, Kogler A, Pelzl B, Huber A (1999) Molecular background of technological properties of selected starches. Starch 51(6):197–211

Przybylski R, Chauhan GS, Eskin NAM (1994) Characterization of quinoa (Chenopodium quinoa) lipids. Food Chem 51(2):187–192

Oakenfull D, Sidhu GS (1990) Could saponins be a useful treatment for hypercholesterolemia. Eur J Clin Nutr 44:79–88

Obaroakpo JU, Nan W, Hao L, Liu L, Zhang S, Lu J, Pang X, Lv J (2020) The hyperglycemic regulatory effect of sprouted quinoa yoghurt in high-fat-diet and streptozotocin-induced type

2 diabetic mice via glucose and lipid homeostasis. Food Funct 11(9):8354–8368. https://doi.org/10.1039/d0fo01575j

Ogungbenle HN (2003) Nutritional evaluation and functional properties of quinoa (Chenopodium quinoa) flour. Int J Food Sci Nutr 54:153–158

Oliveira AC, Reis SMPM, Carvalho EM, Pimenta FMV, Rios KR, Paiva KC, Sousa LM, Almeida M, Arruda SF (2003) Adições crescentes de ácido fitico à dieta não interferiram na digestibilidade da caseína e no ganho de peso em ratos. Rev Nutr 16(2):211–217

Quiroga C, Escalera R, Aroni G, Bonifacio A, González JA, Villca M, Saravia R, Ruiz A (2015) Traditional processes and technological innovations in quinoa harvesting, processing and industrialization. In: Bazile D, Bertero D, Nieto C (eds) State of the art report of quinoa in the World in 2013. FAO & CIRAD, Rome, pp 258–296

Rao V, Poonia A (2022) Process optimization for the development of quinoa based sugar free ice-cream. Indian J Dairy Sci 75(4):338–346. https://doi.org/10.33785/IJDS.2022.v75i04.006

Rao V, Poonia A (2023) Process development and quality characteristics of functional *Kulfi* enriched with malted quinoa flour. Indian J Dairy Sci 77(1):57–63. https://doi.org/10.33785/IJDS.2024.v77i01.008

Reichert RD, Tatarynovich JT, Tyler RT (1986) Abrasive dehulling of quinoa *(Chenopodium quinoa):* effect on saponin content as determined by an adapted hemolytic assay. Cereal Chem 63:471–475

Repo-Carrasco R, Espinoza C, Jacobsen S-E (2003) Nutritional value and use of the Andean crops quinoa (Chenopodium quinoa) and kañiwa (Chenopodium pallidicaule). Food Rev Int 19(1–2):179–189

Repo-Carrasco-Valencia RAM, Serna LA (2011) Quinoa (Chenopodium quinoa, Willd.) as a source of dietary fiber and other functional components. Cienc Tecnol Aliment 31(1):225–230

Risi J, Galwey NW (1984) The *Chenopodium* grains of the Andes: Inca crops for modern agriculture. Adv Appl Biol 10:145–216

Ruales J (1998) Quinoa *(Chenopodium quinoa,* Willd.): nutritional quality and technological aspects as human food. In: Increasing the utilisation of sorghum, buckwheat, grain amaranth and quinoa for improved nutrition. Institute of Food Research, Norwich, pp 49–64

Ruales J, Nair BM (1992) Nutritional quality of the protein in quinoa (Chenopodium quinoa, Willd) seeds. Plant Foods Hum Nutr 42:1–11

Ruales J, Nair BM (1993a) Content of fat, vitamins and minerals in quinoa (Chenopodium quinoa, Willd) seeds. Food Chem 48(2):131–136

Ruales J, Nair BM (1993b) Saponins, phytic acid, tannis and protease inhibitors in quinoa (Chenopodium quinoa Willd) seeds. Food Chem 48(2):137–143

Ryan E, Galvin K, O'Connor TP, Maguire AR, O'Brien NM (2007) Phytosterol, squalene, tocopherol content and fatty acid profile of selected seeds, grains, and legumes. Plant Foods Hum Nutr 62:85–91

Salas S (2003) Quinoa—postharvest and commercialization. Food Rev Int 19:191–201. https://doi.org/10.1081/FRI-120018885

Santos MAT (2006) Efeito do cozimento sobre alguns fatores antinutricionais em folhas de brocou, couve-flor e couve. Ciencia e Agrotecnologia 30(2):294–301

Schlick G, Bubenheim DL (1996) Quinoa: candidate crop for NASA's controlled ecological life support systems. In: Janick J (ed) Progress in new crops. ASHS Press, Arlington, pp 632–640

Schulte-auf'm-Erley G, Kaul HP, Kruse M, Aufhammer W (2005) Yield and nitrogen utilization efficiency of the pseudocereals amaranth, quinoa and buckwheat under differing nitrogen fertilization. Eur J Agron 22(1):95–100

Shi ZX, Hao YQ, Teng C, Yao Y, Ren GX (2019) Functional properties and adipogenesis inhibitory activity of protein hydrolysates from quinoa (Chenopodium quinoa Willd.). Food Sci Nutr 7(6):2103–2112. https://doi.org/10.1002/fsn3.1052

Simmonds NW (1965) The grain chenopods of the tropical American highlands. Econ Bot 19:223–235

Souza LAC, Spehar CR, Santos RLB (2004) Análise de imagem para determinação do teor de saponina em quinoa. Pesq Agrop Brasileira 39(4):397–401

Spehar CR (2007) Quinoa: alternativa para a diversificação agrícola e alimentar. Ed. Técnico. Planaltina, DF, Brasil, Embrapa Cerrados, 103 p
Tagle MB, Planella MT (2002) La quinoa en la zona central de Chile. In: Supervivencia de una tradicion prehispana. Editorial IKU, Santiago, pp 25–29
Tang Y, Tsao R (2017) Phytochemicals in quinoa and amaranth grains and their antioxidant, anti-inflammatory, and potential health beneficial effects: a review. Mol Nutr Food Res 61(7):1600767. https://doi.org/10.1002/mnfr.201600767
Tang H, Watanabe K, Mitsunaga T (2002) Characterization of storage starches from quinoa, barley and adzuki seeds. Carbohydr Polym 49(1):13–22
Tari TA, Annapure US, Singhal RS, Kulkarni PR (1990) Starch-based spherical aggregates: screening of small granule sized starches for entrapment of a model flavouring compound, vanillin. Carbohydr Polym 53(1):45–51
Tapia ME (1979) Historia y Distribucion geographica. Quinua y Kaniwa. Cultivos Andinos. In: Tapia ME (ed) Serie Libros y Materiales Educativos, vol 49. Instituto Interamericano de Ciencias Agricolas, Bogota, pp 11–15
Tapia ME (2001) Zonificación agro ecológica del cultivo de quinua (Chenopodium quinoa willd). In: Primer taller internacional sobre quinua: recursos genéticos y sistemas de producción. International Potato Center, Lima, Peru, pp 17–27
USDA (2011). https://fdc.nal.usda.gov/fdc-app.html#/food-details/168917/nutrients
Valencia-Chamorro SA (2003) Quinoa. In: Caballero B (ed) Encyclopedia of food science and nutrition, vol 8. Academic Press, Amsterdam, pp 4895–4902
Vega-Gálvez A, Miranda M, Vergara J, Uribe E, Puente L, Martínez EA (2010) Nutrition facts and functional potential of quinoa (*Chenopodium quinoa* willd.), an ancient Andean grain: a review. J Sci Food Agric 90(15):2541–2547. https://doi.org/10.1002/jsfa.4158
Villacr'es E, P'astor G, Quelal MB, Zambrano I, Morales SH (2013) Effect of processing on the content of fatty acids, tocopherols and sterols in the oils of quinoa (Chenopodium quinoa Willd.), lupine (Lupinus mutablis Sweet), amaranth (Amaranthus caudatus L.) and sangorache (Amaranthus quitensis L.). Glob J Food Sci Technol 2(4):44–53
Weber EJ (1978) The Inca's ancient answer to food shortage. Nature 272:486
Whistler RL, BeMiller JN (1997) Carbohydrate chemistry for food scientists. Eagan Press, St Paul, MN
Wilson HD (1990) Quinua and relatives (Chenopodium sect.Chenopodium subsect.Celluloid). Econ Bot 44(Suppl 3):92–110. https://doi.org/10.1007/BF02860478
Wright KH, Pike OA, Fairbanks DJ, Huber CS (2002) Composition of Atriplex hortensis, sweet and bitter Chenopodium quinoa seeds. J Food Sci 67(4):1383–1385
Xu XJ, Luo ZG, Yang QY, Xiao ZG, Lu XX (2019) Effect of quinoa flour on baking performance, antioxidant properties and digestibility of wheat bread. Food Chem 294:87–95. https://doi.org/10.1016/j.foodchem.2019.05.037
Yao Y, Yang XS, Shi ZX, Ren GX (2014) Anti-inflammatory activity of saponins from quinoa (Chenopodium quinoa Willd.) seeds in lipopolysaccharide-stimulated RAW 264.7 macrophages cells. J Food Sci 79(5):H1018–H1023. https://doi.org/10.1111/1750-3841.12425
Zamudio FV, Segura Campos MR (2022) Amaranth, quinoa and chia bioactive peptides: a comprehensive review on three ancient grains and their potential role in management and prevention of Type 2 diabetes. Crit Rev Food Sci Nutr 62(10):2707–2721. https://doi.org/10.1080/10408398.2020.1857683
Zamudio FV, Hidalgo-Figueroa SN, Ortiz Andrade RR, Hernandez Alvarez AJ, Segura Campos MR (2022) Identification of antidiabetic peptides derived from in silico hydrolysis of three ancient grains: Amaranth, quinoa and chia. Food Chem 394:133479. https://doi.org/10.1016/j.foodchem.2022.133479
Zapana F, Bruijn J, Vidal L, Melin P, Gonzalez ME, Cabrera G, Williams P, Borquez R (2020) Physical, chemical and nutritional characteristics of puffed quinoa. Int J Food Sci Technol 55(1):313–322
Zheng Y, Wang X, Zhuang Y, Li Y, Tian H, Shi P, Li G (2019) Isolation of novel ACE-inhibitory and antioxidant peptides from quinoa bran albumin assisted with an in silico approach: char-

acterization, in vivo antihypertension, and molecular docking. Molecules 24(24):4562. https://doi.org/10.3390/molecules24244562
Zhang H, Tsao R (2016) Dietary polyphenols, oxidative stress and antioxidant and anti-inflammatory effects. Curr Opin Food Sci 8:33–42. https://doi.org/10.1016/j.cofs.2016.02.002
Zhang D, Wei X, Liu Z, Wu X, Bao C, Sun Y, Su N, Cui J (2021) Transcriptome analysis reveals the molecular mechanism of GABA accumulation during quinoa (Chenopodium quinoa Willd.) germination. J Agric Food Chem 69(41):12171–12186. https://doi.org/10.1021/acs.jafc.1c02933
Zhang J, Zhang ML, Bai X, Zhang YK, Wang C (2022) The impact of high hydrostatic pressure treatment time on the structure, gelatinization and thermal properties and in vitro digestibility of oat starch. Grain & Oil Sci Technol 5(1):1–12. https://doi.org/10.1016/j.gaost.2022.01.002
Zhu N, Sheng S, Sang S, Jhoo JW, Bai N, Karwe MV et al (2002) Triterpene saponins from debittered quinoa (Chenopodium quinoa) seeds. J Agric Food Chem 50(4):865–867

Amaranth

11

Nikhil Dnyaneshwar Patil, Aarti Bains, and Prince Chawla

Abstract

This chapter provides a comprehensive examination of this ancient grain (Amaranth), covering its history, cultivation, nutritional benefits, culinary uses, and future prospects. Amaranth, revered in ancient civilizations, has regained popularity as a superfood due to its rich nutritional profile and versatility. The chapter delves into the grain's origins in Mesoamerica, highlighting its cultural and spiritual significance, and the impact of the Spanish conquest on its cultivation. Modern cultivation practices, including sustainable farming methods and challenges such as pests and diseases, are discussed. Nutritionally, amaranth is noted for its high protein content, essential vitamins, minerals, and bioactive compounds. Its health benefits, including cardiovascular health, blood sugar regulation, and anti-inflammatory properties, are explored in detail. The chapter also addresses anti-nutrients and toxins present in amaranth, offering strategies for their management through processing techniques like soaking and cooking. Processing methods such as milling, puffing, toasting, and fermentation enhance amaranth's culinary appeal and nutritional value. The chapter highlights the potential for value addition, with amaranth being used in products like flour, snacks, protein powder, and beverages. Safety considerations in production and consumption are emphasized to ensure product quality and consumer health. The chapter concludes by acknowledging the challenges and opportunities in the amaranth industry, emphasizing the grain's role in promoting food security and sustainable agriculture. Collaboration and continued research are deemed essen-

N. D. Patil · P. Chawla (✉)
Department of Food Technology and Nutrition, Lovely Professional University, Phagwara, Punjab, India

A. Bains
Department of Microbiology, Lovely Professional University, Phagwara, Punjab, India

J. Singh et al. (eds.), *Cereals and Nutraceuticals*,
https://doi.org/10.1007/978-981-97-2542-7_11

tial for unlocking amaranth's full potential as a nutritious and versatile food source.

Keywords

Amaranth · Cultivation · Nutritional profile · Application · Health benefits

11.1 Introduction

In the ever-evolving landscape of agriculture and nutrition, the humble amaranth plant has emerged as a remarkable and versatile grain. Its story spans thousands of years, from its origins as a staple food in ancient civilizations to its resurgence as a sought-after superfood in contemporary diets (Barrette et al. 2020). This comprehensive chapter explores the multifaceted world of amaranth, delving into its history, cultivation practices, nutritional composition, culinary applications, health implications, and future prospects. From the intriguing origins and historical significance of amaranth to the modern challenges and opportunities it presents, each section of this chapter offers a glimpse into the diverse and dynamic journey of this resilient grain (Wesley and Koyiyumptewa 2021). We will unravel the intricate chemical composition that makes amaranth a nutritional powerhouse, examining the vitamins, minerals, and macronutrients that have earned it a place in health-conscious diets.

While amaranth's nutritional benefits are abundant, we will also explore the presence of anti-nutrients and potential toxins, shedding light on how these compounds can be managed to maximize the health benefits of amaranth consumption (Nyonje 2022). Additionally, we will uncover the bioactive and functional components that contribute to amaranth's positive effects on human health, highlighting its potential as a functional food (Najafian 2023). Drawing from scientific research and studies, we will outline the health implications of incorporating amaranth into one's diet, from its impact on cardiovascular health to its role in weight management. We will also explore the myriad ways in which amaranth is consumed worldwide, from traditional dishes to innovative culinary creations that reflect its adaptability.

Processing techniques play a pivotal role in making amaranth more palatable and versatile, and we will examine various methods, from milling to puffing, that enhance amaranth's culinary appeal (Yousaf et al. 2021). Moreover, we will delve into the realm of value addition, exploring how amaranth can be transformed into an array of products to meet diverse consumer needs (Gunarathne et al. 2022).

Ensuring the safety of amaranth products, from production to consumption, is paramount, and we will discuss safety measures and guidelines for mitigating potential risks (Vinke et al. 2022). Finally, we will gaze into the future of amaranth, acknowledging the challenges it faces while embracing the opportunities that lie ahead (Zitcer 2021). As we conclude this chapter, we reflect on amaranth's role as a sustainable and nutritious food source and its potential to shape a healthier and more food-secure world.

11.2 Origin and Background

Amaranth, often referred to as the "ancient grain," has a rich history that spans thousands of years. This chapter explores the origins and historical significance of amaranth, shedding light on its cultural and agricultural importance.

11.2.1 The Ancient Origins of Amaranth

Amaranth's story begins in the Americas, where it was cultivated and revered by ancient civilizations long before the arrival of Europeans. Evidence of amaranth cultivation dates back over 8000 years, with its roots in Mesoamerica, primarily in what is now Mexico and Central America (Schubart et al. 2022). The Aztecs, in particular, held amaranth in high regard, considering it a staple crop alongside maize and beans. They called it "huautli" and used it in various dishes, including a sweet, popped amaranth treat mixed with honey known as "alegría" (Balakrishnan and Schneider 2022).

11.2.2 Cultural and Spiritual Significance

Amaranth played a crucial role in the cultural and spiritual practices of indigenous peoples. It was used in religious ceremonies and offerings, often symbolizing immortality due to its remarkable ability to thrive in diverse environmental conditions (Sandstrom and Sandstrom 2023). Amaranth was also utilized as a source of sustenance for warriors and travelers due to its nutrient-dense nature and the ability to produce bountiful yields even in challenging terrain (Christopher 2023).

11.2.3 The Spanish Conquest and Suppression

Despite its cultural significance, the cultivation and consumption of amaranth faced severe suppression during the Spanish conquest of the Americas. The conquistadors, who aimed to convert indigenous populations to Christianity, perceived amaranth as a threat to their religious goals (Rinke 2023). This was because the indigenous people used amaranth in ceremonies and rituals. As a result, the Spanish authorities banned amaranth cultivation and even punished those found growing it (García 2021).

11.2.4 Resurgence in the Modern Era

The suppression of amaranth continued for centuries, leading to a decline in its cultivation and consumption among indigenous communities. However, amaranth did not vanish entirely (Turner et al. 2021). In the late twentieth century, there was a resurgence of interest in amaranth as a nutritious and resilient crop. Researchers and agriculturalists recognized its potential to combat malnutrition and contribute to food security (Ram et al. 2023). Consequently, efforts were made to reintroduce amaranth cultivation, both in its native regions and around the world (Ruth et al. 2021).

11.2.5 A Global Perspective

Today, amaranth is cultivated not only in the Americas but also in various parts of the world, including Africa, Asia, and Europe. Its adaptability to diverse climates and its remarkable nutritional profile make it a valuable crop for addressing global food security challenges (Chandra et al. 2021; Imathiu 2021).

In conclusion, amaranth's journey from its ancient origins in the Americas, through suppression during the Spanish conquest, to its modern resurgence and global cultivation, is a testament to its historical and cultural significance. Understanding this rich history provides valuable context for appreciating the role of amaranth in today's agricultural and culinary landscape.

11.3 Production and Cultivation Practices

In this section, we delve into the techniques and practices involved in the cultivation of amaranth. From ideal growing conditions to sustainable farming methods, readers will gain insight into how this resilient crop is produced.

11.3.1 Choosing the Right Growing Conditions

Cultivating amaranth successfully begins with selecting the right growing conditions. While amaranth is known for its adaptability, certain factors can influence its growth and yield (Waselkov et al. 2020).

11.3.1.1 Climate

Amaranth thrives in a variety of climates, but it is essential to choose the right variety for your region. Some varieties are better suited to tropical climates, while others are more cold-tolerant (Ruth et al. 2021). Adequate sunlight and warmth are generally essential for optimal growth (Kumar et al. 2019).

11.3.1.2 Soil

Amaranth prefers well-draining soils with a pH level between 6.0 and 7.5. It can grow in a range of soil types, but sandy loam or loamy soil is often ideal. Soil preparation, including proper fertilization and organic matter incorporation, is crucial (Komlaga et al. 2019; Panagea et al. 2021).

11.3.1.3 Water

Amaranth has moderate water requirements but can tolerate drought better than some other crops. Irrigation should be consistent with an emphasis on not allowing the soil to become waterlogged (Norain et al. 2021).

11.3.2 Sustainable Farming Methods

Sustainability is a growing concern in modern agriculture, and amaranth can be cultivated using environmentally friendly practices.

11.3.2.1 Crop Rotation

Like other crops, amaranth benefits from crop rotation. It helps prevent soil depletion and reduces the risk of pests and diseases. Alternating amaranth with other crops in a rotation plan can enhance soil health (Schmidt et al. 2023).

11.3.2.2 Organic Farming

Many amaranth growers opt for organic farming practices. This means avoiding synthetic fertilizers and pesticides and relying instead on natural alternatives. Organic farming promotes soil health and minimizes environmental impact (Boateng and Donkoh 2023; Swarnalakshmi and Sridevi 2023; Abolarin 2021).

11.3.2.3 Interplanting and Companion Planting

Amaranth can be interplanted with other crops to maximize space and resources. Companion planting with compatible species can also help deter pests and improve overall crop health (Kumpf 2023).

11.3.2.4 Integrated Pest Management (IPM)

IPM strategies involve monitoring for pests and diseases and using targeted, environmentally friendly methods to control them (Zayan 2018). This reduces the need for chemical pesticides (Saeed Ben Youssef 2023).

11.3.2.5 Efficient Harvesting Techniques

Timing is critical when harvesting amaranth. Leaves are typically harvested when they are young and tender, while seeds are collected when they are fully mature. Proper harvesting techniques can optimize yields (Beam et al. 2019; Emmanuel and Babalola 2022).

11.3.3 Propagation and Seed Selection

Amaranth can be propagated through seeds. Selecting the right seeds is crucial for a successful crop (Buddenhagen et al. 2023). Farmers often consider factors like seed purity, germination rates, and disease resistance. Saving seeds from healthy, productive plants can also help maintain desirable traits in future crops (Gebeyehu et al. 2019).

11.3.4 Challenges in Amaranth Cultivation

While amaranth is generally resilient, it does face certain challenges:

11.3.4.1 Weeds

Weeding is essential in amaranth cultivation, as it can be competitive with weeds. Mulching and proper spacing can help reduce weed pressure (Pavlović et al. 2022).

11.3.4.2 Pests and Diseases

Common pests include aphids and leafhoppers, while fungal diseases like powdery mildew can affect amaranth (Umetsu and Shirai 2020). Vigilant monitoring and early intervention are key to managing these issues (Patrícia et al. 2020).

11.3.4.3 Seed Shattering

Some amaranth varieties have seeds that shatter easily, making harvesting more challenging. Selecting non-shattering varieties or using appropriate harvesting techniques can mitigate this issue (Schwartz-Lazaro et al. 2021).

In conclusion, successful amaranth cultivation requires careful consideration of growing conditions, sustainable farming practices, and the selection of appropriate seeds. By addressing these factors, farmers can maximize the yield and quality of this resilient and nutritious crop while minimizing environmental impact.

11.4 Consumption Patterns

Amaranth, often celebrated as a nutrient-rich superfood, has experienced a resurgence in popularity in recent years. This section explores the diverse consumption patterns of amaranth, from traditional dishes to modern culinary innovations, shedding light on its versatility and growing global appeal.

11.4.1 Traditional Uses

11.4.1.1 Porridges and Grains

One of the most common traditional uses of amaranth is in porridge or as a cooked grain. In many cultures, amaranth seeds are boiled or simmered until they reach a soft, porridge-like consistency (Anitha et al. 2022). This basic preparation can be consumed with sweet or savory accompaniments (Sachithra and Chamila 2020).

11.4.1.2 Flatbreads

Amaranth flour is used to make flatbreads in various parts of the world. In India, for example, it is used to create roti, a type of unleavened bread. These flatbreads are often enjoyed with vegetables, curries, or chutneys (Piga et al. 2021).

11.4.1.3 Snacks and Sweets

Amaranth is frequently used to create snacks and sweets. In Mexico, for instance, amaranth seeds are mixed with honey to make "alegría," a traditional sweet treat. In South America, popped amaranth is mixed with molasses to make "sweets" called "dulce de Alegria" (Ruth et al. 2021; Rojas-Rivas et al. 2019).

11.4.1.4 Beverages

Amaranth can also be used to make beverages. For instance, in parts of Africa, amaranth leaves are used to prepare nutritious, leafy green tea (Shembe et al. 2023).

11.4.2 Modern Culinary Innovations

11.4.2.1 Gluten-Free Baking

Amaranth flour is a valuable ingredient in gluten-free baking. It adds a nutty flavor and helps improve the texture of gluten-free bread, muffins, and pancakes (Dragomir and Bahaciu 2019).

11.4.2.2 Salad Greens

Amaranth leaves, particularly young and tender ones, are increasingly used in salads. They provide a fresh, slightly peppery flavor and a nutrient boost to green salads (Rai et al. 2022).

11.4.2.3 Grain Bowls

Amaranth grains can be used as the base for grain bowls, similar to quinoa or rice bowls. They serve as a nutritious foundation for a variety of toppings, including vegetables, proteins, and flavorful sauces (Howard 2021; Mackinnon 2023).

11.4.2.4 Soups and Stews

Amaranth leaves and grains can be added to soups and stews for an added nutritional punch. They contribute both flavor and thickness to these dishes (Holdsworth et al. 2020).

11.4.2.5 Breakfast Cereals

Amaranth flakes or puffs can be used in breakfast cereals, providing a crunchy texture and nutty flavor. They are often combined with other grains, nuts, and dried fruits (Barr 2019).

11.4.2.6 Smoothies

Amaranth can be incorporated into smoothies, either in the form of cooked grains or amaranth flour. It adds thickness and nutrition to these blended beverages (Mccormick 2020).

11.4.3 Global Appeal and Health Benefits

The diverse culinary applications of amaranth have contributed to its growing global appeal. Its nutritional profile, which includes high levels of protein, fiber, vitamins, and minerals, makes it an attractive addition to a wide range of diets (Pereira et al. 2020; Shembe et al. 2023).

11.4.3.1 Gluten-Free and Plant-Based

Amaranth is naturally gluten-free, making it an excellent choice for those with gluten sensitivities or celiac disease. It is also a plant-based source of complete protein, meaning it contains all the essential amino acids (Rao and Poonia 2023).

11.4.3.2 Nutrient Density

Amaranth is rich in essential nutrients, such as iron, magnesium, calcium, and B vitamins, making it a valuable addition to diets aiming to address nutritional deficiencies (Olusanya et al. 2023).

11.4.3.3 Versatile and Sustainable

Amaranth's adaptability to diverse climates and its ability to thrive with minimal water make it a sustainable crop. Its versatility in the kitchen, from sweet to savory dishes, ensures its place on the tables of many cultures (Neumann et al. 2023).

In conclusion, amaranth's consumption patterns are a testament to its adaptability and appeal across various culinary traditions. As it continues to gain recognition for its health benefits and versatility, amaranth is likely to become an even more prominent ingredient in global cuisine.

11.5 Chemical Composition

Amaranth, often referred to as an "ancient grain," boasts a remarkable nutritional profile that has garnered attention from health-conscious consumers and researchers alike. In this section, we will take a deep dive into the chemical composition of amaranth, uncovering the vitamins, minerals, and macronutrients that make it a nutritional powerhouse.

11.5.1 Macronutrients

Macronutrients are the primary components of our diet, providing the energy and building blocks our bodies need for various physiological processes. Amaranth contains three essential macronutrients (Pérez-Beltrán et al. 2022). Amaranth is a carbohydrate-rich grain. The majority of its carbohydrates come from starch, which serves as an energy source. The carbohydrate content in amaranth is similar to that of rice or wheat (Olagunju et al. 2021). However, what sets it apart is its significant fiber content. Amaranth contains both soluble and insoluble fiber, promoting digestive health, reducing the risk of digestive disorders, and potentially aiding in weight management. The high fiber content also contributes to a gradual release of energy, helping to stabilize blood sugar levels (Singhania et al. 2023). Amaranth is renowned for its protein content, particularly among grains. It contains approximately 13–14% protein by weight, which is higher than most other grains. What makes amaranth's protein particularly valuable is its amino acid profile (Szabóová et al. 2020). It contains all the essential amino acids, including lysine, which is often limited in other plant-based foods. This makes amaranth an excellent source of high-quality protein for vegetarians and vegans (Langyan et al. 2022). Additionally, the protein in amaranth is highly digestible. While amaranth is not particularly high in fat, it does contain a modest amount of healthy fats, primarily unsaturated fats. These fats are important for energy storage, as they provide a concentrated source of calories (Singhania et al. 2023). Furthermore, these unsaturated fats, including oleic acid and linoleic acid, are heart-healthy and have been associated with various health benefits, such as reduced risk of cardiovascular diseases (Gülsoy et al. 2023).

11.5.2 Vitamins and Minerals

Amaranth is rich in a variety of essential vitamins and minerals that play crucial roles in maintaining health and well-being. Here are some key nutrients found in amaranth. Furthermore, amaranth is an excellent source of iron, a vital mineral that supports the formation of red blood cells and oxygen transport in the body (Ruth et al. 2021). This is particularly valuable for individuals at risk of iron deficiency, such as pregnant women and those with iron-deficiency anemia (Sharanya et al. 2023). The iron in plant-based foods like amaranth is non-heme iron, which is more readily absorbed when consumed with vitamin C-rich foods. Amaranth is calcium-rich, contributing to strong bones and teeth, as well as muscle function, blood clotting, and nerve signaling (Akbulut and Yeşilkaya 2021). It is an essential mineral, especially for growing children and older adults. The calcium in amaranth is highly bioavailable, meaning the body can absorb it efficiently. Magnesium is abundant in amaranth and plays a role in muscle and nerve function, blood glucose control, and bone health (Marcel et al. 2022). It also supports hundreds of enzymatic reactions in the body, making it a vital mineral for overall health. Amaranth provides a significant amount of phosphorus, which is essential for the formation and maintenance of strong bones and teeth. Phosphorus is also involved in energy metabolism and DNA synthesis (Akbulut and Yeşilkaya 2021). Potassium is important for maintaining proper muscle and nerve function, regulating blood pressure, and supporting overall cardiovascular health. Adequate potassium intake is associated with a reduced risk of stroke, kidney stones, and high blood pressure. Amaranth contains various B vitamins, including folate (B9), riboflavin (B2), niacin (B3), and vitamin B6 (Katayama and Saito 2019). These vitamins are essential for energy metabolism, DNA synthesis, and overall cellular function. Folate, in particular, is crucial during pregnancy as it helps prevent neural tube defects in the developing fetus (Maguolo et al. 2023). Amaranth contains vitamin E, an antioxidant that helps protect cells from oxidative damage. Vitamin E plays a role in immune function, skin health, and vision. While not as high as some other plant sources, amaranth does contain vitamin C, an antioxidant important for immune function and skin health. Vitamin C also enhances the absorption of non-heme iron found in plant-based foods like amaranth (De 2020).

11.5.3 Phytochemicals and Antioxidants

In addition to vitamins and minerals, amaranth contains various phytochemicals and antioxidants, such as flavonoids and polyphenols. These compounds have been associated with a range of health benefits, including reduced inflammation and protection against chronic diseases (Sarker et al. 2020). Phytochemicals can help neutralize harmful free radicals in the body, potentially reducing the risk of conditions like heart disease and cancer (Baliyan et al. 2022).

In summary, amaranth's chemical composition showcases its status as a nutritional powerhouse. Its balanced combination of macronutrients, vitamins, and minerals, along with phytochemicals and antioxidants, makes it a valuable addition to a diverse and health-conscious diet. Whether as a source of complete protein or a provider of essential vitamins and minerals, amaranth continues to earn its reputation as a nutritious and versatile grain, offering numerous health benefits for individuals of all ages and dietary preferences.

11.6 Anti-nutrients and Toxic Compounds in Amaranth

Amaranth, celebrated for its nutritional virtues, is not exempt from the presence of anti-nutrients and potential toxins. In this section, we will conduct an exhaustive exploration of these compounds, elucidating their nature, effects, and strategies for managing them to optimize the health benefits of amaranth consumption.

11.6.1 Anti-nutrients in Amaranth

Amaranth, like many leafy greens and grains, contains oxalates. These naturally occurring organic compounds can form insoluble crystals when they combine with calcium (Popova and Mihaylova 2019). In the human body, high oxalate consumption may increase the risk of kidney stone formation, particularly in individuals predisposed to kidney stones. However, it is important to note that amaranth's oxalate levels are generally not exceptionally high, and most individuals can consume it without any adverse effects (Zainodini et al. 2023). Moreover, cooking amaranth can help reduce oxalate content, making it a safer option for those concerned about kidney stone formation. Phytates are compounds present in many plant-based foods, including amaranth (Sarkar et al. 2023). They have the potential to bind to essential minerals such as iron, zinc, and calcium, forming complexes that are poorly absorbed in the digestive tract (Caetano-Silva et al. 2021). Overconsumption of phytate-rich foods can lead to mineral deficiencies, especially in populations with limited dietary diversity. To mitigate this effect, various food processing techniques, such as soaking, sprouting, or fermenting amaranth, can be employed to reduce phytate levels and enhance mineral bioavailability (Malhotra et al. 2023). Tannins are polyphenolic compounds found in some amaranth varieties. They can interfere with mineral absorption, particularly iron, and may also reduce protein digestibility. However, the levels of tannins in amaranth are generally not considered high, and their impact on nutrient absorption is likely minimal (Manyelo et al. 2020). Tannins are also associated with a characteristic bitter taste, which may be less pronounced in amaranth compared to some other foods. Amaranth contains protease inhibitors, which are compounds that can interfere with protein digestion by inhibiting enzymes responsible for breaking down proteins (Cadenas et al. 2021). Excessive consumption of foods with protease inhibitors could potentially lead to reduced protein absorption. Fortunately, thorough cooking or processing can help reduce the activity of these inhibitors, making the protein in amaranth more digestible (Reyes-Jurado et al. 2023).

11.6.2 Toxic Compounds in Amaranth

Certain leafy varieties of amaranth may accumulate nitrates, particularly when grown in nitrogen-rich soils or with the excessive use of nitrogen-based fertilizers. Nitrates themselves are relatively harmless, but they can be converted into nitrites in the body under specific conditions (O'hara 2020). High levels of nitrites can lead to health issues, including methemoglobinemia, a condition that affects the oxygen-carrying capacity of the blood. To minimize nitrate accumulation in amaranth, it is advisable to select varieties grown in soils with low nitrate levels and to use fertilizers judiciously. Saponins are naturally occurring compounds found in some amaranth varieties (Banerjee et al. 2023). While saponins have diverse biological activities, including potential health benefits, some can have toxic effects when consumed in large quantities. Cooking amaranth thoroughly can reduce saponin content, rendering it safe for consumption (Hussain et al. 2021). Additionally, saponins are often associated with a bitter taste, which can be mitigated through appropriate cooking techniques or by selecting varieties with lower saponin levels. Amaranth, like many crops, can accumulate heavy metals from contaminated soil or water sources (Wang et al. 2022). Cadmium, lead, and arsenic are examples of heavy metals that can be present in amaranth. The consumption of amaranth with elevated levels of heavy metals poses health risks, including toxicity (Li et al. 2021). To mitigate heavy metal exposure, it is essential to source amaranth from reputable suppliers and regions with low heavy metal contamination. Regular testing for heavy metals is advisable for both growers and manufacturers (Aber et al. 2019).

11.6.3 Managing Anti-nutrients and Toxins

Efficient management of anti-nutrients and potential toxins in amaranth is crucial for optimizing its health benefits. Incorporating a diverse range of foods in the diet helps mitigate the impact of anti-nutrients by ensuring that minerals and nutrients are sourced from a variety of foods (Jan et al. 2023). This helps compensate for any reduction in mineral absorption caused by anti-nutrients in amaranth. Certain processing methods, such as soaking, sprouting, fermenting, and cooking, can reduce the levels of anti-nutrients in amaranth (Sahu et al. 2020). For example, soaking and cooking amaranth can lower phytate content and improve mineral absorption. When selecting amaranth varieties, consider their natural content of anti-nutrients and potential toxins (Aderonke Ibidunni Olagunju et al. 2022). Varieties that are low in oxalates, saponins, and nitrates may be preferred if concerns about these compounds exist. Pay attention to the soil and water quality in the cultivation of amaranth (Ruth et al. 2021). Ensuring that amaranth is grown in soils with low heavy metal contamination and using safe irrigation water sources is essential for minimizing heavy metal accumulation. Compliance with food safety and quality regulations, including limits on heavy metal contamination, is essential for both producers and consumers (Obaid et al. 2023). Regular testing of amaranth products for heavy metals and other potential contaminants can help ensure their safety (Laboni et al. 2023).

In conclusion, while amaranth offers a wealth of nutritional benefits, it is prudent to be aware of the presence of anti-nutrients and potential toxins in this grain. By adopting proper processing methods, diversifying the diet, selecting suitable varieties, and monitoring soil and water quality, it is possible to manage these compounds effectively and safely incorporate amaranth into a balanced and health-conscious diet.

11.7 Bioactive and Functional Components in Amaranth

Amaranth, celebrated for its exceptional nutritional profile, houses an array of bioactive and functional components that underpin its numerous health benefits. In this section, we embark on a comprehensive exploration of these compounds, unveiling their nature, mechanisms of action, and their profound positive effects on human health.

11.7.1 Bioactive and Functional Components

Amaranth seeds are a noteworthy source of squalene, a natural hydrocarbon compound that has gained attention for its potential health benefits. Squalene is a precursor to compounds like cholesterol and steroid hormones, and it is also recognized for its antioxidant properties (Gohil et al. 2019). These antioxidants help protect cells from oxidative damage, reducing the risk of chronic diseases such as heart disease and cancer. Amaranth contains beta-glucans, a type of soluble fiber renowned for its ability to reduce cholesterol levels in the blood. Beta-glucans form a viscous gel in the digestive tract, which can bind to cholesterol and prevent its absorption (Mehdi et al. 2020). This mechanism contributes to improved heart health by lowering LDL cholesterol, commonly referred to as "bad" cholesterol. Amaranth is rich in various phytochemicals, including flavonoids and polyphenols (Rao and Poonia 2023). These compounds exert antioxidant and anti-inflammatory effects in the body. By neutralizing harmful free radicals and modulating inflammatory pathways, phytochemicals in amaranth may help reduce the risk of chronic diseases such as cardiovascular disease, diabetes, and certain cancers (Shahbazi et al. 2021). While saponins were discussed in the previous section as potential anti-nutrients, they also possess bioactive properties that can benefit health. Some saponins have demonstrated immune-modulating effects and anti-inflammatory properties (Timilsena et al. 2023). Additionally, saponins may play a role in reducing the risk of certain chronic diseases through their interaction with cholesterol and other lipids. Amaranth contains lignans, which are phytochemicals with antioxidant and estrogenic properties (Shi et al. 2019). These compounds may offer protection against hormone-related conditions, such as breast and prostate cancers. Lignans also play a role in gut health by acting as prebiotics, promoting the growth of beneficial gut bacteria (Pabich and Materska 2019). Amaranthin is a peptide found in amaranth that has shown potential as an anti-inflammatory agent. It may help mitigate inflammation-related disorders and contribute to overall well-being (Rao and Poonia 2023).

11.7.2 Positive Effects on Human Health

The beta-glucans and antioxidants in amaranth contribute to cardiovascular health by lowering LDL cholesterol levels, reducing oxidative stress, and improving blood vessel function. These effects collectively reduce the risk of heart disease and stroke. Amaranth's high fiber content, particularly soluble fiber, can help regulate blood sugar levels by slowing the absorption of glucose (Salvador et al. 2021). This is especially beneficial for individuals with diabetes or those at risk of developing the condition. The fiber-rich nature of amaranth promotes satiety, helping individuals feel full for longer periods (Selby and Taal 2020). This can aid in weight management by reducing overall calorie intake. The phytochemicals, saponins, and amaranthin in amaranth contribute to its anti-inflammatory properties. Chronic inflammation is linked to various diseases, and reducing inflammation can have far-reaching health benefits (Janmohammadi et al. 2023). The antioxidants in amaranth help protect cells from oxidative damage caused by free radicals. This antioxidant defense may reduce the risk of chronic diseases associated with oxidative stress, such as cancer and neurodegenerative disorders (Zhu et al. 2024). Amaranth's fiber content, including both soluble and insoluble fibers, promotes digestive health by preventing constipation, supporting regular bowel movements, and nurturing a healthy gut microbiome (Allai et al. 2022).

In summary, amaranth's bioactive and functional components serve as potent contributors to its diverse array of health benefits. From cardiovascular protection and blood sugar regulation to anti-inflammatory and antioxidant effects, amaranth emerges as a nutritional powerhouse with the potential to enhance overall well-being and reduce the risk of chronic diseases. Its role as a versatile and nutritious grain extends beyond its macronutrient and micronutrient content, making it a valuable addition to a health-conscious diet.

11.8 Health Implications of Amaranth Consumption

11.8.1 Cardiovascular Health

Amaranth's positive impact on cardiovascular health is multifaceted. Its beta-glucans, a soluble fiber, play a crucial role in reducing LDL (low-density lipoprotein) cholesterol levels (Ramakrishna et al. 2019). By binding to cholesterol and preventing its absorption, beta-glucans contribute to the prevention of atherosclerosis and coronary heart disease (Fu et al. 2020). Additionally, amaranth's antioxidants combat oxidative stress and inflammation in blood vessels, enhancing their function and reducing the risk of cardiovascular events such as heart attacks and strokes. The combined effects of reduced LDL cholesterol and improved vascular health make amaranth a valuable addition to heart-healthy diets (Tofas et al. 2019).

11.8.2 Blood Sugar Regulation

Amaranth's high soluble fiber content, specifically beta-glucans, plays a pivotal role in blood sugar regulation. When consumed, soluble fiber forms a gel-like substance in the digestive tract, slowing the absorption of glucose (Galanakis 2019). This gradual release of glucose into the bloodstream helps stabilize blood sugar levels, making amaranth a suitable choice for individuals with diabetes or those at risk of developing the condition. Its ability to prevent rapid spikes in blood sugar is instrumental in managing glycemic control (Okaiyeto et al. 2023; Patlak 2002).

11.8.3 Weight Management

Amaranth's role in weight management is attributed to its satiating nature. The combination of soluble and insoluble fiber in amaranth promotes a sense of fullness and satiety, leading to reduced overall calorie intake (Martínez-Villaluenga et al. 2020). By helping individuals control their appetite and manage portion sizes, amaranth supports weight loss and weight maintenance efforts. This characteristic is particularly valuable for those aiming to achieve or maintain a healthy body weight (Zamudio and Campos 2022).

11.8.4 Anti-inflammatory Effects

Chronic inflammation is a hallmark of numerous chronic diseases, including cardiovascular disease, diabetes, and certain cancers. Amaranth's rich array of phytochemicals, including flavonoids and polyphenols, possess anti-inflammatory properties (Agrawal et al. 2023). They help modulate inflammatory pathways in the body, reducing the risk of chronic inflammation-related conditions. Incorporating amaranth into one's diet may contribute to a more favorable inflammatory profile and overall well-being (Venkatesh et al. 2020).

11.8.5 Antioxidant Protection

Amaranth's antioxidants, including vitamin E and various phytochemicals, play a pivotal role in cellular protection. These compounds neutralize harmful free radicals, preventing oxidative damage to cells and tissues (Michalak 2022). The antioxidant defense provided by amaranth may reduce the risk of chronic diseases associated with oxidative stress, such as cancer and neurodegenerative disorders. It also contributes to the preservation of skin health and immune function (Filleria et al. 2021).

11.8.6 Digestive Health

Amaranth's diverse fiber content, comprising both soluble and insoluble fibers, has a favorable impact on digestive health. Soluble fiber supports regular bowel movements by preventing constipation and promoting gut motility (Gowda et al. 2022). Insoluble fiber adds bulk to stools, aiding in their passage through the digestive tract. Furthermore, the prebiotic properties of certain fibers in amaranth promote the growth of beneficial gut bacteria, fostering a balanced and healthy gut microbiome (Tiwari 2021).

11.8.7 Mineral Bioavailability

While amaranth contains anti-nutrients that can inhibit mineral absorption, appropriate processing techniques, such as soaking or cooking, can significantly reduce these anti-nutrient levels. This enhances the bioavailability of essential minerals like iron and zinc. As a result, amaranth can contribute to the prevention of mineral deficiencies, particularly in populations vulnerable to such deficiencies (Table 11.1).

11.8.8 Immune Modulation

Certain bioactive compounds in amaranth, particularly saponins, exhibit immune-modulating properties. These compounds may bolster the body's immune response, potentially enhancing its ability to defend against infections and diseases. While further research is needed in this area, amaranth's immune-supportive potential is promising (Agrawal et al. 2023; Takaidza 2023).

11.8.9 Hormone-Related Conditions

Amaranth's lignans, which have estrogenic properties and antioxidant effects, hold promise in protecting against hormone-related conditions such as breast and prostate cancers (Gray 2019). These compounds may influence hormonal balance and exhibit protective effects against hormone-driven malignancies (Abancens et al. 2020).

11.8.10 Bone Health

Amaranth's calcium and phosphorus content, essential minerals for bone health, contribute to skeletal integrity (Manyelo et al. 2022). Incorporating amaranth into one's diet can be particularly valuable for preventing osteoporosis and maintaining strong bones and teeth, especially in populations at risk of calcium deficiency (Martínez-Villaluenga et al. 2020).

11.8.11 Nutrient Diversity

Amaranth's rich nutritional profile diversifies the diet by providing essential vitamins, minerals, and high-quality protein. This diversity ensures a well-rounded nutrient intake, supporting overall health and well-being (Sharanagat et al. 2022). It also makes amaranth a valuable dietary choice for individuals with specific nutrient requirements or dietary restrictions (Graziano et al. 2022).

The inclusion of amaranth in a balanced diet offers a multitude of health benefits, ranging from cardiovascular protection and blood sugar regulation to

Table 11.1 Health benefits of the amaranth

Health implication	Description	Citations
Cardiovascular health	Beta-glucans in amaranth reduce LDL cholesterol, lowering the risk of heart disease. Antioxidants improve blood vessel function	Ramakrishna et al. (2019), Fu et al. (2020), Tofas et al. (2019)
Blood sugar regulation	Amaranth's soluble fiber stabilizes blood sugar levels, benefiting individuals with diabetes	Galanakis (2019), Okaiyeto et al. (2023), Patlak (2002)
Weight management	Amaranth's fiber promotes a sense of fullness, aiding in weight loss and control	Martínez-Villaluenga et al. (2020), Zamudio and Campos (2022)
Anti-inflammatory effects	Phytochemicals in amaranth reduce inflammation, lowering the risk of chronic diseases	Agrawal et al. (2023), Venkatesh et al. (2020)
Antioxidant protection	Amaranth's antioxidants combat oxidative stress, reducing the risk of diseases like cancer.	Michalak (2022), Filleria et al. (2021)
Digestive health	Amaranth's fiber supports regular bowel movements and a healthy gut microbiome	Gowda et al. (2022), Tiwari (2021)
Mineral bioavailability	Proper processing enhances mineral absorption, preventing deficiencies	Njoki et al. (2014)
Immune modulation	Amaranth compounds may boost the immune system, aiding in infection defense	Agrawal et al. (2023), Takaidza (2023)
Hormone-related conditions	Amaranth lignans may protect against hormone-related cancers	Gray (2019), Abancens et al. (2020)
Bone health	Amaranth's calcium and phosphorus support strong bones	Manyelo et al.(2022)
Nutrient diversity	Amaranth's rich nutrients make it suitable for diverse dietary needs	Sharanagat et al. (2022), Graziano et al. (2022)

anti-inflammatory and antioxidant effects. Its diverse nutrient profile and potential to enhance mineral bioavailability make it a versatile and nutritious grain that can contribute to overall well-being and reduce the risk of chronic diseases. Incorporating amaranth into one's dietary repertoire aligns with a health-conscious approach to nutrition and can be particularly beneficial for individuals with specific health goals or dietary preferences (Table 11.1).

11.9 Processing Techniques for Amaranth

Processing plays a pivotal role in transforming raw amaranth into edible and versatile products. These techniques not only improve the grain's palatability but also enhance its nutritional quality and expand its culinary applications. In this section, we delve into various processing methods, providing a comprehensive and detailed exploration of each.

11.9.1 Milling and Dehulling

The milling process involves several steps. First, the raw amaranth seeds are cleaned and sorted to remove any impurities or debris. Then, they are milled to produce amaranth flour. Milling may include removing the seed coat (outer layer) to yield a finer, more versatile flour (Kocira et al. 2021). The resulting amaranth flour can vary in particle size, allowing for different culinary applications. Fine flour is suitable for baking, while coarser flour can be used in soups or porridge (Lapčíková et al. 2019). Dehulling is typically performed using mechanical equipment that removes the tough outer hull of some amaranth varieties. This process produces amaranth groats, which are the hulled seeds (Barr 2019). Dehulling makes amaranth easier to cook and digest, expanding its potential uses in both sweet and savory dishes (Gaikwad et al. 2023).

11.9.2 Puffing

Puffing amaranth involves exposing the seeds to rapid and intense heat, typically using specialized equipment like puffing machines. The high temperature causes the moisture inside the seeds to turn into steam, creating pressure (Wang et al. 2023). When the pressure is released, the seeds pop or "puff" into a light and airy form. Puffed amaranth can be used as a cereal, added to baked goods, or incorporated into snacks. Its unique texture and flavor make it an intriguing addition to various recipes (Smith 2021).

11.9.3 Toasting and Roasting

To toast amaranth, the seeds are dry-heated either in a pan on the stovetop or in an oven until they develop a golden-brown color and a nutty aroma. The toasting process can be customized to achieve varying levels of crunchiness and flavor intensity (van Eck and Stieger 2020). Toasted amaranth seeds can be used as toppings for salads, yogurt, or desserts, adding a delightful crunch and a rich, roasted taste. Roasting involves heating amaranth seeds with the addition of oil and seasonings (Green and Hemming 2014). The oil enhances the flavor and helps the seasonings adhere to the seeds. Roasted amaranth seeds can be savory, with herbs and spices, or sweet, with honey or sugar. They make for a flavorful and versatile snack option or a flavorful ingredient in both sweet and savory recipes (Green 2006).

11.9.4 Sprouting

Sprouting amaranth seeds entail soaking them in water for a designated period, allowing them to germinate (Elliott et al. 2022). During this process, enzymes are activated, which can enhance the nutritional value and digestibility of amaranth.

Sprouted amaranth can be used in salads, wraps, sandwiches, or as a garnish, offering a unique texture and a mild, slightly nutty flavor (Aída et al. 2013).

11.9.5 Fermentation

Fermenting amaranth involves the action of beneficial microorganisms, such as lactic acid bacteria, to break down complex carbohydrates and proteins. This process results in changes in flavor, texture, and nutritional content (Cruz-Casas et al. 2023). Fermented amaranth products, like porridge or gruel, can have a tangy flavor and are often enjoyed as a staple food in certain cultures. The fermentation process not only enhances digestibility but may also improve nutrient bioavailability (Qin et al. 2022).

11.9.6 Extrusion

Extrusion is a high-temperature and high-pressure process used to create various amaranth-based products (Wangui 2015). In this method, amaranth flour is mixed with water and exposed to extreme conditions, resulting in the formation of different shapes and sizes, such as flakes, crisps, or pasta-like forms. Extruded amaranth products are convenient, versatile, and easy to incorporate into a variety of dishes, both sweet and savory (Muñoz-Pabon et al. 2022).

11.9.7 Blending and Formulation

Blending amaranth flour with other flours or ingredients allows for the customization of taste, texture, and nutritional content (Dabija et al. 2022). For example, blending amaranth flour with wheat flour can result in baked goods that not only have an improved protein content but also a distinctive flavor profile. This approach is valuable for creating unique and nutritionally balanced products (Siddiqui et al. 2022).

11.9.8 Oil Extraction

Amaranth seeds contain nutritious oil that can be extracted using mechanical pressing or solvent-based methods. Mechanical pressing involves crushing the seeds to release the oil, while solvent-based extraction uses chemical solvents (Martins et al. 2021). The resulting amaranth oil is rich in unsaturated fatty acids, making it suitable for culinary applications such as salad dressings, and cooking oil, or as a nutritious addition to various dishes (Vijayan and Sadanandan 2023).

11.9.9 Packaging and Storage

Proper packaging is crucial to maintain the quality and shelf life of processed amaranth products. Packaging materials should be airtight and moisture-resistant to prevent spoilage and maintain freshness (Reaz et al. 2023). Furthermore, storage conditions are essential; amaranth products should be stored in cool, dry areas away from direct sunlight to prevent degradation (Turner et al. 2020).

In summary, processing techniques for amaranth are diverse and sophisticated, each contributing to the grain's culinary appeal, nutritional quality, and versatility. Whether it is milling for flour, puffing for cereals, toasting or roasting for snacks, sprouting for unique textures, fermenting for enhanced digestibility, extruding for convenient shapes, blending for customized formulations, or extracting oil for culinary use, these methods unlock the full potential of amaranth in modern cuisine. Careful packaging and storage ensure that processed amaranth products maintain their quality, making them accessible and enjoyable for consumers while preserving their valuable nutritional attributes.

11.10 Value Addition in Amaranth

Amaranth, with its rich nutritional profile and versatility, offers numerous opportunities for value addition. This section explores how amaranth can be transformed into a wide range of products that cater to diverse consumer needs and preferences.

11.10.1 Value Addition in Amaranth Products

11.10.1.1 Amaranth Flour

Amaranth flour is a versatile and nutritious addition to various recipes. It can be used in baking, replacing, or supplementing wheat flour to boost protein content and enhance the nutritional profile of bread, muffins, pancakes, and other baked goods (Niccum 2022).

11.10.1.2 Puffed Amaranth

Puffed amaranth is a crunchy and nutritious cereal that can be enjoyed with milk or yogurt. It can also be used as a topping for smoothie bowls, or ice cream, or incorporated into energy bars, adding a delightful texture and nutty flavor (Grescoe 2023).

11.10.1.3 Amaranth Snacks

Toasted or roasted amaranth seeds can be seasoned and packaged as healthy snacks. These snacks can cater to different taste preferences, offering both savory and sweet options (Bitsoie and Fraioli 2021).

11.10.1.4 Amaranth-Based Granola

Amaranth seeds can be combined with oats, nuts, seeds, and dried fruits to create amaranth-based granola. This granola is an excellent source of energy and nutrients, suitable for breakfast or as a snack (Mathur et al. 2020).

11.10.1.5 Amaranth Pasta

Amaranth flour can be used to make gluten-free pasta, catering to individuals with dietary restrictions. Amaranth pasta offers a unique texture and flavor profile, adding variety to the pasta market (Balakrishnan and Schneider 2022).

11.10.1.6 Amaranth Protein Powder

Amaranth seeds can be processed into protein powder, which is rich in essential amino acids. Amaranth protein powder is an excellent choice for vegetarians and vegans seeking a plant-based protein source (Soriano-García and Aguirre-Díaz 2020).

11.10.1.7 Amaranth Oil

The oil extracted from amaranth seeds is a nutritious cooking oil that is rich in unsaturated fatty acids. It can be marketed as a premium, heart-healthy cooking oil (O'brien 2008).

11.10.1.8 Amaranth-Based Energy Bars

Amaranth can be incorporated into energy bars, enhancing their nutritional content. These bars can provide sustained energy and serve as convenient snacks for active individuals (Singh et al. 2022).

11.10.1.9 Amaranth Breakfast Cereals

Amaranth can be used to create a variety of breakfast cereals, including flakes and puffs. These cereals can be fortified with vitamins and minerals, appealing to health-conscious consumers (Hernández-García et al. 2022).

11.10.1.10 Amaranth Beverage

Amaranth can be used to create nutritious beverages, such as amaranth milk or amaranth-based smoothie mixes. These products cater to those seeking dairy alternatives and functional beverages (Table 11.2).

11.10.1.11 Amaranth-Based Soups and Porridge

Amaranth flour can be used to thicken soups and create creamy porridge. These products offer a comforting and nutritious meal option (Souza et al. 2021).

11.10.1.12 Amaranth Snack Bars

Amaranth can be combined with nuts, dried fruits, and natural sweeteners to produce nutrient-dense snack bars. These bars are suitable for on-the-go consumption and can target the health-conscious market (King and Joyce 2021).

11.10.1.13 Amaranth-Based Baby Food

Amaranth's mild flavor and high nutrient content make it an ideal ingredient for baby food. It can be incorporated into purees and cereals to provide essential nutrients for infant growth and development (Jimenez et al. 2020).

Amaranth's adaptability and nutritional richness make it a valuable ingredient for value-added products. From flours and cereals to snacks, pasta, and protein powders, amaranth can be incorporated into a wide variety of culinary creations to cater to different dietary preferences and consumer needs. These value-added products not only expand the market for amaranth but also contribute to its recognition as a versatile and nutritious grain (Table 11.2).

Table 11.2 Application of the amaranth in food products

Value-added product	Description	Citations
Amaranth flour	Used in baking to boost protein and enhance nutrition in bread, muffins, and pancakes	Niccum (2022)
Puffed amaranth	Crunchy cereal is enjoyed with milk or yogurt, or used as a topping for smoothie bowls, ice cream, and energy bars	Grescoe (2023)
Amaranth snacks	Seasoned and packaged roasted seeds as healthy snacks, offering both savory and sweet options	Bitsoie and Fraioli (2021)
Amaranth-based granola	The blend of amaranth seeds, oats, nuts, seeds, and dried fruits, provides energy and nutrients for breakfast or snacks	Mathur et al. (2020)
Amaranth pasta	Gluten-free pasta made from amaranth flour, offering a unique texture and flavor	Balakrishnan and Schneider (2022)
Amaranth protein powder	Plant-based protein source rich in essential amino acids, suitable for vegetarians and vegans	Soriano-García and Aguirre-Díaz (2020)
Amaranth oil	Nutritious cooking oil extracted from amaranth seeds, rich in unsaturated fatty acids	O'brien (2008)
Amaranth-based energy bars	Energy bars incorporate amaranth, providing sustained energy and convenience for active individuals	Singh et al. (2022)
Amaranth breakfast cereals	Various cereals, including flakes and puffs, fortified with vitamins and minerals for health-conscious consumers	Hernández-García et al. (2022)
Amaranth beverage	Nutritious beverages like amaranth milk or smoothie mixes, cater to those seeking dairy alternatives and functional drinks	Manassero et al. (2020)
Amaranth-based soups and porridge	Creamy soups and porridge thickened with amaranth flour, offer comfort and nutrition in a meal	Souza et al. (2021)
Amaranth snack bars	Nutrient-dense bars made with amaranth, nuts, dried fruits, and natural sweeteners are suitable for on-the-go consumption	King and Joyce (2021)
Amaranth-based baby food	Ideal ingredient for baby food, incorporated into purees and cereals to provide essential nutrients for infant growth and development	Jimenez et al. (2020)

11.11 Safety Considerations in Amaranth Production and Consumption

Ensuring the safety of amaranth products, from production to consumption, is of utmost importance. This section discusses various safety considerations and provides guidelines for mitigating potential risks.

11.11.1 Agricultural Practices

11.11.1.1 Pesticide Use

To ensure the safety of amaranth, it is essential to follow integrated pest management practices, minimizing the use of synthetic pesticides. Organic farming methods can be employed to reduce pesticide residues in amaranth products (Ofuya et al. 2023).

11.11.1.2 Soil and Water Quality

Monitoring soil and water quality is crucial to prevent contamination of amaranth crops with heavy metals, pathogens, or harmful chemicals. Regular soil testing can help identify potential issues (Iqbal et al. 2020).

11.11.2 Harvesting and Post-harvest Handling

11.11.2.1 Hygiene

Ensure that workers involved in harvesting and handling amaranth adhere to good hygiene practices, including proper handwashing. Contamination during harvesting and post-harvest stages should be minimized (Kavindu 2022).

11.11.2.2 Storage

Store harvested amaranth in clean, dry, and well-ventilated facilities. Proper storage helps prevent moisture buildup and mold growth (Osei et al. 2020).

11.11.3 Processing and Packaging

11.11.3.1 Food Safety Standards

Implement food safety standards and practices during processing. Maintain a clean and sanitized processing environment to prevent contamination of amaranth products (Malik et al. 2021).

11.11.3.2 Packaging Materials

Use food-grade, airtight packaging materials to protect amaranth products from moisture, pests, and external contaminants. Ensure that packaging is labeled with relevant information, including expiration dates and nutritional content (Gupta et al. 2022).

11.11.4 Allergen Control

11.11.4.1 Cross-Contamination

Prevent cross-contamination with allergens such as gluten, nuts, or soy in processing facilities. Label products accurately to inform consumers of potential allergens (Katidi et al. 2023).

11.11.5 Nutrient Retention

11.11.5.1 Processing Methods

Choose processing methods that retain the nutritional value of amaranth products. Minimize heat exposure and processing times to preserve vitamins, minerals, and bioactive compounds (Aderibigbe et al. 2022).

11.11.6 Microbiological Safety

Train workers in proper hygiene practices to prevent the introduction of harmful microorganisms during processing. Conduct regular microbiological testing to ensure the absence of harmful pathogens in amaranth products (Malik et al. 2021).

11.11.7 Allergen Labeling

Clear Labeling: Clearly label amaranth products with allergen information, especially if they have come into contact with common allergens during processing (Cassin et al. 2022).

11.11.8 Consumer Education

Provide consumers with clear instructions on the safe storage and handling of amaranth products to prevent spoilage and ensure food safety at home (Ukwo et al. 2022).

11.11.9 Regulatory Compliance

Comply with local and international food safety regulations and standards. Regularly update safety practices to align with evolving guidelines (Faour-Klingbeil and Todd 2019).

11.11.10 Traceability

Record-Keeping: Maintain comprehensive records of the production, processing, and distribution of amaranth products. This facilitates traceability in case of safety concerns (Abia et al. 2021).

Safeguarding the safety of amaranth products is a shared responsibility involving producers, processors, and consumers. By adhering to best agricultural practices, implementing stringent food safety measures during processing, and ensuring proper labeling and consumer education, the potential risks associated with amaranth consumption can be effectively minimized. Maintaining safety and quality standards is essential to promote the consumption of this nutritious grain and its value-added products.

11.12 Future Challenges and Opportunities in the Amaranth Industry

11.12.1 Challenges

Amaranth cultivation, despite its resilience, faces several hurdles. Pests, diseases, and adverse weather conditions can pose agronomic challenges, necessitating the development of pest-resistant varieties and sustainable farming practices. Limited market access and awareness, particularly for indigenous amaranth varieties, need to be addressed through efforts aimed at promoting amaranth as a staple crop and increasing consumer awareness. Regulatory standards and certification processes, especially for organic and gluten-free products, can be onerous for small-scale amaranth producers, requiring streamlined procedures for easier market entry. Further research is essential to unlock amaranth's full potential, including understanding specific health benefits, optimizing processing techniques, and addressing safety concerns. Additionally, ensuring a reliable supply chain, from farm to market, is vital and necessitates investment in infrastructure and transportation to minimize post-harvest losses and ensure product availability.

11.12.2 Opportunities

Amaranth's remarkable nutritional profile positions it well to meet the increasing consumer demand for healthier food options. Leveraging this nutritional awareness, there is an opportunity to market amaranth-based products as functional foods. Amaranth's inherent sustainability and resilience align with the growing interest in sustainable and climate-resilient crops. Its versatility in culinary applications, from gluten-free baking to snack bars and beverages, offers opportunities for diversifying product offerings. Both local and global markets present growth potential, with the rising popularity of ethnic and specialty foods. Exporting amaranth-based products can expand its reach. Continued research and innovation in amaranth processing, breeding, and value addition can unlock new possibilities and enhance existing products. Beyond commerce, amaranth cultivation can have a positive social impact, contributing to food security, especially in regions grappling with malnutrition and climate change. Collaborative efforts between stakeholders, including government agencies, research institutions, NGOs, and the private sector, can collectively address challenges and promote the amaranth industry's growth and sustainability.

11.13 Conclusion

In conclusion, amaranth is a remarkable grain with a rich history, exceptional nutritional content, and promising potential. It has shown resilience in diverse agricultural settings and offers a sustainable and nutritious food source for a world facing numerous challenges, including food security and climate change. This chapter has provided a comprehensive exploration of amaranth, from its origins and cultivation practices to its nutritional composition and health implications. By understanding the unique qualities of amaranth and the challenges and opportunities it presents, stakeholders in the amaranth industry can work together to harness its full potential. As we move forward, continued research, innovation, and collaboration will be key to addressing challenges, expanding market access, and ensuring that amaranth plays a vital role in meeting the dietary and nutritional needs of diverse populations. Embracing amaranth as a sustainable and nutritious crop offers a path toward a healthier and more food-secure future for individuals and communities worldwide.

References

Abancens M, Bustos V, Harvey H, McBryan J, Harvey BJ (2020) Sexual dimorphism in colon cancer. Front Oncol 10:607909

Aber H, Mulindwa J, Lungaho M, Nyakundi F, Wambui J, Jager M et al (2019) Hazard analysis and critical control point plan for hazards in Ugandan amaranth vegetable value chain. Afr J Food Agric Nutr Dev 19(2):14458–14482

Abia WA, Montgomery H, Nugent AP, Elliott CT (2021) Tropane alkaloid contamination of agricultural commodities and food products in relation to consumer health: learnings from the 2019 Uganda food aid outbreak. Compr Rev Food Sci Food Saf 20(1):501–525

Abolarin EB (2021) Factor analysis of organic production of spices and vegetables among farmers in Ekiti State, Nigeria. Doctoral dissertation

Aderibigbe OR, Ezekiel OO, Owolade SO, Korese JK, Sturm B, Hensel O (2022) Exploring the potentials of underutilized grain amaranth (Amaranthus Spp.) along the value chain for food and nutrition security: a review. Crit Rev Food Sci Nutr 62(3):656–669

Agrawal RS, Ranveer RC, Rathod NB, Nirmal NP (2023) Phytochemicals as bioactive ingredients for functional foods. In: Recent frontiers of phytochemicals. Elsevier, London, pp 95–108

Aída J, Barrera-Pacheco A, Lara-González S, Montero-Morán GM, Díaz-Gois A, De Mejia EG, Barba De AP, Rosa L (2013) In vitro inhibition of dipeptidyl peptidase IV by peptides derived from the hydrolysis of amaranth (Amaranthus Hypochondriacus L.) proteins. Food Chem 136(2):758–764

Akbulut Ş, Yeşilkaya B (2021) Evaluation of vegan nutrition in regards to health. Sağlık Akademisyenleri Dergisi 8(2):163–168

Allai FM, Azad Z, Gul K, Dar BN (2022) Wholegrains: a review on the amino acid profile, mineral content, physicochemical, bioactive composition and health benefits. Int J Food Sci Technol 57(4):1849–1865

Anitha S, Afari-Sefa V, Kalumikiza Z, Mhango K, Mosha I, Muzanila Y, Mwangwela A, Ochieng J, Okori P, Tsusaka TW (2022) Improved household nutrition through home-grown produce and consumption of nutritious and healthy products. In: Sustainable agricultural intensification: a handbook for practitioners in East and Southern Africa. CABI, Wallingford, pp 119–133

Balakrishnan G, Schneider RG (2022) The role of amaranth, quinoa, and millets for the development of healthy, sustainable food products—a concise review. Foods 11(16):2442

Baliyan S, Mukherjee R, Priyadarshini A, Vibhuti A, Gupta A, Pandey RP, Chang C-M (2022) Determination of antioxidants by DPPH radical scavenging activity and quantitative phytochemical analysis of ficus religiosa. Molecules 27(4):1326

Banerjee P, Garai P, Nimai C, Saha S, Saha P, Sharma AK (2023) A critical review on the effect of nitrate pollution in aquatic invertebrates and fish. Water Air Soil Pollut 234(6):333

Barr S (2019) Technology of cereals, pulses and oilseeds. Scientific e-Resources

Barrette E, Rajchel D, Kambos J, Fiore-Bloom T, Wolf CR, Zaman N, Ress S (2020) Llewellyn's 2021 herbal almanac: a practical guide to growing. Llewellyn Worldwide, Woodbury

Beam SC, Mirsky S, Cahoon C, Haak D, Flessner M (2019) Harvest weed seed control of Italian ryegrass [Lolium Perenne L. Ssp. Multiflorum (Lam.) Husnot], common ragweed (Ambrosia Artemisiifolia L.), and Palmer Amaranth (Amaranthus Palmeri S. Watson). Weed Technol 33(4):627–632

Bitsoie F, Fraioli JO (2021) New native kitchen: celebrating modern recipes of the American Indian. Abrams

Boateng V, Donkoh MTSA (2023) Drivers of smallholder farmers' organic farming adoption and the organic share of the total cropland in Northern Ghana. All Life 16(1):2224526

Buddenhagen CE, Hackell D, Henderson HV, Wynne-Jones B (2023) Factors impacting the detection of weed seed contaminants in seed lots. Pest Manag Sci 79(2):881–890

Cadenas R, Caballero I, Nimubona D, Blanco CA (2021) Brewing with starchy adjuncts: its influence on the sensory and nutritional properties of beer. Foods 10(8):1726

Caetano-Silva ME, Netto FM, Bertoldo-Pacheco MT, Alegría A, Cilla A (2021) Peptide-metal complexes: obtention and role in increasing bioavailability and decreasing the pro-oxidant effect of minerals. Crit Rev Food Sci Nutr 61(9):1470–1489

Cassin A, Devonshire A, Ward S, Mcneill M (2022) Food allergy. In: Pediatric nutrition for dietitians. CRC Press, London, pp 185–202

Chandra AK, Chandora R, Sood S, Malhotra N (2021) Global production, demand, and supply. In: Millets and pseudo cereals. Elsevier, London, pp 7–18

Christopher J (2023) Early agricultural lives: bioarchaeological inferences from neolithic and early copper age tombs in the Central Po Valley, Italy

Cruz-Casas DE, Aguilar CN, Ascacio-Valdés JA, Rodríguez-Herrera R, Chávez-González ML, Flores-Gallegos AC (2023) Bioactive protein hydrolysates obtained from amaranth by fermentation with lactic acid bacteria and Bacillus species. Heliyon 9(2):e13491

Dabija A, Ciocan ME, Chetrariu A, Codină GG (2022) Buckwheat and amaranth as raw materials for brewing, a review. Plan Theory 11(6):756

De LC (2020) Edible seeds and nuts in human diet for immunity development. Int J Recent Sci Res 6(11):38877–38881

Dragomir N, Bahaciu G-V (2019) Specific gluten-based flours recommended in the gluten-free diet. Scientific papers. Series D. Animal Science 62(2):302–308. CERES Publishing House, Bucharest

Elliott H, Woods P, Green BD, Nugent AP (2022) Can sprouting reduce phytate and improve the nutritional composition and nutrient bioaccessibility in cereals and legumes? Nutr Bull 47(2):138–156

Emmanuel OC, Babalola OO (2022) Amaranth production and consumption in South Africa: the challenges of sustainability for food and nutrition security. Int J Agric Sustain 20(4):449–460

Faour-Klingbeil D, Todd ECD (2019) Prevention and control of foodborne diseases in Middle-East North African countries: review of national control systems. Int J Environ Res Public Health 17(1):70
Filleria G, Rodriguez M, Tironi VA (2021) Antioxidant effect of amaranth flour or protein isolate incorporated in high-fat diets fed to wistar rats. Influence of dose and administration duration. J Food Biochem 45(1):e13552
Fu J, Zhang Y, Hu Y, Zhao G, Tang Y, Liang Z (2020) Concise review: coarse cereals exert multiple beneficial effects on human health. Food Chem 325:126761
Gaikwad PS, Reddy NBP, Sandhya K, Yadav BK (2023) Traditional vs. modern usage of nutri-cereals. In: Nutri-cereals. CRC Press, Boca Raton, pp 27–56
Galanakis CM (2019) Dietary fiber: properties, recovery, and applications. Academic Press, Cambridge, MA
García P (2021) Routledge handbook of the archaeology of indigenous-colonial interaction in the Americas. Routledge, New York
Gebeyehu S, Kangile J, Mwakatobe E (2019) Assessment of seed quality along the rice seed value chain in Tanzania. Dev Pract 29(7):854–866
Gohil N, Bhattacharjee G, Khambhati K, Braddick D, Singh V (2019) Corrigendum: engineering strategies in microorganisms for the enhanced production of squalene: advances, challenges and opportunities. Front Bioeng Biotechnol 7:114
Gowda N, Nanje A, Siliveru K, Prasad PVV, Bhatt Y, Netravati BP, Gurikar C (2022) Modern processing of Indian millets: a perspective on changes in nutritional properties. Foods 11(4):499
Gray L (2019) Cancer-free with food: a step-by-step plan with 100+ recipes to fight disease, nourish your body & restore your health. Hay House, Carlsbad
Graziano S, Agrimonti C, Marmiroli N, Gullì M (2022) Utilisation and limitations of pseudocereals (quinoa, amaranth, and buckwheat) in food production: a review. Trends Food Sci Technol 125:154–165
Green A (2006) Field guide to herbs & spices: how to identify, select, and use virtually every seasoning on the market. Quirk Books, Philadelphia
Green P, Hemming C (2014) Grain power: over 100 delicious gluten-free ancient grain & superblend recipe: a cookbook. Penguin, Toronto
Grescoe T (2023) The lost supper: searching for the future of food in the flavors of the past. Greystone Books, Vancouver
Gülsoy E, Kaya ED, Türkhan A, Bulut M, Koyuncu M, Güler E, Sayın F, Muradoğlu F (2023) The effect of altitude on phenolic, antioxidant and fatty acid compositions of some Turkish Hazelnut (Coryllus Avellana L.) cultivars. Molecules 28(13):5067. https://doi.org/10.3390/molecules28135067
Gunarathne A, Hemmerling S, Labonte N, Zühlsdorf A, Spiller A (2022) Passionate about food: exploring 'foodie' segmentation by nutritional knowledge. J Consum Mark 39(7):678–690
Gupta V, Biswas D, Roy S (2022) A comprehensive review of biodegradable polymer-based films and coatings and their food packaging applications. Materials 15(17):5899
Hernández-García Y, Melgar-Lalanne G, Téllez-Medina DI, Ruiz-May E, de la Paz Salgado-Cruz M, Andrade-Velásquez A, Dorantes-Álvarez L, López-Hernández D, Gómez MPS (2022) Scavenging peptides, antioxidant activity, and hypoglycemic activity of a germinated Amaranth (Amaranthus Hypochondriacus L.) beverage fermented by lactiplantibacillus plantarum. J Food Biochem 46(7):e14139
Holdsworth M, Pradeilles R, Tandoh A, Green M, Wanjohi M, Zotor F, Asiki G et al (2020) Unhealthy eating practices of city-dwelling Africans in deprived neighbourhoods: evidence for policy action from Ghana and Kenya. Glob Food Sec 26:100452
Howard MA (2021) High-protein plant-based diet for beginners: quick and easy recipes for everyday meals. Zeitgeist.
Hussain MI, Farooq M, Syed QA, Ishaq A, Al-Ghamdi AA, Hatamleh AA (2021) Botany, nutritional value, phytochemical composition and biological activities of quinoa. Plan Theory 10(11):2258

Imathiu S (2021) Neglected and underutilized cultivated crops with respect to indigenous African leafy vegetables for food and nutrition security. J Food Sec 9(3):115–125
Iqbal MS, Iqbal Z, Ansari MI, Singh SP, Pandey B, Srivastava JK, Singh AK (2020) Impacts of soil pollution on human health with special reference to human physiognomy and physiology. In: Plant responses to soil pollution. Springer, Singapore, pp 163–177
Jan N, Hussain SZ, Naseer B, Bhat TA (2023) Amaranth and quinoa as potential nutraceuticals: a review of anti-nutritional factors, health benefits and their applications in food, medicinal and cosmetic sectors. Food Chem X 18:100687
Janmohammadi H, Hosseintabar-Ghasemabad B, Oliyai M, Alijani S, Gorlov IF, Slozhenkina MI, Mosolov AA et al (2023) Effect of dietary amaranth (Amaranthus hybridus chlorostachys) supplemented with enzyme blend on egg quality, serum biochemistry and antioxidant status in laying hens. Antioxidants 12(2):456. https://doi.org/10.3390/antiox12020456
Jimenez D, Miraballes M, Gámbaro A, Lobo M, Samman N (2020) Baby purees elaborated with andean crops. Influence of germination and oils in physico-chemical and sensory characteristics. Food Sci Technol 124:108901
Katayama K, Saito M (2019) Muscle sympathetic nerve activity during exercise. J Physiol Sci 69(4):589–598
Katidi A, Xanthopoulou S, Vlassopoulos A, Noutsos S, Priftis K, Kapsokefalou M (2023) Food allergens in ultra-processed foods according to the NOVA classification system: a Greek branded food level analysis. Nutrients 15(12):2767. https://doi.org/10.3390/nu15122767
Kavindu MJ (2022) Assessment of the safety of selected locally processed baobab products and HACCP development for baobab pulp value chain in Kenya. PhD diss, JKUAT-AGRICULTURE
King J, Joyce M (2021) Too busy to diet. Jacqueline King, Arlington Height
Kocira A, Kozłowicz K, Panasiewicz K, Staniak M, Szpunar-Krok E, Hortyńska P (2021) Polysaccharides as edible films and coatings: characteristics and influence on fruit and vegetable quality—a review. Agronomy (Basel, Switzerland) 11(5):813
Komlaga GA, Oduro I, Essel EM (2019) Utilization of selected tropical crops (cocoa, kola nuts, sorghum, millet, and shea butter). In: Byproducts from agriculture and fisheries. Wiley, Chichester, pp 563–580
Kumar L, Hasanuzzaman M, Rahim NA (2019) Global advancement of solar thermal energy technologies for industrial process heat and its future prospects: a review. Energy Convers Manag 195:885–908
Kumpf D (2023) The organic monocrop experimenting with green growth on Indian tea plantations. Environ Human 15(2):1–18
Laboni FA, Ahmed MW, Kaium A, Alam MK, Parven A, Jubayer MF, Rahman MA, Meftaul IM, Khan MSI (2023) Heavy metals in widely consumed vegetables grown in industrial areas of Bangladesh: a potential human health Hazard. Biol Trace Elem Res 201(2):995–1005
Langyan S, Yadava P, Belwal T, Kaul T (2022) Editorial: sustaining protein nutrition through plant-based foods: a paradigm shift. Front Sustain Food Syst 6:1037086. https://doi.org/10.3389/fsufs.2022.1037086
Lapčíková B, Burešová I, Lapčík L, Dabash V, Valenta T (2019) Impact of particle size on wheat dough and bread characteristics. Food Chem 297:124938
Li Y, Dong X, Zhao L (2021) Application of magnetic chitosan nanocomposites modified by graphene oxide and polyethyleneimine for removal of toxic heavy metals and dyes from water. Int J Biol Macromol 192:118–125
Mackinnon M (2023) The hormone type cookbook: nourishing recipes for balancing the 7 different hormone types-recipes for healthy cycles, PMS, PCOS, fertility, and menopause. Fair Winds Press, Beverly
Maguolo A, Gabbianelli R, Maffeis C (2023) Micronutrients in early life and offspring metabolic health programming: a promising target for preventing non-communicable diseases. Eur J Clin Nutr 77(12):1105–1112. https://doi.org/10.1038/s41430-023-01333-4
Malhotra U, Roy M, Sontakke M, Choudhary P (2023) A recent paradigm on iron absorption, prevalence, and emerging dietary approaches to eradicate iron deficiency. Food Bioeng 2(1):53–63

Malik S, Krishnaswamy K, Mustapha A (2021) Hazard analysis and risk-based preventive controls (HARPC): current food safety and quality standards for complementary foods. Foods 10(9):2199

Manassero CA, Añón MC, Speroni F (2020) Development of a high protein beverage based on amaranth. Plant Foods for Human Nutrition 75:599–607

Manyelo TG, Sebola NA, van Rensburg EJ, Mabelebele M (2020) The probable use of genus Amaranthus as feed material for monogastric animals. Animals 10(9):1504

Manyelo, TG, Sebola NA, Mabelebele M (2022). Effect of amaranth leaf meal on performance, meat, and bone characteristics of Ross 308 broiler chickens. Plos one 17, no. 8 (2022): e0271903

Marcel MR, Chacha JS, Ofoedu CE (2022) Nutritional evaluation of complementary porridge formulated from orange-fleshed sweet potato, amaranth grain, pumpkin seed, and soybean flours. Food Sci Nutr 10(2):536–553

Martínez-Villaluenga C, Peñas E, Hernández-Ledesma B (2020) Pseudocereal grains: nutritional value, health benefits and current applications for the development of gluten-free foods. Food Chem Toxicol 137:111178

Martins FCOL, Sentanin MA, De Souza D (2021) Categories of food additives and analytical techniques for their determination. In: Innovative food analysis. Elsevier, London, pp 123–156

Mathur M, Kumari A, Grewal RB, Panghal A, Rani M, Kargwal R (2020) Development and optimization of ingredients for multigrain fibre and protein enriched composite bars using sensory evaluation. Pharma Innov J 9(11):143–149

Mccormick M (2020) A southern culinary cure: 130+ crazy delicious, gluten-free recipes to reduce inflammation and make your gut happy. Simon and Schuster, Noida

Mehdi NV, Kumar P, Zucca EM, Varoni L, Dini E, Panzarini J (2020) Lifestyle, oxidative stress, and antioxidants: back and forth in the pathophysiology of chronic diseases. Front Physiol 11:694

Michalak M (2022) Plant-derived antioxidants: significance in skin health and the ageing process. Int J Mol Sci 23(2):585

Muñoz-Pabon KS, Parra-Polanco AS, Roa-Acosta DF, Hoyos-Concha JL, Bravo-Gomez JE (2022) Physical and paste properties comparison of four snacks produced by high protein quinoa flour extrusion cooking. Front Sustain Food Syst 6:852224. https://doi.org/10.3389/fsufs.2022.852224

Najafian L (2023) A Review of bioactive peptides as functional food ingredients: mechanisms of action and their applications in active packaging and food quality improvement. Food Funct 14(13):5835–5857

Njoki JW, Sila DN, Onyango AN (2014) Impact of processing techniques on nutrient and antinutrient content of grain amaranth (A. albus). Food Science and Quality Management 25:10–17

Neumann CE, Parks LL, Parks JG (2023) Global dishes: favorite meals from around the world. Bloomsbury Publishing USA

Niccum C (2022) Can pea and white hemp proteins be used as a more sustainable protein source to improve nutritional and sensory results in gluten-free baked products?

Norain FJ, Massawe S, Mayes WK, Ho A, Singh RC (2021) Physiological screening for drought tolerance traits in vegetable amaranth (Amaranthus Tricolor) Germplasm. Agriculture 11(10):994

Nyonje WA (2022) Phenotypic based characterization of nutritional and bioactive traits of amaranth vegetable for improved iron bioavailability and acceptability. PhD diss, JKUAT-AGRICULTURE

O'brien RD (2008) Fats and oils: formulating and processing for applications. CRC Press, London

O'Hara B (2020) No-till intensive vegetable culture: pesticide-free methods for restoring soil and growing nutrient-rich, high-yielding crops. Chelsea Green Publishing, London

Obaid H, Ma L, Nader MH, Hashimi S, Sharifi H, Kakar J, Ni C (2023) Heavy metal contamination status of water, agricultural soil, and plant in the semiarid region of Kandahar, Afghanistan. ACS Earth Space Chem 7(7):1446–1458

Ofuya TI, Okunlola AI, Mbata GN (2023) A review of insect pest management in vegetable crop production in Nigeria. Insects 14(2):111

Okaiyeto K, Kerebba N, Rautenbach F, Singh SK, Dua K, Oguntibeju OO (2023) UPLC-ESI-QTOF-MS phenolic compounds identification and quantification from ethanolic extract of: in vitro antioxidant and antidiabetic potentials. Arab J Chem 16:104447

Olagunju AI, Oluwajuyitan TD, Oyeleye SI (2021) Multigrain bread: dough rheology, quality characteristics, in vitro antioxidant and antidiabetic properties. J Food Meas Charact 15(2):1851–1864

Olagunju AI, Arigbede TI, Makanjuola SA, Oyebode ET (2022) Nutritional compositions, bioactive properties, and in-vivo glycemic indices of amaranth-based optimized multigrain snack bar products. Meas Food 7(100039):100039

Olusanya RN, Kolanisi U, Ngobese NZ (2023) Mineral composition and consumer acceptability of amaranthus leaf powder supplemented ujeqe for improved nutrition security. Foods 12(11):2182. https://doi.org/10.3390/foods12112182

Osei MK, Adjebeng-Danquah J, Egbadzor KF, Osuman AS, Adjei EA, Abebrese SO, Melomey LD, Agyare RY (2020) Vegetable seed production: prospects and challenges: the case of Ghana. In: Advances in seed production and management. Springer, Singapore, pp 91–117

Pabich M, Materska M (2019) Biological effect of soy Isoflavones in the prevention of civilization diseases. Nutrients 11(7):1660

Panagea IS, Berti A, Čermak P, Diels J, Elsen A, Kusá H, Piccoli I et al (2021) Soil water retention as affected by management induced changes of soil organic carbon: analysis of long-term experiments in Europe. Land 10(12):1362

Patlak M (2002) New weapons to combat an ancient disease: treating diabetes. FASEB J 16(14):1853

Patrícia KA, Viana DP, Raggio MT, Hosey LR (2020) Recommended procedures for the management of early childhood caries lesions—a scoping review by the Children Experiencing Dental Anxiety: Collaboration on Research and Education (CEDACORE). BMC Oral Health 20:1–11

Pavlović D, Vrbničanin S, Anđelković A, Božić D, Rajković M, Malidža G (2022) Non-chemical weed control for plant health and environment: ecological integrated weed management (EIWM). Agronomy (Basel, Switzerland) 12(5):1091

Pereira AG, Fraga-Corral M, García-Oliveira P, Jimenez-Lopez C, Lourenço-Lopes C, Carpena M, Paz O, Gullón P, Prieto MA, Simal-Gandara J (2020) Culinary and nutritional value of edible wild plants from Northern Spain rich in phenolic compounds with potential health benefits. Food Funct 11(10):8493–8515

Pérez-Beltrán YE, Tovar J, Sáyago-Ayerdi SG (2022) Influence of the food matrix on the digestibility and metabolic responses to starchy foods. In: Starch and starchy food products. CRC Press, Boca Raton, pp 291–322

Piga A, Conte P, Fois S, Catzeddu P, Del Caro A, Sanguinetti AM, Fadda C (2021) Technological, nutritional and sensory properties of an innovative gluten-free double-layered flat bread enriched with amaranth flour. Foods 10(5):920. https://doi.org/10.3390/foods10050920

Popova A, Mihaylova D (2019) Antinutrients in plant-based foods: a review. Open Biotechnol J 13:68–76

Qin H, Wu H, Shen K, Liu Y, Li M, Wang H, Qiao Z, Zhixin M (2022) Fermented minor grain foods: classification, functional components, and probiotic potential. Foods 11(20):3155

Rai U, Rai S, Datta S (2022) Microgreens: a super food for nutritional security in the 21st century—a review. Environ Ecol 40(4A):2239–2244

Ram A, Thattantavide A, Kumar A (2023) Re-emergence of pseudocereals as superfoods for food security and human health: current progress and future prospects. In: Wild food plants for zero hunger and resilient agriculture. Springer, Singapore, pp 207–236

Ramakrishna R, Sarkar D, Shetty K (2019) Functional bioactives from barley for human health benefits. In: Functional foods and biotechnology, Food biotechnology series. Taylor & Francis, Boca Raton, pp 61–85

Rao V, Poonia A (2023) Protein characteristics, amino acid profile, health benefits and methods of extraction and isolation of proteins from some pseudocereals—a review. Food Product Process Nutr 5(1):37. https://doi.org/10.1186/s43014-023-00154-z

Reaz AH, Abedin MJ, Abdullah ATM, Satter MA, Farzana T (2023) Physicochemical and structural impact of CMC-hydrocolloids on the development of gluten-free foxtail millet biscuits. Heliyon 9(6):e17176

Reyes-Jurado F, Soto-Reyes N, Dávila-Rodríguez M, Lorenzo-Leal AC, Jiménez-Munguía MT, Mani-López E, López-Malo A (2023) Plant-based milk alternatives: types, processes, benefits, and characteristics. Food Rev Intl 39(4):2320–2351

Rinke S (2023) Conquistadors and aztecs: a history of the fall of tenochtitlan. Oxford University Press, Oxford

Rojas-Rivas E, Espinoza-Ortega A, Thomé-Ortíz H, Moctezuma-Pérez S, Cuffia F (2019) Understanding consumers' perception and consumption motives towards amaranth in Mexico using the Pierre Bourdieu's theoretical concept of habitus. Appetite 139(August):180–188

Ruth ON, Unathi K, Nomali N, Chinsamy M (2021) Underutilization versus nutritional-nutraceutical potential of the amaranthus food plant: a mini-review. Appl Sci 11(15):6879

Sachithra JK, Chamila VL (2020) Indigenous and traditional foods of Sri Lanka. J Ethnic Foods 7(1):1–19

Sahu P, Tripathy B, Rout S (2020) Significance of anti-nutritional compounds in vegetables. Agric Rural Dev Spat Issues Chall Approaches 98:98–109

Salvador M, Rodrigo-Garcia J, Escudero-Feliu M, Garcia-Costela E, Lima-Cabello A, Carazo-Gallego S, Morales-Santana J, Leon JC (2021) Current advances research in nutraceutical compounds of legumes, pseudocereals and cereals. In: Grain and seed proteins functionality. IntechOpen, London

Sandstrom AR, Sandstrom PE (2023) Pilgrimage to broken mountain: Nahua sacred journeys in Mexico's huasteca veracruzana. University Press of Colorado, Denver

Sarkar T, Salauddin M, Roy S, Chakraborty R, Rebezov M, Shariati MA, Thiruvengadam M, Rengasamy KRR (2023) Underutilized green leafy vegetables: frontier in fortified food development and nutrition. Crit Rev Food Sci Nutr 63(33):11679–11733

Sarker U, Hossain MN, Iqbal MA, Oba S (2020) Bioactive components and radical scavenging activity in selected advance lines of salt-tolerant vegetable amaranth. Front Nutr 7(November):587257

Schmidt D, Verruma-Bernardi MR, Forti VA, Borges MTMR (2023) Quinoa and amaranth as functional foods: a review. Food Rev Intl 39(4):2277–2296

Schubart CD, Deli T, Mancinelli G, Cilenti L, Fernández AG, Falco S, Berger S (2022) Phylogeography of the Atlantic blue crab Callinectes sapidus (Brachyura: Portunidae) in the Americas versus the Mediterranean Sea: determining origins and genetic connectivity of a large-scale invasion. Biology 12(1):35

Schwartz-Lazaro LM, Shergill LS, Evans JA, Bagavathiannan MV, Beam SC, Bish MD, Bond JA et al (2021) Seed-shattering phenology at soybean harvest of economically important weeds in multiple regions of the United States. Part 1: broadleaf species. Weed Sci 69(1):95–103

Selby NM, Taal MW (2020) An updated overview of diabetic nephropathy: diagnosis, prognosis, treatment goals and latest guidelines. Diabet Obes Metabol 22 Suppl 1(S1):3–15

Shahbazi R, Sharifzad F, Bagheri R, Alsadi N, Yasavoli-Sharahi H, Matar C (2021) Anti-inflammatory and immunomodulatory properties of fermented plant foods. Nutrients 13(5):1516. https://doi.org/10.3390/nu13051516

Sharanagat VS, Singh L, Nema PK (2022) Approaches for development of functional and low gluten bread from sorghum: a review. J Food Process Preserv 46(11):e17089. https://doi.org/10.1111/jfpp.17089

Sharanya H, Shivaranjani KS, Divya S, Aishwarya R (2023) Comparison of different iron formulations for treating iron deficiency anemia in pregnancy—a systemic review. Int J Acad Med Pharm 5(4):1885–1890

Shembe PS, Ngobese NZ, Siwela M, Kolanisi U (2023) The potential repositioning of South African underutilised plants for food and nutrition security: a scoping review. Heliyon 9(6):e17232

Shi Z-Y, Zeng J-Z, Wong AST (2019) Chemical structures and pharmacological profiles of ginseng saponins. Molecules (Basel, Switzerland) 24(13):2443

Siddiqui SA, Chayan Mahmud MM, Abdi G, Wanich U, Farooqi MQU, Settapramote N, Khan S, Wani SA (2022) New alternatives from sustainable sources to wheat in bakery foods: science, technology, and challenges. J Food Biochem 46(9):e14185

Singh A, Kumari A, Chauhan AK (2022) Formulation and evaluation of novel functional snack bar with amaranth, rolled oat, and unripened banana peel powder. J Food Sci Technol 59(9):3511–3521

Singhania N, Kumar R, Pramila S, Bishnoi AB, Ray A (2023) Bioactive properties and health benefits of amaranthus. In: Harvesting food from weeds. Wiley, Hoboken, pp 351–383

Smith AF (2021) Popped culture: a social history of popcorn in America. University of South Carolina Press, Columbia

Soriano-García M, Aguirre-Díaz IS (2020) Nutritional functional value and therapeutic utilization of amaranth. In: Nutritional value of amaranth. IntechOpen, London

Souza D, Charmaine C, Souza D, Souza SD (2021) The good health always cookbook: recipes and nutrition secrets for a better you. Penguin Random House India Private Limited, Gurugram

Swarnalakshmi R, Sridevi D (2023) Organic cultivation of trigreens with vermicompost. Lampyrid 13:83–89

Szabóová M, Záhorský M, Gažo J, Geuens J, Vermoesen A, D'Hondt E, Hricová A (2020) Differences in seed weight, amino acid, fatty acid, oil, and squalene content in γ-irradiation-developed and commercial amaranth varieties (Amaranthus Spp.). Plants 9(11):1412

Takaidza S (2023) Indigenous South African food: nutrition and health benefits. IntechOpen, London

Timilsena YP, Phosanam A, Stockmann R (2023) Perspectives on saponins: food functionality and applications. Int J Mol Sci 24(17):13538. https://doi.org/10.3390/ijms241713538

Tiwari A (2021). Formulation of lito with mustard green leaves (Brassica juncea L.). PhD diss., Department of Nutrition & Dietetics Central Campus of Technology Institute of Science and Technology Tribhuvan University, Nepal. 2021, 2023

Tofas T, Draganidis D, Deli CK, Georgakouli K, Fatouros IG, Jamurtas AZ (2019) Exercise-induced regulation of redox status in cardiovascular diseases: the role of exercise training and detraining. Antioxidants (Basel, Switzerland) 9(1):13

Turner ER, Luo Y, Buchanan RL (2020) Microgreen nutrition, food safety, and shelf life: a review. J Food Sci 85(4):870–882

Turner MI, Adams KR, Berkebile JN, Dockter AR (2021) Ancient grains: new evidence for ancestral puebloan use of domesticated amaranth. Am Antiq 86(4):815–832

Ukwo SP, Udo II, Ndaeyo N (2022) Food additives: overview of related safety concerns. Food Sci Nutr Res 5(1):1–10. https://doi.org/10.33425/2641-4295.1052

Umetsu N, Shirai Y (2020) Development of novel pesticides in the 21st century. J Pestic Sci 45(2):54–74

van Eck A, Stieger M (2020) Oral processing behavior, sensory perception and intake of composite foods. Trends Food Sci Technol 106:219–231

Venkatesh HN, Ravish H, Silvia CRWD, Srinivas H (2020) Molecular signature of the immune response to yoga therapy in stress-related chronic disease conditions: an insight. Int J Yoga 13(1):9–17

Vijayan V, Sadanandan L (2023) Salvia hispanica L.–a promise towards food security. Biodiversity challenges and threats; current scenario (2023): 59.

Vinke S, Rais I, Millett P (2022) The dual-use education gap: awareness and education of life science researchers on nonpathogen-related dual-use research. Health Sec 20(1):35–42

Wang Y, Tuccillo F, Lampi A-M, Knaapila A, Pulkkinen M, Kariluoto S, Coda R et al (2022) Flavor challenges in extruded plant-based meat alternatives: a review. Compr Rev Food Sci Food Saf 21(3):2898–2929

Wang Y, Lyu B, Hongling F, Li J, Ji L, Gong H, Zhang R, Liu J, Hansong Y (2023) The development process of plant-based meat alternatives: raw material formulations and processing strategies. Food Res Int 167:112689

Wangui J (2015) Impact of different processing techniques on nutrients and antinutrients content of grain amaranth (Amaranthus albus). PhD diss. 2015

Waselkov KE, Regenold ND, Lum RC, Olsen KM (2020) Agricultural adaptation in the native North American weed waterhemp, Amaranthus tuberculatus (Amaranthaceae). PLoS One 15(9):e0238861

Wesley SB, Koyiyumptewa G (2021) In: Kuwanwisiwma LJ (ed) Becoming Hopi: a history. University of Arizona Press, Tucson

Yousaf L, Hou D, Liaqat H, Shen Q (2021) Millet: a review of its nutritional and functional changes during processing. Food Res Int 142:110197

Saeed Ben Youssef A (2023) Modern trends for the application of biological control and modern technologies in agricultural projects. Int J Modern Agric Environ 1:26–53

Zainodini N, Dousdampanis P, Ahmadi Z, Mohamadi M, Nazari A (2023) Association of oxalate consumption and some individual habits with the risk of kidney stones. Chin Med Sci J 38(3):250–256

Zamudio V, Campos MRS (2022) Amaranth, quinoa and chia bioactive peptides: a comprehensive review on three ancient grains and their potential role in management and prevention of type 2 diabetes. Crit Rev Food Sci Nutr 62(10):2707–2721

Zayan S (2018) Impact of climate change on plant diseases and IPM strategies. Arab J Plant Protect 36(1):75–79

Zhu Y, Wang K, Jia X, Caili F, Yu H, Wang Y (2024) Antioxidant peptides, the guardian of life from oxidative stress. Med Res Rev 44(1):275–364. https://doi.org/10.1002/med.21986

Zitcer A (2021) Practicing cooperation: mutual aid beyond capitalism. University of Minnesota Press, Minneapolis